D1747362

Gunter Neubert

Dictionary of Fluid Power

Wörterbuch Fluidtechnik

parat

VCH

©VCH Verlagsgesellschaft mbH, D-6940 Weinheim (Federal Republic of Germany), 1993

Distribution:

VCH, P. O. Box 10 1161, D-6940 Weinheim (Federal Republic of Germany)

Switzerland: VCH, P. O. Box, CH-4020 Basel (Switzerland)

United Kingdom and Ireland: VCH (UK) Ltd., 8 Wellington Court, Cambridge CB1 1HZ (England)

USA and Canada: VCH, 220 East 23rd Street, New York, NY 10010–4606 (USA)

ISBN 3-527-29013-3 (VCH, Weinheim)   ISSN 0930-6862
ISBN 1-56081-713-5 (VCH, New York)

Gunter Neubert

parat

Dictionary of Fluid Power

English/German
German/English

Wörterbuch Fluidtechnik

Englisch/Deutsch
Deutsch/Englisch

VCH

Weinheim · New York · Basel · Cambridge

Prof. Dr.-Ing. habil. G. Neubert
Bundschuhstraße 5
O-8019 Dresden
Federal Republic of Germany

Editor of series "parat"
Dr. Hans-Dieter Junge
Cavaillonstraße 78/I
D-6940 Weinheim
Federal Republic of Germany

This book was carefully produced. Nevertheless, author and publisher do not warrant the information contained therein to be free of errors. Readers are advised to keep in mind that statements, data, illustrations, procedural details or other items may inadvertently be inaccurate.

Published jointly by
VCH Verlagsgesellschaft mbH, Weinheim (Federal Republic of Germany)
VCH Publishers, Inc., New York, NY (USA)

Editorial Director: Dr. Hans-Dieter Junge
Production Manager: Dipl.-Wirt.-Ing. (FH) Hans-Jochen Schmitt

Library of Congress Card No. applied for

A catalogue record for this book is available from the British Library

Die Deutsche Bibliothek – CIP-Einheitsaufnahme
Neubert, Gunter:
Dictionary of fluid power : English/German,
German/English = Wörterbuch Fluidtechnik / Gunter Neubert.
– Weinheim ; New York ; Basel ; Cambridge : VCH, 1993
   (Parat)
   ISBN 3-527-29013-3 (Weinheim ...)
   ISBN 1-56081-713-5 (New York)
NE: HST

© VCH Verlagsgesellschaft mbH, D-6940 Weinheim (Federal Republic of Germany), 1993

Printed on acid-free and low-chlorine paper
Gedruckt auf säurefreiem und chlorarm gebleichtem Papier

All rights reserved (including those of translation into other languages). No part of this book may be reproduced in any form – by photoprinting, microfilm, or any other means – nor transmitted or translated into a machine language without written permission from the publishers. Registered names, trademarks, etc. used in this book, even when not specifically marked as such, are not to be considered unprotected by law.
Alle Rechte, insbesondere die der Übersetzung in andere Sprachen, vorbehalten. Kein Teil dieses Buches darf ohne schriftliche Genehmigung des Verlages in irgendeiner Form – durch Photokopie, Mikroverfilmung oder irgendein anderes Verfahren – reproduziert oder in eine von Maschinen, insbesondere von Datenverarbeitungsmaschinen, verwendbare Sprache übertragen oder übersetzt werden. Die Wiedergabe von Warenbezeichnungen, Handelsnamen oder sonstigen Kennzeichen in diesem Buch berechtigt nicht zu der Annahme, daß diese von jedermann frei benutzt werden dürfen. Vielmehr kann es sich auch dann um eingetragene Warenzeichen oder sonstige gesetzlich geschützte Kennzeichen handeln, wenn sie nicht eigens als solche markiert sind.
Composition: Filmsatz Unger & Sommer, D-6940 Weinheim
Printing: Strauss Offsetdruck GmbH, D-6945 Hirschberg 2
Bookbinding: Großbuchbinderei J. Schäffer, D-6718 Grünstadt
Printed in the Federal Republic of Germany

Vorwort

Mit ihren bekannten Vorzügen ist die Fluidtechnik – Entwicklung und Einsatz hydraulischer und pneumatischer Geräte, Anlagen und Systeme – ein unverzichtbares Glied sowohl konventioneller handbedienter als auch hochgradig automatisierter Steuerungen und Regelungen über das ganze Feld der Technik hinweg: vom herkömmlichen Ackerbaugerät bis zum Raumfahrzeug, von der Theatermaschine bis zum Schweißroboter.

Das Wörterbuch bietet den speziellen Wortschatz der Fluidtechnik und berücksichtigt ausdrücklich den oft verwirrenden Reichtum an Wörtern gleicher oder ähnlicher Bedeutung in jeder der beiden Sprachen, den ihre Anwendung durch die Fachleute so vieler verschiedener technischer Gebiete mit sich bringt.

Abkürzungen	ED	DE
Allgemeines	nicht markiert	
Fluidstatische und fluiddynamische Grundbegriffe	THEOR	THEOR
Erzeugung von Fluiddruckströmen (FDS):		
Pumpen	PU/MOT	PU/MOT
Kompressoren	COMPR	KOMPR
Verbrauch von FDS:		
Rotations- und Schwenkmotoren	PU/MOT	PU/MOT
Arbeitszylinder	CYL	ZYL
hydrostatische Getriebe	PU/MOT	PU/MOT
Leitung von FDS:		
Fluide	FLUID	FLUID
Rohr- und Schlauchleitungen	CONDUI	LEIT
Verbindungen	FITT	VERBIN
Verkettungstechnik (für Ventile)		
Aufbereitung von Fluiden (Filter, Öler, Trockner, Wärmetauscher, Entlüfter, Schalldämpfer)	PREPAR	AUFBER
Isolierung von FDS (Dichtungen, Abstreifer)	SEAL	DICHT
Steuerung und Regelung von FDS:		
Meß- und Prüfgeräte	MEASUR	MESS
Wegeventile	D-VALV	WEGEV
Druckventile	P-VALV	DRUCKV
Stromventile	F-VALV	STROMV
Sperrventile	CH-VALV	SPERRV
Ventile allgemein	VALV	VENTIL
Logikelemente	LOGEL	LOGEL
Stell- und Regeleinrichtungen	VALVOP	STELL
Speicherung von Fluiden:		
Behälter	TANK	BEHAEL
Speicher, Druckbehälter	ACC	SPEICH
Wandlung zwischen FDS, Druckverstärker, Druckumsetzer	INTENS	WANDL
Grundschaltungen, Anwendungsfälle, Darstellung, Entwurf, Instandhaltung, Fertigung	CIRC APPL	KREISL ANWEND

English / German
Englisch / Deutsch

A

ability of air separation *FLUIDS* Luftausscheidevermögen *n*, Luftabscheidevermögen *n*

ability of contaminant separation *FLUIDS* Schmutzabscheidevermögen *n*, Schmutzabsetzvermögen *n*

ability of gas separation *FLUIDS* Gasausscheidevermögen *n*, Gasabscheidevermögen *n*

abrupt contraction *THEOR* plötzliche Verengung *f*

abrupt enlargement *THEOR* plötzliche Erweiterung *f*

absolute filter fineness (rating) absolute Filterfeinheit *f*

absolute filtration fineness (rating) *s.* absolute filter fineness

absolute pressure absoluter Druck *m*, Absolutdruck *m*

absolute pressure gauge Absolutdruckmesser *m*

absolute pressure transducer Absolutdruckwandler *m*

absolute viscosity *FLUIDS* dynamische Viskosität *f*

absorb *v* *motor* schlucken *Motor*

absorber *s.* absorptive muffler

absorption dryer *PREPAR* Absorptionstrockner *m*

absorptive muffler *PREPAR* Absorptionsschalldämpfer *m*, Schalldämpfer *m* mit schallschluckendem Material

abutment motor Sperrtrommelmotor *m*

abutment pump Sperrtrommelpumpe *f*

acceleration head (pressure) *THEOR* Beschleunigungsdruck *m*

acceleration tolerance Beschleunigungsverträglichkeit *f*

accessories *pl* Zubehör *n*

AC coarse test dust *air cleaner coarse test dust FLUIDS* Luftfilter-Prüfgrobstaub *m*

accordian boot *SEALS* Faltenbalg *m*

ACCTD *FLUIDS* *s.* AC coarse test dust

acoustic filter akustisches Filter *n*

accumulator *hydraulic accumulator* Druckflüssigkeitsspeicher *m*, Flüssigkeitsspeicher *m*, Hydraulikspeicher *m*, Hydrospeicher *m*

accumulator bag (bladder) Speicherblase *f*

accumulator capacity Speichernutzvolumen *n*

accumulator circuit Speicherkreislauf *m*

AC fine test dust *air cleaner fine test dust FLUIDS* Luftfilter-Prüffeinstaub *m*

ACFTD *FLUIDS* *s.* AC fine test dust

acid value *FLUIDS* Neutralisationszahl *f*, NZ *f*, Säurezahl *f*, SZ *f*

acoustic velocity *THEOR* Schallgeschwindigkeit *f*

acoustic wave sensor *fluidic ear LOG EL* pneumatisch-akustischer Sensor *m*

A. C. solenoid *alternating current solenoid VALV OP* Wechselstrommagnet *m*

act *v* **on (upon)** *pressure* beaufschlagen *Druck beaufschlagt eine Fläche*, wirken auf *eine Fläche*, anliegen an *z. B. einem Ventil*

activated alumina desiccant *PREPAR* Aktivtonerde-Trockenmittel *n*

activated carbon filter Aktivkohlefilter *m,n*
actual motor inlet (input, intake) flow rate effektiver (tatsächlicher) Schluckstrom *m*
actual pump delivery (discharge rate, flow rate, output) effektiver (tatsächlicher) Förderstrom *m*
actuate *v* *VALV OP* betätigen, stellen, steuern, verstellen
actuated position *PCV* Betätigungsstellung *f*
actuating force *VALV OP* Stellkraft *f*, Betätigungskraft *f*, Verstellkraft *f*, Steuerkraft *f*, Schaltkraft *f*
actuating piston *VALV OP* Stellkolben *m*, Betätigungskolben *m*
actuating pressure *VALV OP* Stelldruck *m*, Betätigungsdruck *m*, Verstelldruck *m*, Steuerdruck *m*
actuating torque *VALV OP* Stellmoment *n*, Betätigungsmoment *n*, Verstellmoment *n*, Steuermoment *n*
actuator Verbraucher *m*, Druckstromverbraucher *m*, Motor *m*; *CYL* s. linear actuator; *PU/MOT* s. rotary actuator; *VALV OP* s. valve actuator
actuator control [circuit, system] *CIRC* Motorsteuerkreislauf *m*, Verbrauchersteuerkreislauf *m*, Motorsteuersystem *n*, Verbrauchersteuersystem *n*
actuator port *DCV* Verbraucheranschluß *m*
adaptor *FITT* Anpaßstück *n*
add-fluid point Nachfüllstelle *f*
additive *FLUIDS* Additiv *n*, Wirkstoff *m*, Zusatz *m*
adding *of gas bottles ACCUM* Nachschaltung *f* *von Gasflaschen*

adhesive fitting Klebverbindung *f*
adjustable-angle cam plate *PU/MOT* Schwenkscheibe *f*
adjustable cushion *CYL* einstellbare Endlagenbremse (Hubendebremse) *f*
adjustable elbow richtungseinstellbare Winkelverschraubung *f*
adjustable fitting richtungseinstellbare Verschraubung *f*, Schwenkverschraubung *f*
adjustable restriction *FCV* Verstelldrossel *f*, einstellbare Drossel *f*
adjustable relief valve einstellbares Druckbegrenzungsventil *n*
adjustable-stroke cylinder Zylinder *m* mit verstellbarem Anschlag
adjustment sensitivity *of an adjustable restriction FCV* Auflösungsvermögen *n* *einer Verstelldrossel*
admission Zufuhr *f*, Speisung *f*, Beaufschlagung *f*, Lieferung *f*, Versorgung *f*
admittance *flow/pressure drop THEOR* Admittanz *f* *Volumenstrom/Druckabfall*
adsorption filter Adsorptionsfilter *m,n*
adsorptive dryer *PREPAR* Adsorptionstrockner *m*
aerate *v: be aerated* *fluid is aerated* Luft *f* aufnehmen *Flüssigkeit nimmt Luft auf*; *RESERVOIRS* belüften, entlüften
aerated fluid lufthaltige Flüssigkeit *f*
aeration *of the fluid* Luftaufnahme *f* *durch das Öl*; *RESERVOIRS* Belüftung *f*, Entlüftung *f*
aeration opening *RESERVOIRS* Belüftungsöffnung *f*, Entlüftungsöffnung *f*, Lüftungsöffnung *f*

aerodynamic aerodynamisch, gasdynamisch
aerodynamics Aerodynamik *f*, Gasdynamik *f*
aeromechanics Aeromechanik *f*, Gasmechanik *f*
aerosol *FLUIDS* Aerosol *n*
aerostatics Aerostatik *f*
AF hydraulics *alternating-fluid hydraulics* Wechselstromhydraulik *f*
aftercool *v* *PREPAR* nachkühlen
aftercooler *PREPAR* Nachkühler *m*
aftercooling *PREPAR* Nachkühlung *f*
after-warmer *PREPAR* Nachwärmer *m*
age *v* *FLUIDS* altern
ageing *FLUIDS* Alterung *f*
 ageing rate *FLUIDS* Alterungsgeschwindigkeit *f*
 ageing resistance (stability) *FLUIDS* Alterungsbeständigkeit *f*, Alterungsstabilität *f*
agent *FLUIDS* Additiv *n*, Wirkstoff *m*, Zusatz *m*
agricultural hydraulics *APPL* Landmaschinenhydraulik *f*
air Luft *f*, *Weiterbildungen s auch unter* pneumatic, pneumo-, compressed air
 air actuator *VALV OP* pneumatische Stelleinheit (Betätigungseinrichtung) *f*, Druckluftstelleinheit *f*
 air beam *LOG EL* Luftstrahl *m*
 air bearing *APPL* Luftlager *m*
 air bleeder *for entrained air* Entlüftungseinrichtung *f*, Entlüfter *m*, Entlüftung *f* *für ungelöste Luft*
 air bottle *ACCUM* Luftflasche *f*
 air breather *RESERVOIRS* Belüftungsorgan *n*, Entlüftungsorgan *n*, Lüftungsorgan *n*
 air breather filter *RESERVOIRS* Belüftungsfilter *m,n*, Entlüftungsfilter *m,n*, Luftfilter *m,n*
 air bubble *FLUIDS* Luftblase *f*
 air-charged accumulator luftbelasteter (hydropneumatischer) Speicher *m*, Luftdruckspeicher *m*
 air circuit Pneumatikschaltung *f*, Pneumatiksystem *f*, Druckluftschaltung *f*, Druckluftsystem *f*, Luftsystem *n*
 air circulation Luftumlauf *m*, Luftzirkulation *f*
 air cleaner coarse test dust *FLUIDS* Luftfilter-Prüfgrobstaub *m*
 air cleaner fine test dust *FLUIDS* Luftfilter-Prüffeinstaub *m*
 air clutch *APPL* pneumatische Kupplung *f*, Druckluftkupplung *f*
 air compressor Luftverdichter *m*, Luftkompressor *m*
 air conditioner unit *filter-regulator-lubricator* *PREPAR* Druckluft-Wartungseinheit *f*, Luftaufbereitungseinheit *f*, Luftaufbereiter *m*
 air conditioning Druckluftaufbereitung *f*, Luftaufbereitung *f*
 air connection Pneumatikanschluß *m*, Druckluftanschluß *m*, Luftanschluß *m*
 air consumption Druckluftverbrauch *m*, Luftverbrauch *m*
 air consumption rating Nennluftverbrauch *m*
 air content *FLUIDS* Luftgehalt *m*
 air control Pneumatiksteuerung *f*, pneumatische Steuerung *f*, Drucklufsteuerung *f*
 air control *VALV OP* pneumatische Stelleinheit (Betätigungseinrichtung) *f*, Druckluftstelleinheit *f*

air-controlled air-pilot operator *VALV OP* pneumopneumatische Stelleinheit *f*
air-controlled hydraulic pilot operator *VALV OP* pneumohydraulische Stelleinheit *f*
air conveying *APPL* pneumatisches Fördern *n*
air-cooled heat exchanger Öl-Luft-Wärmeübertrager *m*
air-cooled oil cooler luftgekühlter Ölkühler *m*, Luft-Öl- Kühler *m*
air-cooling Luftkühlung *f*
air current Luftstrom *m*
air cylinder Pneumatikzylinder *m*, Druckluftzylinder *m*, Luftzylinder *m*
air demand Druckluftbedarf *m*, Luftbedarf *m*
air de-trainer plate *RESERVOIRS* Entlüftungsblech *n*
air distribution system *LINES* Druckluftverteilungssystem *n*, Luftverteilungssystem *n*
air drier *s.* air dryer
air drive *APPL* Pneumatikantrieb *m*, pneumatischer Antrieb *m*, Druckluftantrieb *m*
air dryer *PREPAR* Drucklufttrockner *m*
air emulsion *FLUIDS* Aeroemulsion *f*
air film handling *APPL* Luftfilmtransport *m*, Luftfilmförderung *f*
air filter *HY* Luftfilter *m,n*; *PN* Druckluftfilter *m,n*, Luftfilter *m,n*, Pneumatikfilter *m,n*
air flow Luftstrom *m*; *s.* air flow [rate]
air flow rate Luftvolumenstrom *m*, Luftstrom *m*
air gap Luftspalt *m*

air-gap solenoid *VALV OP* luftschaltender Magnet *m*, Trockenmagnet *m*
air-gap torque motor *VALV OP* trockener Torque-Motor (Drehmomentmotor) *m*
air hose Druckluftschlauch *m*, Luftschlauch *m*, Pneumatikschlauch *m*
air-hydraulic hydropneumatisch, pneumohydraulisch
air-hydraulic accumulator luftbelasteter (hydropneumatischer) Speicher *m*, Luftdruckspeicher *m*
air-hydraulic intensifier pneumohydraulischer Druckübersetzer *m*
air-hydraulic pump hydropneumatische Pumpe *f*, Drucklufthydraulikpumpe *f*
air-hydraulics Hydropneumatik *f*, Pneumohydraulik *f*
air inclusion Luftsack *m*, Lufteinschluß *m*
air-in-oil emulsion *FLUIDS* Aeroemulsion *f*
air in solution *FLUIDS* gelöste Luft *f*
air jet *LOG EL* Luftstrahl *m*
air leak in a hydraulic system Lufteindringstelle *f* in einer Hydraulikanlage
air leakage Luftleckage *f*, Luftleckverlust *m*
air line Druckluftleitung *f*, Luftleitung *f*, Pneumatikleitung *f*
air line drain Luftleitungsablaß *m*
air line filter Druckluftfilter *m,n*, Luftfilter *m,n*, Pneumatikfilter *m,n*
air line lubricator *PREPAR* Druckluftölgerät *n*, Druckluftöler *m*
air-loaded accumulutor luftbelaste-

ter (hydropneumatischer) Speicher *m*, Luftdruckspeicher *m*
air logic control *LOG EL* pneumatische Logik *f* (Digitaltechnik *f*)
air lubricator *PREPAR* Druckluftölgerät *n*, Druckluftöler *m*
air mains Druckluftnetz *n*, Luftnetz *n*, Pneumatiknetz *n*
air motor Druckluftmotor *m*, Luftmotor *m*
air-oil accumulator luftbelasteter (hydropneumatischer) Speicher *m*, Luftdruckspeicher *m*
air-oil heat exchanger Öl-Luft-Wärmeübertrager *m*
air-oil intensifier pneumohydraulischer Druckübersetzer *m*
air-operated pneumatisch angetrieben, druckluftgetrieben, druckluftbetrieben, luftgetrieben
air-operated pump hydropneumatische Pumpe *f*, Drucklufthydraulikpumpe *f*
air operator *VALV OP* pneumatische Stelleinheit (Betätigungseinrichtung) *f*, Druckluftstelleinheit *f*
air-over-oil tank Behälter *m* ohne Trennmittel
air-piloted *VALV OP* druckluftvorgesteuert, pneumatisch vorgesteuert
air port Pneumatikanschluß *m*, Druckluftanschluß *m*, Luftanschluß *m*
air-powered pneumatisch angetrieben, druckluftgetrieben, druckluftbetrieben, luftgetrieben
air-powered pump hydropneumatische Pumpe *f*, Druckluftpumpe *f*
air receiver Druckluftspeicher *m*, Druckluftbehälter *m*, Windkessel *m*
air release *ability of air separation*
FLUIDS Luftausscheidevermögen *n*, Luftabscheidevermögen *n*
air sampling Luftprobenahme *f*
air solubility *FLUIDS* Luftlösungsvermögen *n*
air spring Luftfeder *f*
air stream Luftstrom *m*; Luftstrahl *m*
air supply Druckluftzufuhr *f*, Luftzufuhr *f*
air system Pneumatikschaltung *f*, Pneumatiksystem *n*, Druckluftschaltung *f*, Druckluftsystem *n*, Luftsystem *n*; Pneumatikanlage *f*, Pneumatiksystem *n*, Pneumatik *f*, Druckluftanlage *f*
air-to-air intensifier reinpneumatischer Druckübersetzer *m*
air tool *APPL* Druckluftwerkzeug *n*
air-tool fitting Werkzeugverschraubung *f*
air trap Luftsack *m*, Lufteinschluß *m*
air under pressure Druckluft *f*, Luft *f* unter Druck
air valve Pneumatikventil *n*, Druckluftventil *n*, Luftventil *n*
airborne aus der Luft stammend, in der Luft enthalten, Luft ...
aircraft fluid Luftfahrtflüssigkeit *f*, Flugzeugflüssigkeit *f*
aircraft hydraulics *APPL* Luftfahrthydraulik *f*, Flugzeughydraulik *f*
airdraulic *s.* air-hydraulic
airdraulics *s.* air-hydraulics
airline Druckluftleitung *f*, Luftleitung *f*, Pneumatikleitung *f*
airtight luftdicht
AIT *autoignition temperature*
FLUIDS Selbstentzündungstemperatur *f*

ALC *air logic control* *LOG EL* pneumatische Logik (Digitaltechnik) *f*
all-hydraulic vollhydraulisch, reinhydraulisch
all-pneumatic vollpneumatisch, reinpneumatisch
alternating current solenoid *VALV OP* Wechselstrommagnet *m*
alternating-fluid hydraulics Wechselstromhydraulik *f*
aluminium tubing *GB* Aluminiumrohr *n*, Alu-Rohr *n*
aluminum tubing *USA* *s.* aluminium tubing
ambient pressure Umgebungsdruck *m*
ambient temperature Umgebungstemperatur *f*
amplifier *s.* hydraulic amplifier
amplify *v* *pressure* verstärken Druck
analog fluidic device *LOG EL* analoges Fluidikelement *n*
angle check valve Eck-Rückschlagventil *n*
angle mount *CYL* Winkelbefestigung *f*
angle of overlap *DCV* Überdeckungswinkel *m*
angle of underlap *DCV* Unterdeckungswinkel *m*
angle plate *PU/MOT* Steuerscheibe *f*, Hubscheibe *f*
angle valve Eckventil *n*
angled piston motor Axialkolbenmotor *m* mit abgewinkelter Hauptachse, Schrägtrommelmotor *m*, Schwenkkopfmotor *m*
angled piston pump Axialkolbenpumpe *f* mit abgewinkelter Hauptachse, Schrägtrommelpumpe *f*, Schwenkkopfpumpe *f*

anhydrous fluid nicht-wasserhaltige (wasserfreie) Flüssigkeit *f*
aniline point *FLUIDS* Anilinpunkt *m*, AP *m*
annular gap *THEOR* Ringspalt *m*
annular groove *DCV* Ringnut *f*, Ringkanal *m*
annular orifice *FCV* Ringdrossel *f*
annular piston *CYL* Ringkolben *m*
annulus *CYL* Ringraum *m*; *CYL* Ringraumfläche *f*, Ringraumquerschnitt *m*; *s.* annular gap; *s.* annular groove
annulus [area] *CYL* Ringraumfläche *f*, Ringraumquerschnitt *m*
annulus side *CYL* Ringraumseite *f*
ANSI = American National Standards Institute
anti-cavitation valve Kavitationsschutzventil *n*
anti-corrosion additive *FLUIDS* Korrosionsinhibitor *m*, Korrosionsschutzadditiv *n*
anti-corrosive power *FLUIDS* Korrosionsschutzfähigkeit *f*, Korrosionsschutzvermögen *n*
anti-extrusion ring *SEALS* Stützring *m*, Back-up-Ring *m*, Kammerungsring *m*
anti-foaming additive *FLUIDS* Anti-Schaum-Additiv *n*, Entschäumadditiv *n*, Schaumhemmer *m*
anti-freeze [additive] *FLUIDS* Frostschutzadditiv *n*, Gefrierschutzadditiv *n*, Frostschutz *m*, Gefrierschutz *m*
anti-freeze lubricant *PREPAR* Vereisungsschutzschmierstoff *m*, Enteisungsschmierstoff *m*
anti-leak additive *FLUIDS* Leckageinhibitor *m*

anti-oxidant *FLUIDS* Oxidationsinhibitor *m*, Oxidationshemmer *m*
anti-rotation device *CYL* Verdrehsicherung *f*
anti-rust additive *FLUIDS* Rostinhibitor *m*, Rostschutzadditiv *n*
anti-void valve Unterdruckbegrenzungsventil *n*
anti-wear additive *FLUIDS* Verschleißinhibitor *m*
anti-wear oil Verschleißschutzöl *n*
anticav *s.* anti-cavitation valve
APC *automatic particle counter* automatischer Partikelzähler *m*, APZ *m*, APC *m*
apply *v* **pressure** *eg to a chamber, at an area* Druck *m* zuführen *z. B. einem Raum, einer Fläche*, mit Druck *m* beaufschlagen *z. B. einen Raum, eine Fläche, als Adj.* auch zusammengesetzt: druckbeaufschlagt
apply *v* **pressure fluid** *eg to an actuator, at an area* Druckflüssigkeit *f* zuführen *z. B. einem Verbraucher, einer Fläche*, mit Druckflüssigkeit *f* beaufschlagen *z. B. einen Verbraucher, eine Fläche*
aqueous fluid wasserhaltige (wäßrige, wasserbasische) Flüssigkeit *f*
area characteristics *of a restriction THEOR* Drosselcharakteristik *f*, Drosselverhalten *n*, Öffnungscharakteristik *f*
area ratio *CYL* Flächenverhältnis *n*
armature *s.* solenoid armature
ash content *FLUIDS* Aschegehalt *m*
aspirator *of a pressure regulator* Kompensationsdüse *f*, Saugrohr *n* im Druckminderventil
ATF *automatic transmission fluid* ATF-Öl *n* *für Drehmomentwandler*

atmosphere: open to atmosphere *HY* mit dem Behälter *m* verbunden, drucklos; *PN* mit der Atmosphäre *f* verbunden, nach außen entlüftend
atmospheric *nonpressurized* drucklos, nicht unter Druck *m*, nicht vorgespannt
atmospheric air atmosphärische Luft *f*, Luft *f* im Ansaugzustand
atmospheric pressure atmosphärischer Druck *m*, Atmosphärendruck *m*, Umgebungsluftdruck *m*
atmospheric return line drucklose Rücklaufleitung *f*
attached jet *LOG EL* Wandstrahl *m*
attenuate *v* *vibrations* dämpfen *Schwingungen*
auto[genous]-ignition temperature *FLUIDS* Selbstentzündungstemperatur *f*
automatic drain *PREPAR* automatischer (selbsttätiger) Ablaß *m*
automatic particle counter automatischer Partikelzähler *m*, APZ *m*, APC *m*
automatic seal selbstwirkende (druckgespannte) Dichtung *f*
automatic shut-off valve *FCV* Leitungsbruchventil *f*, Rohrbruchventil *f*
automatic transmission fluid ATF-Öl *n* *für Drehmomentwandler*
automotive hydraulics *APPL* Kraftfahrzeughydraulik *f*, Kfz-Hydraulik *f*, Fahrzeughydraulik *f*
auxiliary circuit Hilfskreislauf *m*
auxiliary control *VALV OP* Hilfsbetätigung *f*, Zusatzbetätigung *f*
auxiliary pump Hilfspumpe *f*
auxiliary valve Hilfsventil *n*
A. V. *acid value FLUIDS* Neutralisationszahl *f*, NZ *f*, Säurezahl *f*, SZ *f*

average bulk modulus *FLUIDS* Sekantenkompressionsmodul *m*, mittlerer Kompressionsmodul *m*
axial[-flow] compressor Axialverdichter *m*, Axialkompressor *m*
axial notch *FCV* Längskerbe *f*, Axialkerbe *f*
axial piston motor Axialkolbenmotor *m*
axial piston pump Axialkolbenpumpe *f*
axial piston pump *or* **motor** Axialkolbenmaschine *f*, Axialkolbeneinheit *f*, Axialkolbengerät *n*
axial piston transmission Axialkolbengetriebe *n*
axial seal axiale Dichtung *f*

B

back-flush *v* *PREPAR* durch Stromumkehr *f* säubern, rückspülen
back pressure Gegendruck *m*, Staudruck *m*
back-pressure valve *PCV* Vorspannventil *n*, Gegendruckventil *n*, Widerstandsventil *n*
back pressuring Gegendruckerzeugung *f*
back support ring *female adaptor of a V-ring assembly* Druckring *m*, Sattelring *m* *eines Dichtungssatzes*
back-to-back check valves *CIRC* antiparallel geschaltete Rückschlagventile *npl*
back-up ring *SEALS* Stützring *m*, Back-up-Ring *m*, Kammerungsring *m*
backflow Rückstrom *m*
backlash valve Spielausgleichsventil *n*

backwash *v* *PREPAR* durch Stromumkehr *f* säubern, rückspülen
baffle *RESERVOIRS* Leitblech *n*, Lenkblech *n*, Umlenkblech *n*, Leitplatte *f*, Lenkplatte *f*, Umlenkplatte *f*
baffle *v* *RESERVOIRS* mit Leitblech[en], ... versehen, unterteilen
baffle plate *s.* baffle
bag *ACCUM* Speicherblase *f*
bag accumulator Blasenspeicher *m*
bag protection valve *ACCUM* Blasenschutzventil *n*
bag-type strainer *PREPAR* Siebkorb *m*, Filterkorb *m*
balance *v* **pressure** Druck *m* ausgleichen, kompensieren, vom Druck *m* entlasten
balance line Entlastungsleitung *f*, Ausgleichsleitung *f*
balanced pump druckausgeglichene (druckentlastete, druckkompensierte, ausgeglichene, entlastete) Pumpe *f*
balanced relief valve ablaufdruckentlastetes Druckbegrenzungsventil *n*, Druckbegrenzungsventil *n* mit äußerer Leckflüssigkeitsrückführung
balanced-rotor vane pumpe Flügelzellenpumpe *f* mit druckentlastetem Rotor
balancing groove *DCV* Zentriernut *f*, Druckausgleichsnut *f*
balancing valve *FCV* Gleichlaufventil *n*, Synchronisierventil *n*
ball check valve Kugelrückschlagventil *n*
ball element *LOG EL* Kugelelement *n*
ball-end piston *in axial piston pumps or motors* Kugelkolben *m* *in Axialkolbenmaschinen*

ball piston *PU/MOT* kugelförmiger Kolben *m*, Kugelkolben *m*
ball-piston motor Kugelkolbenmotor *m*
ball-piston pump Kugelkolbenpumpe *f*
ball screw rotary actuator *PU/MOT* Kugelumlaufschwenkmotor *m*, Schwenkmotor *m* mit Kugelumlaufmutter
ball-type piston *PU/MOT* kugelförmiger Kolben *m*, Kugelkolben *m*
ball-type poppet *CHECKS* Ventilkugel *m*
ball valve Kugelventil *n*; *mit kugelförmigem Steuerelement* Kugelhahn *m*, Kugeldrehschieberventil *n*
ball viscometer *MEASUR* Kugelfallviskosimeter *n*
band cylinder Bandzylinder *m*
band seal Liniendichtung *f*
banjo richtungseinstellbare Verschraubung *f*, Schwenkverschraubung *f*
bank *valve bank* Ventilbatterie *f*, *wenn höhenverkettet* Steuersäule *f*
bank *v* *valves* verketten *Ventile*
banking *valves* verkettungsfähig, verkettbar *Ventile*
barb fitting Stecknippelverbindung *f*, Wulstnippelverbindung *f*
barbed nipple *FITT* Wulstnippel *n*
barrel *CYL* Zylinderrohr *n*, Zylindermantel *m*
barrel slot Zylindermantelschlitz *m*
barrel wall Zylinderwand *f*
base-mounted *VALV* unterflächenmontiert, *s auch* surface-mounted
basic circuit Grundschaltung *f*, Grundkreislauf *m*
basic symbol Grundsymbol *n*, Grundschaltzeichen *n*

bayonet-type coupling *FITT* Bajonettverschlußkupplung *f*
beaded insert *FITT* Wulstnippel *n*
beam Strahl *m*
beam deflection element *LOG EL* Strahlablenkelement *n*
bearing length *CYL* Führungslänge *f*
bearing ring *CYL* Führungsring *m*, Gleitring *m*
bellows *SEALS* Faltenbalg *m*
bellows accumulator Faltenbalgspeicher *m*, Balgspeicher *m*
bellows cylinder Balgzylinder *m*
bellows pressure gauge Balgfedermanometer *n*
bellows pressure transducer Balgdruckwandler *m*
bellows pressure switch Balgdruckschalter *m*
bend *pipe bend* Rohrkrümmer *m*, Rohrbogen *m*
bend loss *THEOR* Krümmerverlust *m*, Bogenverlust *m*, *s auch DE* Knieverlust *m*
bend radius *LINES* Biegeradius *m*
bender Rohrbiegevorrichtung *f*
bent-axis axial piston motor Axialkolbenmotor *m* mit abgewinkelter Hauptachse, Schrägtrommelmotor *m*, Schwenkkopfmotor *m*
bent-axis axial piston pump Axialkolbenpumpe *f* mit abgewinkelter Hauptachse, Schrägtrommelpumpe *f*, Schwenkkopfpumpe *f*
Bernoulli's equation *THEOR* Bernoulli-Gleichung *f*
Bernoulli's law *of energy conservation in a flowing fluid* Bernoullisches Gesetz *n* Energieerhaltung bei strömendem Medium
beta ratio *(β for X μm particle size*

= *particles > X μm upstream / particles > X μm downstream)* Beta-Wert *m*, β-Wert *m*, Beta-Verhältnis *n* ein Maß für den Filterwirkungsgrad
bevel washer seal Lippenscheibe *f*
bidirectional filter Umkehrfilter *m,n* automatischer Elementewechsel bei Stromumkehr
bidirectional motor Umsteuermotor *m*
bidirectional pump Pumpe *f* für umkehrbare Drehrichtung, Umsteuerpumpe *f*
bimetallic thermometer Bimetallthermometer *n*
binding *of the valve spool DCV* hydraulisches Verklemmen *n* (Klemmen *n*, Verkleben *n*, Kleben *n*) *des Steuerkolbens*
biodegradable *FLUIDS* biologisch (biochemisch) abbaubar
birotational motor Umsteuermotor *m*
birotational pump Pumpe *f* für umkehrbare Drehrichtung, Umsteuerpumpe *f*
bite fitting Schneidringverschraubung *f*
bladder *ACCUM* Speicherblase *f*
bladder accumulator Blasenspeicher *m*
bladder protection valve *ACCUM* Blasenschutzventil *n*
bladder rotary actuator Blasenschwenkmotor *m*
blank end *CYL* Kolbenseite *f*, Bodenseite *f*; kolbenseitiges (bodenseitiges) Zylinderende *n*; *s auch* blank end chamber
blank end chamber *CYL* Kolbenraum *m*

blank end pressure *CYL* kolbenseitiger (bodenseitiger) Druck *m*
blanking plate *of a valve stack* Sperrplatte *f*, Abschlußplatte *f*, Dichtungsplatte *f* einer Steuersäule
Blasius' law *of fluid resistance* Blasiussches Gesetz *n* Strömungswiderstand
bleed *v* im Nebenschluß *m* ableiten, umleiten; Abluft *f* ableiten, entlüften; entlüften *ungelöste Luft entfernen*
bleed line Entlüftungsleitung *f*
bleed-off *s.* bleed *v*; *s.* bleed-off fluid
bleed-off circuit Nebenstromkreislauf *m*, Nebenschlußkreislauf *m*, Bypassschaltung *f*
bleed-off filter Filter *m,n* im Abzweig zum Behälter
bleed-off flow control Nebenstromsteuerung *f*, Bypass-Stromsteuerung *f*, Abzweigstromsteuerung *f*
bleed-off fluid im Nebenschluß abgeleitete Flüssigkeit *f*
bleed-off system *LOG EL* Entlüftungssteuerung *f*
bleed-off valve Überströmventil *n*
bleed-on system *LOG EL* Belüftungssteuerung *f*
bleed-operated *VALV OP* druckentlastungsgesteuert
bleed throttle *for exhaust air* Entlüftungsdrossel *f*, Abluftdrossel *f*
bleeder *for entrained air* Entlüftungseinrichtung *f*, Entlüfter *m*, Entlüftung *f* *für ungelöste Luft*
bleeder hole (port) *for entrained air* Entlüftungsöffnung *f*, Entlüftungsbohrung *f* *für ungelöste Luft*

bleeding *of entrained air* Entlüftung *f* Entfernung ungelöster Luf
bleeding valve Entlüftungsventil *n*
blind end *CYL* Kolbenseite *f*, Bodenseite *f*; kolbenseitiges (bodenseitiges) Zylinderende *n*; *s.* blind end chamber
blind end chamber *CYL* Kolbenraum *m*
blind end pressure *CYL* kolbenseitiger (bodenseitiger) Druck *m*
blind run *of a pipeline* Blindstück *n*, Blindabschnitt *m einer Rohrleitung*
block *VALV* Blockventil *n*, Ventilblock *m*, Steuerblock *m*
block *v eg flow, line* absperren, sperren, blockieren *z. B. Strom, Leitung*
block diagram *CIRC* Blockschaltbild *n*
block-head cylinder Zylinder *m* mit Quadratflansch *bzw.* Quadratkopf
block off *v eg flow, line* absperren, sperren,
blockieren *z. B. Strom, Leitung*
blow [off] *v PCV* ansprechen, sich öffnen, öffnen
blow-out disk *PCV* Berstscheibe *f*, Berstmembran *f*, Reißscheibe *f*
blowby air innerer Leckverlust *m* Luft
X-bodied *CYL* mit Körper aus *X*
body *CYL* Zylinderkörper *m*; *FILTERS* Filtergehäuse *n*, Filterkörper *m*; *VALV* Ventilgehäuse *n*, Ventilkörper *m*
boiling point *FLUIDS* Siedepunkt *m*, Siedetemperatur *f*
boundary layer *THEOR* Grenzschicht *f*

bonded fabric filter Vliesfilter *m,n*
bonded seal (washer) Verbundstoffdichtung *f*
boost *v pump* vorfüllen, füllen *Pumpe*; *pressure* verstärken *Druck*
boost pressure ratio *INTENS* Druckübersetzungsverhältnis *n*
boost pump *in a closed circuit* Spülpumpe *f*, Hilfspumpe *f in einem geschlossenen Kreislauf*
booster Druckübersetzer *m*
booster-operated druckübersetzergespeist
booster pump Speisepumpe *f*, Füllpumpe *f*, Vorfüllpumpe *f*, Zuförderpumpe *f*, Ladepumpe *f*
boot *SEALS* Faltenbalg *m*
bore Innendurchmesser *m*, ID; *s auch* cylinder bore
bore area *CYL* Zylinderquerschnittsfläche *f*, Zylinderfläche *f*
bore size Innendurchmesser *m*, ID; *s auch* cylinder bore size
bottom *CYL* Zylinderboden *m*, Zylinderfuß *m*
bottom out *v CYL* Zylinderende *n* [vollständig] erreichen
bottom valve Ablaßventil *n*, Bodenventil *n*, Fußventil *n*
Bourdon (bourdon) pressure gauge Rohrfedermanometer *n*
Bourdon pressure switch Rohrfederdruckschalter *m*
Bourdon pressure transducer Rohrfederdruckwandler *m*
Bourdon spring (tube) *MEASUR* Rohrfeder *f*
bowl filter Glockenfilter *m,n*
b. p. *boiling point FLUIDS* Siedepunkt *m*, Siedetemperatur *f*

bracket mounting *CYL, VALV*
Konsolbefestigung *f*
braking duty *PU/MOT* Bremsbetrieb *m*
braking valve Bremsventil *n*
branch Zweig *m* *eines Kreislaufs*; Abzweig *m* *einer T-Verschraubung*
branch *v* *a flow* abzweigen *einen Strom*
branch circuit Zweigkreislauf *m*
branch flow [rate] Zweig[volumen]strom *m*
branching loss *THEOR* Verzweigungsverlust *m*
brazed fitting Lötverbindung *f*, Lötverschraubung *f*
brazeless fitting lötlose Verbindung (Verschraubung) *f*
breakaway force *CYL* Losbrechkraft *f*
breakaway torque *PU/MOT* Losbrechmoment *n*
breakaway-type quick-disconnect coupling Schlauchkupplung *f* mit Abreißsicherung
breathe *v* *RESERVOIRS* belüften, entlüften
breather *RESERVOIRS* Belüftungsorgan *n*, Entlüftungsorgan *n*, Lüftungsorgan *n*
breather-filler *RESERVOIRS* Einfüll-Belüftungs- Kombination *f*
breather hole *RESERVOIRS* Belüftungsöffnung *f*, Entlüftungsöffnung *f*, Lüftungsöffnung *f*
breathing *RESERVOIRS* Belüftung *f*, Entlüftung *f*, Lüftung *f*
bridge circuit Brückenschaltung *f*
Briggs thread *s.* NPT thread
British Standard Pipe thread *FITT* BSP-Rohrgewinde *n* *keglig bzw. keglig/gerade*

BSP thread *s.* British Standard Pipe thread
bubble leakage *immersion testing of seals* Blasenleckage *f*, Blasenstrom *m* *bei der Tauchprüfung von Dichtungen*
bubble-point test *of filters* Blasendruckprüfung *f*, Blasenpunkttest *m* *von Filtern*
bubble test *s.* bubble-point test
bubble-tight gasdicht; druckdicht
buckling lenght *CYL* Knicklänge *f*
buckling resistance (strength) *CYL* Knickfestigkeit *f*
buffer Stoßdämpfer *m*, Stoßfänger *m*, Prallfänger *m*
built-in check valve eingebautes Rückschlagventil *n*
built-in contamination *FLUIDS* Montageverschmutzung *f*, eingebaute Verschmutzung *f*
bulk filtration *PREPAR* Außenfilterung *f*, Spülung *f*
bulk liquid Flüssigkeitskörper *m*
bulk modulus of elasticity *FLUIDS* Kompressionsmodul *m*, Elastizitätsmodul *m*
bulkhead branch tee T-Verschraubung *f* mit Schottabzweig
bulkhead connector Schottverschraubung *f*
bulkhead elbow Winkel-Schottverschraubung *f*
bulkhead fitting Schottverschraubung *f*
bulkhead side tee T-Verschraubung *f* mit Schottabzweig
bulkhead union Schottverschraubung *f* mit zweiseitigem Rohranschluß

Bunsen absorption coefficient
FLUIDS Bunsenscher Lösungskoeffizient *m*, Bunsenkoeffizient *m*
burst filter element geborstenes Filterelement *n*
burst pressure *FILTERS* Berstdruck *m*, Platzdruck *m*
butt-joint fitting Stoßverschraubung *f*
butt welding elbow *FITT* Verschweißwinkel *m*
butt welding fitting Verschweißverschraubung *f*
button-actuated *VALV OP* druckknopfbetätigt
bypass Nebenschluß *m*, Umgehung *f*, Bypass *m* (*pl* Bypässe), Nebenstrom *m*
bypass *vt* im Nebenschluß *m* ableiten, umleiten; *vi* im Nebenschluß *m* fließen, umgehen
bypass circuit Nebenstromkreislauf *m*, Nebenschlußkreislauf *m*, Bypassschaltung *f*
bypass cracking pressure *of a filter when clogged* Öffnungsdruck *m* des Umgehungsventils *des Filters bei Verschmutzung*
bypass filter Filter *m,n* mit Umgehung, Umgehungsfilter *m,n*; *s.* bypass-flow filter
bypass filtration Nebenstromfilterung *f*
bypass-flow control Nebenstromsteuerung *f*, Bypass- Stromsteuerung *f*, Abzweigstromsteuerung *f*
bypass-flow control valve Dreiwege-Stromregelventil *n*, Dreiwege-Strombegrenzungsventil *n*
bypass-flow filter Filter *m,n* im Nebenstrom, Nebenstromfilter *m,n*

bypass-flow filtration Nebenstromfilterung *m,n*
bypass flow regulator Dreiwege-Stromregelventil *n*, Dreiwege-Strombegrenzungsventil *n*
bypass installation Einbau *m* (Anordnung *f*) im Nebenstrom
bypass line Umgehungsleitung *f*
bypass port Entlastungsanschluß *m*
bypass valve *of a filter* Kurzschlußventil *n*, Umgehungsventil *n*, Bypassventil *n* *eines Filters*; *s auch* bleed-off valve

C

C-ring *SEALS* C-Ring *m*
C-type Bourdon tube *MEASUR* C-förmige Rohrfeder *f*, Kreisformrohrfeder *m*
cable cylinder Zugseilzylinder *m*, Kabelzylinder *m*
calibration fluid Kalibrierfluid *n*, Eichfluid *n*
cam-actuated (-operated) valve Nockenventil *n*
cam plate *PU/MOT* Steuerscheibe *f*, Hubscheibe *f*
cam plate actuated (operated) valve kurvenscheibenbetätigtes Ventil *n*
cam ring *of a vane pump or motor* Leitring *m*, Gehäusering *m*, Führungsring *m* *in einer Flügelzellenmaschine*
cam valve Nockenventil *n*
can filter Glockenfilter *m,n*
capacitance *THEOR* Kapazität *f*, kapazitiver Widerstand *m*
capacitance pressure transducer *MEASUR* kapazitiver Druckwandler *m*

capacity *flow capacity* Durchflußkapazität *f*, Durchflußleistung *f*; *ACCUM* Speichernutzvolumen *n*
capacity rating *PU/MOT* Nennverdrängungsvolumen *n*; Nennvolumenstrom *m*, Nenndurchflußstrom *m*, Nenndurchfluß *m*, Nennstrom *m*; Nennleistung *f* abgegebene Leistung
capacity-to-size ratio Leistungs-Abmessungs-Verhältnis *n*
cap end *CYL* Kolbenseite *f*, Bodenseite *f*; kolbenseitiges (bodenseitiges) Zylinderende *n*; *s auch* cap end chamber
cap end chamber *CYL* Kolbenraum *m*
cap end pressure *CYL* kolbenseitiger (bodenseitiger) Druck *m*
cap mount *CYL* bodenseitige Befestigung *f*
capillary Kapillarrohr *n*, Kapillare *f*
capillary-action lubricator *PREPAR* Kapillarnebelöler *m*
capillary viscometer Auslaufviskosimeter *n*, Kapillarviskosimeter *n*, Ausflußviskosimeter *n*
capped seal Dichtung *f* mit Verschleißring
capsule pressure gauge Kapselfedermanometer *n*
capsule pressure transducer Kapselfederdruckwandler *m*
captive seal [ein]gefaßte Dichtung *f*
carry-over *of oil PREPAR* Öleintrag *m*
cartridge *FILTERS* Filtereinsatz *m*, Filterpatrone *f*; *SEALS* Einbaudichtsatz *m*, Einbaudichtung *f*
cartridge check [valve] Rückschlagventil *n* für Bohrungseinbau, Einbaurückschlagventil *n*
cartridge flow valve Stromventil *n* für Bohrungseinbau, Einbaustromventil *n*
cartridge insert valve Ventil *n* für Bohrungseinbau, Einbauventil *n*
cartridge logic valve Logikventil *n*, Zweiwege-Einbauventil *n*, Wegesitzventil *n*, Cartridge *n*, Hydrologikventil *n*
cartridge pump Einbaupumpe *f*, Steckpumpe *f*
cartridge valve *s.* cartridge insert valve; *s.* cartridge logic valve
cartridge valve cavity Ventilaufnahmebohrung *f*, Ventilaufnahme *f*
cartridge valve manifold Ventileinbauplatte *f*
cascade *v* hintereinanderschalten, in Serie (Reihe) *f* schalten
case drain *PU/MOT* Gehäuseleckanschluß *m*, Gehäuseablaß *m*
cased seal [ein]gefaßte Dichtung *f*
cast iron cylinder Graugußzylinder *m*, Gußzylinder *m*
cast steel cylinder Stahlgußzylinder *m*, Gußzylinder *m*
castor-base fluid rizinusbasische Flüssigkeit *f*
cast reservoir (tank) Gußbehälter *m*
cavitate *v* kavitieren
cavitation Kavitation *f*
  cavitation loss Kavitationsverlust *m*
celerity Druckwellengeschwindigkeit *f*
cellulose filter Zellulosefilter *m,n*
centered position *DCV* Mittelstellung *f*, *s auch* neutral position
  centered spool zentrisch angeordneter Kolben *m*

centering groove *DCV* Zentriernut *f*, Druckausgleichnut *f*
centering spring *VALV OP* Zentrierfeder *f*
centerline lugs mount *CYL* Befestigung *f* an Laschen in Höhe der Zylinderachse
centerline mount *CYL* Befestigung *f* in einer Ebene mit der Zylinderachse
center position *DCV* Mittelstellung *f*, *s auch* neutral position
central compressed-air source *APPL* zentrale Druckluftversorgung *f*
central hydraulic system *APPL* Zentralhydraulik *f*
centrally guided radial-piston pump Radialkolbenpumpe *f* mit innerer Kolbenabstützung (mit innenliegender Hubkurve), Exzentermaschine *f*
centrally ported radial-piston pump innenbeaufschlagte Radialkolbenpumpe *f*
centrifugal compressor Radialverdichter *m*
centrifugal dryer *PREPAR* Zentrifugaltrockner *m*, Zentrifugalabscheider *m*
centrifugal filter Zentrifugalfilter *m,n*
centrifugal pump Strömungspumpe *f*, Zentrifugalpumpe *f*, Kreiselpumpe *f*, Turbopumpe *f*
ceramic filter Keramikfilter *m,n*
ceramic valve Keramikventil *f*
CETOP = Comité Européen des Transmissions Oléohydrauliques et Pneumatiques
cfm *cubic feet per minute* Einheit des Volumenstroms: 1 cfm = $472 \cdot 10^{-6}$ m^3/s = 28,32 L/min

chain rotary actuator Kettenkolbenschwenkmotor *m*
chain-return cylinder Kettenrückzugzylinder *m*
chamber *of a vane pump or motor* Zelle *f* einer Flügelzellenmaschine
chamber configuration *DCV* Kanalsystem *n*, Kammersystem *n*
change of delivery direction *PU/MOT* Förderrichtungswechsel *m*, Förderrichtungsumkehr *f*
change of direction of rotation *PU/MOT* Drehrichtungswechsel *m*, Drehrichtungsumkehr *f*
change of discharge (output) direction *PU/MOT* Förderrichtungswechsel *m*, Förderrichtungsumkehr *f*
change of sectional area *THEOR* Querschnittsänderung *f*
channel *LINES* Strömungsweg *m*, Kanal *m*
characteristic diameter *THEOR* hydraulischer Durchmesser *m*
characteristic radius *THEOR* hydraulischer Radius *m*
charge *v* *with fluid, gas ACCUM* aufladen, laden, füllen *mit Flüssigkeit, Gas*; *v* with pressure Druck *m* zuführen, mit Druck *m* beaufschlagen, *als Adj. auch zusammengesetzt:* druckbeaufschlagt
charge pressure relief valve *in a closed circuit* Speisedruckbegrenzungsventil *n* *in einem geschlossenen Kreislauf*
charge pump Speisepumpe *f*, Füllpumpe *f*, Vorfüllpumpe *f*, Zuförderpumpe *f*, Ladepumpe *f*
charging *ACCUM* Aufladung *f*, Ladung *f*, Füllung *f*

charging line Fülleitung *f*, Vorfüllleitung *f*, Zuförderleitung *f*
charging pressure *ACCUM* Fülldruck *m*, Aufladedruck *m*, Vorspanndruck *m*
charging pump *s.* charge pump
chatter *v* *eg valves* rattern, schnarren, flattern *z. B. Ventile*
check [valve] Sperrventil *n*; *nonreturn valve* Rückschlagventil *n*
check-valve pump druckgesteuerte (sitzventilgesteuerte) Pumpe *f*
chevron packing *SEALS* Dachringsatz *m*, Dachmanschettensatz *m*
chevron ring *SEALS* Dachring *m*, Dachmanschette *f*
chlorinated hydrocarbon fluid Flüssigkeit *f* auf Basis chlorierter Kohlenwasserstoffe
choke Drossel[stelle] *f*, Strömungswiderstand *m*; *laminar-type restriction THEOR* Laminarwiderstand *m*, Drossel *f*
choke *v* *flow* drosseln *Volumenstrom*
choke block *of a valve stack* Drosselplatte *f* *einer Steuersäule*
choke length *THEOR* Drossellänge *f*, Drosselstrecke *f*
choked flow *critical airflow through restriction THEOR* kritische Geschwindigkeit *f*, Laval-Geschwindigkeit *f* *Luftstrom durch Drosselstelle*
CHP *corner horsepower PU/MOT* Eckleistung *f*
churning oil *in an tank* Schwallöl *n* in einem Behälter
cim *cubic inches per minute* Einheit des Volumenstroms: $1\ cim = 0{,}273 \cdot 10^{-6}\ m^3/s = 16{,}39 \cdot 10^{-3}\ L/min$

CIP *cold isostatic pressing APPL* isostatisches Kaltpressen *n*
circuit Kreislauf *m*, System *n*, Schaltung *f*
circuit design Kreislaufprojektierung *f*, Kreislaufentwurf *m*
circuit diagram (drawing) Funktionsschaltplan *m*, Schaltplan *m*
circuit manifold *VALV* maschinengebundene (kreislaufgebundene) Unterplatte *f*, Batterieplatte *f*, Sammelanschlußplatte *f*
circuit pressure Kreislaufdruck *m*
circuit tool Zeichenschablone *f* *für Schaltzeichen*
circuitry Kreislauf *m*, System *n*, Schaltung *f*; *s auch* circuit diagram
circulate *v* *fluid* umlaufen, zirkulieren *Flüssigkeit*; umlaufen (zirkulieren) lassen *Flüssigkeit*
circumferential groove *DCV* Ringnut *f*, Ringkanal *m*; *FCV* Umfangskerbe *f*
cis *cubic inches per second* Einheit des Volumenstroms: $1\ cis = 16{,}39 \cdot 10^{-6}\ m^3/s = 0{,}98\ L/min$
clamp-type hose fitting Klemm-Schlauchverbindung *f*
clamping circuit *APPL* Spannkreislauf *m*
clamping cylinder *APPL* Spannzylinder *m*
claw coupling *FITT* Klauenkupplung *f*
clean air *PREPAR* Reinluft *f*
clean face *FILTERS* Reinseite *f*
clean pressure drop *FILTERS* Druckabfall *m* im Neuzustand
cleanable [filter element] reinigungsfähiges (regenerierbares) Filterelement *n*

cleaning fluid   Reinigungsflüssigkeit *f*, *s aber* flushing fluid
cleanliness code   *FLUIDS* Reinheitsgraduierung *f*
cleanout opening   *RESERVOIRS* Reinigungsöffnung *f*
clearance compensation   *eg in a gear pump* Spielausgleich *m*, Spaltausgleich *m*   *z. B. in der Zahnradpumpe*
clearance loss   *THEOR* Spaltverlust *m*
clearance seal   Spaltdichtung *f*
clearance volume   *CYL* schädlicher (toter) Raum *m*, Totraum *m*
clevis mount   *CYL* Schwenkgabelbefestigung *f*, Gabelbefestigung *f*
clog *v*   *small orifice* verstopfen, [sich] zusetzen, zuwachsen   *enge Drosselquerschnitte*; *FILTERS* verstopfen, zusetzen, *bei Oberflächenfiltern auch* verlegen
clogging   *FILTERS* Verstopfung *f*, Zusetzen *n*, *bei Oberflächenfiltern auch* Verlegung *f*
clogging indicator   *FILTERS* Verstopfungsanzeige *f*
clogging sensitivity   *FCV* Verstopfungsempfindlichkeit *f*, Zusetzempfindlichkeit *f*
close *vi*   *valve* [sich] schließen   *Ventil*
closed: normally closed   *DCV* in Ruhestellung *f* geschlossen
closed-center circuit   Kreislauf *m* mit Sperrstellung
closed-center position   *DCV* Sperrstellung *f*
closed-center valve   Ventil *n* mit Sperrstellung
closed circuit   geschlossener Kreislauf *m*

closed-circuit hydrostatic transmission   hydrostatisches Getriebe *n* mit geschlossenem Kreislauf
closed-circuit relief valve   Kreislaufsicherheitsventil *n*
closed-crossover valve   Ventil *n* mit positiver Schaltüberdeckung
closed end   *of a conduit THEOR* festes Leitungsende *n*
closed-loop circuit   geschlossener Kreislauf *m*
closed-loop tester   *torque and power* Verspannungsprüfstand *m*
closed position   *DCV* gesperrte Stellung *f*, Sperrstellung *f*
closed termination   *of a conduit THEOR* festes Leitungsende *n*
closing pressure   *PCV* Schließdruck *m*
closing pressure surge   Schließdruckstoß *m*, Schließdruckschlag *m*
closing shock   *s.* closing pressure surge
closing stroke   *PCV* Schließweg *m*
closing time   *VALV* Schließzeit *f*
cloud point   *FLUIDS* Trübungspunkt *m*
cluster fitting   Rohrgruppenverbindung *f*, Multikupplung *f*, Kupplungsträger *m*
coalescing filter   Koaleszenzfilter *m,n*
Coanda effect   *LOG EL* Coanda-Effekt *m*, Wandhafteffekt *m*
Coanda effect element   *LOG EL* Wandstrahlelement *n*, Haftstrahlelement *n*, Coanda-Element *n*
coarse filter   Grobfilter *m,n*
coarse filtration   Grobfilterung *f*
coarse-meshed   grobmaschig
cocoon   *sound-proof enclosure PREPAR* schalldichtes Gehäuse *n*, Schallschutzhaube *f*

coil *VALV OP* Magnetspule *f*
coil tube *LINES* Spiralrohr *n*
cold isostatic pressing *APPL* isostatisches Kaltpressen *n*
cold start *s.* cold-temperature start
cold-start viscosity *PU/MOT* Startviskosität *f*
cold-temperature start Kaltanlauf *m*, Kaltstart *m*
collapse-burst test *FILTERS* Zusammenbruch-/Berstfestigkeitsprüfung *f*, Kollaps-/Berstfestigkeitsprüfung *f*
collapsed filter element zusammengebrochenes Filterelement *n*
collapsed length *CYL* Einbaulänge *f*, Einfahrlänge *f*
collapsed pressure *FILTERS* Zusammenbrechdruck *m*, Kollapsdruck *m*
collar seal Hutmanschette *f*, Innenlippenring *m*
collection chamber *LOG EL* Fangraum *m*
collection pressure *LOG EL* Fangdruck *m*
collector tube *LOG EL* Fangdüse *f*, Fangrohr *n*
column length *CYL* Knicklänge *f*
combined seal kombinierte Dichtung *f*, Zweistoffdichtung *f*
combiner valve Stromvereinigungsventil *n*
commercial pressure gauge Betriebsmanometer *n*
commutator [valve] *PU/MOT* Kommutatorventil *n*
compensate *v* *for leakage* ausgleichen, kompensieren *Leckverluste*; *pressure Druck* ausgleichen, kompensieren, *vom Druck* entlasten

compensated flow-control valve Stromregelventil *n*, Stromregler *m*, Strombegrenzungsventil *n*
compensated pump druckausgeglichene (druckentlastete, druckkompensierte, ausgeglichene, entlastete) Pumpe *f*
compensation line Entlastungsleitung *f*, Ausgleichleitung *f*
compensator *s.* pressure compensator
compensator control *s.* pressure-compensator control
compensator spool *of a flow regulator* Regelkolben *m im Stromregelventil*
complete symbol vollständiges Symbol (Schaltzeichen) *n*
component *s.* hydraulic component; *s.* pneumatic component
component life Gerätelebensdauer *f*, Bauteillebensdauer *f*
composite seal kombinierte Dichtung *f*, Zweistoffdichtung *f*
composite media filter Kombinationsfilter *m,n*
compound liquid spring Gegenkolbenflüssigkeitsfeder *f*
compress *v* verdichten, komprimieren
compressed air Druckluft *f*, *in der Werkstattumgangssprache auch* Preßluft *f*, *Zusammensetzungen s auch unter* pneumatic, pneumo-, air
compressed air circuit Pneumatikschaltung *f*, Pneumatiksystem *n*, Druckluftschaltung *f*, Druckluftsystem *n*, Luftsystem *n*
compressed air connection Pneumatikanschluß *m*, Druckluftanschluß *m*, Luftanschluß *m*
compressed air consumption Druckluftverbrauch *m*, Luftverbrauch *m*

compressed air control Pneumatiksteuerung *f*, pneumatische Steuerung *f*, Druckluftsteuerung *f*
compressed air cylinder Pneumatikzylinder *m*, Druckluftzylinder *m*, Luftzylinder *m*
compressed air demand Druckluftbedarf *m*, Luftbedarf *m*
compressed air distribution system Druckluftverteilungssystem *n*, Luftverteilungssystem *n*
compressed air drive *APPL* Pneumatikantrieb *m*, pneumatischer Antrieb *m*, Druckluftantrieb *m*
compressed air hose Druckluftschlauch *m*, Luftschlauch *m*, Pneumatikschlauch *m*
compressed air line Druckluftleitung *f*, Luftleitung *f*, Pneumatikleitung *f*
compressed air mains Druckluftnetz *n*, Luftnetz *n*, Pneumatiknetz *n*
compressed air motor Druckluftmotor *m*, Luftmotor *m*
compressed air preparation Druckluftaufbereitung *f*
compressed air supply Druckluftzufuhr *f*, Luftzufuhr *f*
compressed air system Pneumatikanlage *f*, Pneumatiksystem *n*, pneumatische Anlage *f*, Pneumatik *f*, Druckluftanlage *f*, Druckluftsystem *n*; Pneumatikschaltung *f*, Pneumatiksystem *n*, Druckluftschaltung *f*, Druckluftsystem *n*, Luftsystem *n*
compressed air valve Pneumatikventil *n*, pneumatisches Ventil *n*, Druckluftventil *n*, Luftventil *n*
compressed water Druckwasser *n*, in der Werkstattumgangssprache auch Preßwasser *n*; *Zusammensetzungen s auch unter* water
compressed water pump Druckwasserpumpe *f*, Preßwasserpumpe *f*, Preßpumpe *f*
compressibility *FLUIDS* Kompressibilität *f*
compressibility flow *THEOR* Kompressibilitätsstrom *m*, Kompressionsstrom *m*
compressibility value *FLUIDS* Preßzahl *f*, Kompressibilitätszahl *f*
compressible kompressibel, zusammendrückbar
compressible flow *THEOR* kompressible Strömung *f*
compressible-fluid shock absorber *Stoßdämpfer, bei dem eine kompressible Flüssigkeit eingesaugt und zusammengedrückt wird*
compression fitting Klemmringverschraubung *f*, Keilringverschraubung *f*, Druckringverschraubung *f*
compression seal Weichdichtung *f*
compression work *THEOR* Kompressionsarbeit *f*
compressional viscosity *FLUIDS* Druckviskosität *f*
compressor Kompressor *m*, Verdichter *m*
compressor room Kompressorraum *m*, Verdichterraum *m*
compressor stage Kompressorstufe *f*, Verdichterstufe *f*
condensate discharge (drain) Kondensatablaß *m*, Kondensatableiter *m*
conditioner *s.* air conditioner unit
conditioning *s.* air conditioning
conductance *THEOR* Leitwert *m* reeller Leitwert, Konduktanz *f*
conductor *s.* conduit

conduit Leitung *f*
confinement-controlled seal anzugsbegrenzte Dichtung *f*
conical-seal union fitting Schweißkegelverschraubung *f*, Schweißnippelverschraubung *f*
connect *v* eg to a cylinder, with port A verbinden z. B. mit dem Zylinder, anschließen z. B. an Anschluß A
connect *v* **in parallel** parallelschalten
connect *v* **in series** hintereinanderschalten, in Serie (Reihe) *f* schalten
connecting line Verbindungsleitung *f*, Verbindung *f*
***n*-connection valve** s. *n*-port valve
connector Verbindungsleitung *f*, Verbindung *f*; Endstück *n*
connector plate *VALV* Verbindungsplatte *f*, Anschlußplatte *f*
constant-deceleration cushion *CYL* Endlagenbremse (Hubendebremse) *f* mit konstanter Verzögerung, Konstantbremse *f*
constant-delivery pump nicht stellbare Pumpe *f*, Konstantpumpe *f*
constant extension speed telescopic cylinder Gleichlauf- Teleskopzylinder *m*
constant-flow circuit Kreislauf *m* mit Volumenstromquelle, Kreislauf *m* ohne Volumenstrombeeinflussung
constant-flow source (supply) Volumenstromquelle *f*, Stromquelle *f*
constant-force liquid spring Gleichkraft-Flüssigkeitsfeder *f*
constant-horsepower control of a pump Leistungsregelung *f* einer Pumpe
constant power transmission Konstantleistungsgetriebe *n*

constant-pressure source (supply) Druckquelle *f*
constant torque transmission Konstantmomentgetriebe *n*
constant-volume motor nicht stellbarer Motor *m*, Konstantmotor *m*
constant-volume pump nicht stellbare Pumpe *f*, Konstantpumpe *f*
contact pressure gauge Kontaktmanometer *n*
contact seal Berührungsdichtung *f*
contaminant *FLUIDS* Schmutz *m*, Verschmutzung *f*
contaminant-carrying capacity *FLUIDS* Schmutztragevermögen *n*, Schmutztransportvermögen *n*
contaminant concentration *FLUIDS* Schmutzkonzentration *f*, Verschmutzungskonzentration *f*
contaminant content *FLUIDS* Schmutzgehalt *m*
contaminant-free *FLUIDS* verschmutzungsfrei, schmutzfrei
contaminant-holding capacity *FILTERS* Schmutztragevermögen *n*, Schmutzaufnahmevermögen *n*
contaminant ingress Eindringen *n* von Schmutz
contaminant level *FLUIDS* Verschmutzungsgrad *m*, Schmutzveau *n*
contaminant particle *FLUIDS* Schmutzteilchen *n*, Schmutzpartikel *n*, Verschmutzungsteilchen *n*
contaminant release *FLUIDS* Schmutzabscheidevermögen *n*, Schmutzabsetzvermögen *n*
contaminant-sensitive *FLUIDS* schmutzempfindlich, verschmutzungsempfindlich
contaminant sensitivity *FLUIDS*

Schmutzempfindlichkeit *f*, Verschmutzungsempfindlichkeit *f*
contaminant sensitivity class (grade) of a component *FLUIDS* Schmutzempfindlichkeitsklasse *f* eines Geräts
contaminate *v* verunreinigen, verschmutzen
contaminated face *FILTERS* Schmutzseite *f*
contamination *FLUIDS* Schmutz *m*, Verschmutzung *f*
contamination characteristic *FLUIDS* Schmutzkennwert *m*, Verschmutzungskennwert *m*
contamination class *FLUIDS* Schmutzklasse *f*, Verschmutzungsklasse *f*
contamination code *for hydraulic fluids* Reinheitsgraduierung *f* von Hydraulikflüssigkeiten
contamination level *FLUIDS* Verschmutzungsgrad *m*, Schmutzniveau *n*
contamination-loaded *FLUIDS* schmutzhaltig, verschmutzt
contamination monitor *MEASUR* Verschmutzungsanzeiger *m*
contamination source *FLUIDS* Verschmutzungsquelle *f*, Schmutzquelle *f*
contraction loss *THEOR* Verengungsverlust *m*, Kontraktionsverlust *m*
continuity equation *THEOR* Kontinuitätsgleichung *f*
continuous intensifier Dauerstromdruckverstärker *m*
continuous operating pressure Dauerbetriebsdruck *m*, Dauerarbeitsdruck *m*
continuous pressure signal Dauerdrucksignal *n*
continuous working pressure Dauerbetriebsdruck *m*, Dauerarbeitsdruck *m*
contour ring *of a vane pump or motor* Leitring *m*, Gehäusering *m*, Führungsring *m* einer Flügelzellenmaschine
control *PU/MOT* Stelleinrichtung *f*, Stelleinheit *f*, Stellkopf *m*; *s auch* valve control
control *v VALV OP* betätigen, stellen, steuern, verstellen
control area *fluid dynamics* Kontrollgebiet *n*, Bezugsgebiet *n* Fluiddynamik
control chamber *DCV* Schieberkammer *f*, Steuerkammer *f*
control circuit Steuerkreislauf *m*, Steuerkreis *m*, Steuerschaltung *f*
control duct Steuerkanal *m*
control edge *DCV* Steuerkante *f*
control element *DCV* Ventilelement *n*, Steuerelement *n*
control flow [rate] *VALV OP* Steuervolumenstrom *m*, Steuerdurchflußstrom *m*, Steuerstrom *m*
control fluid *VALV OP* Steuerflüssigkeit *f*
control force *VALV OP* Stellkraft *f*, Betätigungskraft *f*, Verstellkraft *f*, Steuerkraft *f*, Schaltkraft *f*
control input Steuereingang *m*
control jet *LOG EL* Steuerstrahl *m*
control mechanism *s.* valve control mechanism
control nozzle *LOG EL* Steuerdüse *f*, Steuerrohr *n*
control piston *of the displacement control PU/MOT* Stellkolben *m* der Verdrängervolumenverstellung; *VALV OP* Stellkolben *m*, Betätigungskolben *m*

control plane *DCV* Steuerebene *f*, Schaltebene *f*
control port Steueranschluß *m*
control position *DCV* Schaltstellung *f*, Schieberstellung *f*
control pressure Steuerdruck *m*; *VALV OP* Stelldruck *m*, Betätigungsdruck *m*, Steuerdruck *m*
control pressure ratio *LOG EL* Steuerdruckverhältnis *n*
control return fluid Steuerrückflüssigkeit *f*
control surface *fluid dynamics* Kontrollfläche *f*, Bezugsfläche *f Fluiddynamik*
control torque *VALV OP* Stellmoment *n*, Betätigungsmoment *n*, Verstellmoment *n*, Steuermoment *n*
control tube *LOG EL* Steuerdüse *f*, Steuerrohr *n*
control valve Stromventil *n*; *auch allgemein* Ventil *n*
control volume *fluid dynamics THEOR* Kontrollvolumen *n*, Bezugsvolumen *n*
controlled-force proportional solenoid *VALV OP* kraftgesteuerter Proportionalmagnet *m*
controlled orifice *in a compensated flow-control valve* Regeldrossel *f*, Aktivdrossel *f im Stromregelventil*
controlled-stroke proportional solenoid *VALV OP* hubgesteuerter Proportionalmagnet *m*
controller *PU/MOT* Stelleinrichtung *f*, Stelleinheit *f*, Stellkopf *m*
convoluted diaphragm pressure transducer Wellmembrandruckwandler *m*
coolant *PREPAR* Kühlmittel *n*
coolant injection *COMPR* Kühlmitteleinspritzung *f*

cooling Kühlung *f*
cooling medium Kühlmittel *n*
copper-strip test *of corrosivity FLUIDS* Kupferstreifenprüfung *f* auf Korrosivität
copper tubing Kupferrohr *n*
copy *v* kopieren, nachformen
copying valve *APPL* Kopierventil *n*
corner horsepower *PU/MOT* Eckleistung *f*
corrodibility *FLUIDS* Korrosionsempfindlichkeit *f*, Korrosionsanfälligkeit *f*
corrosion inhibitor *FLUIDS* Korrosionsinhibitor *m*, Korrosionsschutzadditiv *n*
corrosive *FLUIDS* korrosiv, korrosionswirksam, korrodierend wirkend
corrosiveness *FLUIDS* Korrosionsempfindlichkeit *f*, Korrosionsanfälligkeit *f*; *s auch* corrosivity
corrosivity *FLUIDS* korrodierende Wirkung *f*, Korrosivität *f*
corrugated metal hose *LINES* Wellrohrschlauch *m*, Metallschlauch *m*
corrugated tube *LINES* Wellrohr *n*
cotton filter Baumwollfilter *m,n*
counterbalance valve Vorspannventil *n*, Gegendruckventil *n*, Widerstandsventil *n*
counterflow exchanger Gegenstromwärmeübertrager *m*
coupler *FITT* Kupplungsdose *f*
coupling *hose fitting* Schlauchverbindung *f*, Schlauchverschraubung *f*; *quick-disconnect coupling* Schlauchkupplung *f*, Schnellkupplung *f*
coupling half *FITT* Kupplungshälfte *f*
coupling nipple *s.* coupling plug

coupling plug *FITT* Kupplungsstecker *m*, Kupplungsnippel *m*
coupling socket *FITT* Kupplungsdose *f*
cover *cylinder end cover* Zylinderdeckel *m*; *hose cover* Schlauchmantel *m*, Schlauchumhüllung *f*, Schlauchdecke *m*; *reservoir cover* Behälterdeckel *m*, Behälterabdeckplatte *f*
cover-removed fitting *hose fitting* Schälverbindung *f*, Schärfverbindung *f* *Schlauchverbindung*
CPR *control pressure ratio LOG EL* Steuerdruckverhältnis *n*
crack *v PCV* ansprechen, [sich] öffnen
cracking characteristics *PCV* Ansprechverhalten *n*
cracking flow *PCV* Ansprech-[volumen]strom *m*, Ansprechdurchflußstrom *m*
cracking pressure *PCV* Ansprechdruck *m*, Öffnungsdruck *m*
crank rotary actuator Kurbelschwenkmotor *m*
creeping speed *PU/MOT* Kriechdrehzahl *f*
crescent *s.* crescent-shaped separator
crescent motor Innenzahnradmotor *m* mit Trennsichel
crescent pump Innenzahnradpumpe *f* mit Trennsichel
crescent-seal internal gear motor *s.* crescent motor
crescent-seal internal gear pump *s.* crescent pump
crescent-shaped separator *in an gear pump or motor* Trennsichel *f*, Dichtkeil *m*

crimper Verpreßmaschine *f*, Schlauchpresse *f*
crimp fitting Verpreßverbindung *f*, Quetschverbindung *f*, Preßverbindung *f*
crimping device *s.* crimper
critical flow *airflow through restriction THEOR* kritische Geschwindigkeit *f*, Laval-Geschwindigkeit *f* *Luftstrom durch Drosselstellen*
critical Reynolds' number *THEOR* kritische Reynolds-Zahl *f* (Re-Zahl *f*)
cross *FITT* Kreuzverschraubung *f*, Kreuzstück *n*
cross-bleed *DCV* Überschneidungsentlüftung *f*, Schaltentlüftung *f*
cross-drilled manifold *VALV* maschinengebundene (kreislaufgebundene) Unterplatte *f*, Batterieplatte *f*, Sammelanschlußplatte *f*
cross fitting Kreuzverschraubung *f*, Kreuzstück *n*
cross-piston leakage *CYL* innerer Leckverlust *m*
cross-port leakage *PU/MOT* Leckstrom *m* zwischen den Anschlüssen
cross-port relief valve Umsteuerentlastungsventil *n*; Kreislauf-Druckbegrenzungsventil *n*
crossover bleed *DCV* Überschneidungsentlüftung *f*, Schaltentlüftung *f*
crossover characteristics *DCV* Mittelstellungsverhalten *n*, Nulldurchgangsverhalten *n*
crossover lap *DCV* Schaltüberdeckung *f*
crossover leakage *DCV* Schaltleckstrom *m*
crossover plate *of a valve stack* Umlenkplatte *f* *einer Steuersäule*

crossover position *DCV* Mittelstellung *f*, *s auch* neutral position
crossover relief valve Umsteuerentlastungsventil *n*
crossover zone *from suction to pressure side PU/MOT* Umsteuerbereich *m* zwischen Saug- und Druckseite
cruciform ring *SEALS* X-Ring *m*
cubic feet per minute Einheit des Volumenstroms: 1 cfm = $472 \cdot 10^{-6}$ m³/s = 28,32 L/min
cubic inches per minute Einheit des Volumenstroms: 1 cim = $0,273 \cdot 10^{-6}$ m³/s = $16,39 \cdot 10^{-3}$ L/min
cubic inches per second Einheit des Volumenstroms: 1 cis = $16,39 \cdot 10^{-6}$ m³/s = 0,98 L/min
cup ring (seal) Topfmanschette *f*, Außenlippenring *m*
current Flüssigkeitsstrom *m*, Fluidstrom *m*, Strom *m*
current-to-pressure transducer Stromstärke-Druck-Wandler *m*
cushion *CYL* Endlagenbremse *f*, Hubendebremse *f*
cushion *v CYL* bremsen *am Hubende*
cushion area *CYL* Bremsfläche *f*, Bremsquerschnitt *m*
cushion cavity *s.* cushion chamber
cushion chamber *CYL* Bremsraum *m*
cushion dashpot *s.* cushion chamber
cushion plunger *CYL* Bremsansatz *m*
cushion seal *CYL* Bremsdichtung *f*
cushion spear *CYL* Bremsansatz *m*
cushion stroke *CYL* Bremsweg *m*
cushion valve Dämpfungsventil *n*

cushioning *CYL* Endlagenbremsung *f*, Hubendebremsung *f*
cushioning force *CYL* Bremskraft *f*
custom-filled fluid vom Kunden eingefüllte Flüssigkeit *f*
custom[-made] manifold *VALV* maschinengebundene (kreislaufgebundene) Unterplatte *f*, Batterieplatte *f*, Sammelanschlußplatte *f*
custom seal kundenspezifische Dichtung *f*, Dichtung *f* in Sonderausführung, Spezialdichtung *f*
cutaway diagram Schnittbild *n*
cutoff pressure *of a pressure-compensator control PU/MOT* Ansprechdruck *m* einer Nullhubsteuerung
cutting edge *FITT* Schneidlippe *f*, Schneidkante *f*
cutting ferrule (sleeve) *FITT* Schneidring *m*
cycle diagram (plot) Arbeitszyklusdiagramm *n*, Zyklusdiagramm *n*
cycle sequence diagram (plot) Funktionsablaufplan *m*, Ablaufdiagramm *n*
cylinder Arbeitszylinder *m*, Zylinder *m*; Luftflasche *f*, Gasflasche *f*, Druckbehälter *m*
cylinder action *auch* Wirkungsweise *f* eines Zylinders
cylinder alignment Zylinderausrichtung *f*
cylinder area Zylinderquerschnittsfläche *f*, Zylinderfläche *f*
cylinder barrel Zylinderrohr *n*, Zylindermantel *m*
cylinder block *PU/MOT* Zylinderkörper *m*, Zylinderblock *m*, Kolbenträger *m*, *bei Radialkolbenmaschinen auch* Zylinderstern *m*, *bei*

Axialkolbenmaschinen auch Zylindertrommel *f*, Kolbentrommel *f*
cylinder body Zylinderkörper *m*
cylinder bore Zylinderinnendurchmesser *m*, Zylinderbohrungsdurchmesser *m*, *in lockerem Stil auch* Zylinderbohrung *f*; Zylinderinnenwand *f*, Zylinderbohrung *f*
cylinder bore area Zylinderquerschnittsfläche *f*, Zylinderfläche *f*
cylinder bore size Zylinderinnendurchmesser *m*, Zylinderbohrungsdurchmesser *m*, *in lockerem Stil auch* Zylinderbohrung *f*
cylinder bottom Zylinderboden *m*, Zylinderfuß *m*
cylinder chamber Zylinderraum *m*
cylinder connection Zylinderanschluß *m*
cylinder cover *s.* cylinder end cap
cylinder cushion Endlagenbremse *f*, Hubendebremse *f*; *Weiterbildungen s unter* cushion
cylinder design Zylinderbauform *f*, Zylindertyp *m*
cylinder end Zylinderende *n*
cylinder end cap (closure, cover) Zylinderdeckel *m*, Zylinderboden *m*
cylinder end seal Zylinderenddichtung *f*, Zylinderkopfdichtung *f*
cylinder flange Zylinderflansch *m*
cylinder front cover Zylinderkopfdeckel *m*, Zylinderkopf *m*, Dichtungsgehäuse *n*
cylinder head Zylinderkopf *m*; Zylinderende *n*
cylinder head cap (cover) Zylinderkopfdeckel *m*, Zylinderkopf *m*, Dichtungsgehäuse *n*
cylinder head seal Zylinderenddichtung *f*, Zylinderkopfdichtung *f*

cylinder ID *s.* cylinder inner diameter
cylinder inner diameter Zylinderinnendurchmesser *m*, Zylinderbohrungsdurchmesser *m*, *in lockerem Stil auch* Zylinderbohrung *f*
cylinder inner wall Zylinderinnenwand *f*, Zylinderbohrung *f*
cylinder inside diameter *s.* cylinder inner diameter
cylinder mount (mounting) Zylinderbefestigungselement *n*, Zylinderbefestigung *f*
cylinder port *CYL, DCV* Zylinderanschluß *m*
cylinder pressure port *DCV* Zylinderdruckanschluß *m*, Zylinderzulauf[anschluß] *m*
cylinder return port *DCV* Zylinderrücklaufanschluß *m*
cylinder rod *s.* piston rod
cylinder sensor Positionsgeber *m*, Stellungsgeber *m*, Lagegeber *m*, Zylindersensor *m*
cylinder type Zylinderbauform *f*, Zylindertyp *m*
cylinder wall Zylinderwand *f*
cylindrical plug valve, rotary spool valve *DCV* Kolbendrehschieberventil *n*

D

D-ring *SEALS* D-Ring *m*
damp[en] *v vibrations* dämpfen *Schwingungen*
damper Stoßdämpfer *m*, Stoßfänger *m*, Prallfänger *m*; *pulsation damper* Druckpulsationsdämpfer *m*, Druckschwingungsdämpfer *m*

damping hose Dämpfungsschlauch *m*
damping plug Dämpfungseinsatz *m*
damping valve Dämpfungsventil *n*
Darcy-Weisbach formula *THEOR* Formel für den Druckverlust in Rohrleitungen
dashpot Endlagenbremse *f*, Hubendebremse *f*; *shock absorber* Stoßdämpfer *m*, Pralldämpfer *m*, Prallfänger *m*
dash size *hose size: number of 1/16-in increments in inside diameter; USA Maß für Schlauchinnendurchmesser; tubing: number of 1/16-in increments in outside diameter; USA Maß für Rohraußendurchmesser*
D. C. solenoid *direct current solenoid VALV OP* Gleichstrommagnet *m*
DCV *directional control valve* Wegeventil *n*
dead center *CYL* Totpunkt *m*, Umkehrpunkt *m*
dead end *of a conduit* festes Leitungsende *n*
dead point *CYL* Totpunkt *m*, Umkehrpunkt *m*
dead volume *CYL* schädlicher Raum *m*, toter Raum *m*, Totraum *m*
dead water *THEOR* Totwasser *n*
dead-weight pressure gauge tester gewichtsbelastete Manometerprüfvorrichtung *f*
deadband *DCV* Totzone *f*, inaktiver Bereich *m*
deadhead *v* mit Nullförderstrom laufen *Nullhubpumpe*
deadhead pressure *PU/MOT* Druck *m* bei Nullförderung, Nullförderdruck *m*
deadzone *DCV* Totzone *f*, inaktiver Bereich *m*

deaerate *v* *entrained air* ungelöste Luft *f* entfernen
deaeration *of entrained air* Entlüftung *f* *(Entfernung ungelöster Luft)*
deaerator *for entrained air* Entlüftungseinrichtung *f*, Entlüfter *m*, Entlüftung *f* *(für ungelöste Luft)*
deaerator valve Entlüftungsventil *n*
decay *v* *eg pressure* abfallen, sinken *z. B. Druck*
deceleration valve *FCV* Verzögerungsventil *n*
decelerator Stoßdämpfer *m*, Pralldämpfer *m*, Prallfänger *m*; *CYL* Endlagenbremse *f*, Hubendebremse *f*
decomposition temperature *FLUIDS* Zersetzungstemperatur *f*, Zerfallstemperatur *f*
decompression allmählicher Druckabbau *m*, Dekompression *f*; Vorauslaß *m*, Voröffnung *f* *(bei entsperrbaren Rückschlagventilen)*
decompression pressure surge Entspannungsdruckstoß *m*, Entspannungsdruckschlag *m*
decompression shock *s.* decompression pressure surge
decompression valve *DCV* Vorauslaßventil *n*; *PCV* Entspannungsventil *n*
de-energize *v* *solenoid VALV OP* stromlos machen *(den Magneten)*
de-energized *solenoid VALV OP* entregt, stromlos *Magnet*
deflection loss *THEOR* Ablenkverlust *m*
defoam *v* *FLUIDS* entschäumen
defoamer *FLUIDS* Anti-Schaum-Additiv *n*, Entschäumadditiv *n*, Schaumhemmer *m*

degree Engler *conventional unit of kinematic viscosity* Engler-Grad *m* *(veraltete Einheit der kinematischen Viskosität)*
dehumidifier *s.* dehydrator
dehydrator *PREPAR* Trockner *m*
deliquescent dryer *PREPAR* Absorptionstrockner *m*
deliver *v* fördern, liefern, abgeben *Flüssigkeit in (an) das System,* beaufschlagen *Komponente mit Flüssigkeit,* speisen, beliefern, versorgen *System mit Flüssigkeit*
delivered air *COMPR* geförderte Luft *f*, Förderluft *f*
delivering stroke *PU/MOT* Förderhub *m*
delivery Zufuhr *f*, Speisung *f*, Beaufschlagung *f*, Lieferung *f*, Versorgung *f*; *PU/MOT* Förderstrom *m*; *PU/MOT* Förderung *f*
delivery chamber Förderraum *m*, Druckraum *m*, Auslaßkammer *f*, Austrittskammer *f*, Ausgangsraum *m*
delivery channel Förderkanal *m*, Druckkanal *m*, Auslaßkanal *m*, Austrittskanal *m*, Abflußkanal *m*
delivery characteristics *PU/MOT* Förderkennlinie *f*, Förderverhalten *n*
delivery control *PU/MOT* Förderstromverstellung *f*
delivery direction *PU/MOT* Förderrichtung *f*
delivery duct Förderkanal *m*, Druckkanal *m*, Auslaßkanal *m*, Austrittskanal *m*, Abflußkanal *m*
delivery line Förderleitung *f*, Druckleitung *f*, Auslaßleitung *f*, Austrittsleitung *f*, Ausgangsleitung *f*, Abflußleitung *f*, Ablaufleitung *f*, Abführleitung *f*, Ableitung *f*

delivery per cycle *PU/MOT* Fördervolumen *n* je Umdrehung
delivery port Förderanschluß *m*, Druckanschluß *m*, Auslaßanschluß *m*, Austrittsanschluß *m*, Abführanschluß *m*
delivery pressure Förderdruck *m*, Auslaßdruck *m*, Austrittsdruck *m*, Ausgangsdruck *m*, Ablaufdruck *m*
delivery range *PU/MOT* Förderstrombereich *m*, Volumenstrombereich *m*
delivery rate *COMPR* Liefermenge *f*; *PU/MOT* Förderstrom *m*
delivery rating Nennförderstrom *m*
delta ring *(SEALS* Deltaring *m*, Dreieckring *m*
demand flow circuit Schaltung *f* für Volumenstrombeeinflussung
demand oil volume *ACCUM* gefordertes Ölvolumen *n*
demulsibility *FLUIDS* Demulgierbarkeit *f*, Dismulgierbarkeit *f*; Wasserabscheidevermögen *n*
demulsible *FLUIDS* demulgierbar, dismulgierbar
demulsification *FLUIDS* Demulgierung *f*, Dismulgierung *f*
demulsifier *FLUIDS* Demulgator *m*, Dismulgator *m*
demulsify *v* *FLUIDS* demulgieren, dismulgieren
demulsifying agent *FLUIDS* Demulgator *m*, Dismulgator *m*
density-pressure characteristics *FLUIDS* Dichte-Druck-Verhalten *n*, d-p-Verhalten *n*
depression Drucksenkung *f*, Druckverminderung *f*, Druckverringerung *f*
depression wave *THEOR* Depressionswelle *f*, Unterdruckwelle *f*

depressurize

depressurize *v* drucklos machen
depth filter Tiefenfilter *m,n*
depth filtration Tiefenfilterung *f*
desiccant *PREPAR* Trockenmittel *n*
desiccant dryer Adsorptionstrockner *m*
desiccant-style breather *RESERVOIRS* kombinierter Trockner-Belüfter *m*
design pressure Auslegungsdruck *m*, Bemessungsdruck *m*
destroke *v* *variable pump* einschwenken, zurückverstellen *Verstellpumpe*
destroking characteristics *of a pressure-compensated pump* Einschwenkcharakteristik *f einer Nullhubpumpe*
desurger Druckstoßdämpfer *m*
detachable fitting lösbare Verbindung (Verschraubung) *f*
detent position *VALV OP* Raststellung *f*
detent-positioned valve *VALV OP* rastgesichertes Ventil *n*
detergent *FLUIDS* Detergens *n*, absetzverhinderndes Mittel *n*
detrainer plate *RESERVOIRS* Entlüftungsblech *n*
develop *v* *flow THEOR* sich ausbilden *Strömung*
developed pipeline length abgewickelte (gestreckte) Rohrlänge *f*
development of flow *THEOR* Strömungsausbildung *f*
dewpoint *FLUIDS* Taupunkt *m*
diaphragm *ACCUM* Trennglied *n*, Trennelement *n*, Trennwand *f*
diaphragm accumulator Membranspeicher *m*
diaphragm-actuated valve membranbetätigtes Ventil *n*, Membranventil *n*
diaphragm-ball element *LOG EL* Membran-Kugel-Element *n*
diaphragm chamber *VALV OP* Membrankammer *f*
diaphragm compressor Membranverdichter *m*
diaphragm element *LOG EL* Membranelement *n*
diaphragm motor *an air motor design* Membranmotor *m* *ein Druckluftmotor*
diaphragm-operated valve membranbetätigtes Ventil *n*, Membranventil *n*
diaphragm piston *CYL* Membrankolben *m*
diaphragm pressure gauge Plattenfedermanometer *n*, Membranmanometer *n*
diaphragm pressure switch Membrandruckschalter *m*
diaphragm pressure transducer Membrandruckwandler *m*, Druckmeßdose *f*
diaphragm-sphere element *LOG EL* Membran-Kugel-Element *f*
diaphragm stack *LOG EL* Membran[en]paket *n*, Membran[en]satz *m*
diaphragm stroke Membranhub *m*
diaphragm valve *DCV* Membranventil *n* *mit Membran als Ventilelement*; *s.* diaphragm-operated valve
dibasic acid ester fluid *FLUIDS* Flüssigkeit *f auf Basis von Dicarbonsäureestern*
differential accumulator Differentialkolbenspeicher *m*
differential-area piston *CYL* Differentialkolben *m*
differential circuit Umströmungsschaltung *f*, Differentialschaltung *f*, Rückspeiseschaltung *f*, Eilgangschal-

tung *f*, Eilvorlaufschaltung *f*   *mit Differentialzylinder*
differential cylinder   Differentialzylinder *m*
differential-piston intensifier   Differentialkolben-Druckübersetzer *m*
differential pressure   *pressure difference*   Differenzdruck *m*   *Meßgröße*
differential pressure gauge   Differenzdruckmanometer *n*, Druckdifferenzmanometer *n*
differential pressure regulator   *PCV*   Druckdifferenzventil *n*, *für die Differenz zwischen Ein- und Ausgangsdruck auch* Druckgefälleventil *n*
differential pressure stability   *of filtration rating* Differenzdruckstabilität *f   der Filterfeinheit*
differential pressure switch   Differenzdruckschalter *m*
differential pressure transducer   Druckdifferenzwandler *m*
differential spool   *in a flow regulator* Stufenkolben *m   im Stromregelventil*
diffuser   *in a reservoir* Diffusor *m*, Rückstromteiler *m   in einem Behälter*
digital cylinder   Mehrpositionszylinder *m*, Mehrstellungszylinder *m*, Digitalzylinder *m*
digital fluidics   *LOG EL* Digitalfluidik *f*
digital pressure gauge   Digitalmanometer *n*
digital valve   *DCV* Digitalventil *n*
digitally controlled valve   *DCV* Digitalventil *n*
dilatational viscosity   *FLUIDS* Dilatationsviskosität *f*, Sekundärviskosität *f*
DIM   *direct impact modulator LOG EL* Gegenstrahlelement *n* mit axialer Steuerdüse, direkter Impaktmodulator *m*
DIN flange   *FITT* DIN-Flansch *m*
direct *v*   *eg flow to actuator* leiten *z. B. Flüssigkeit zum Verbraucher*
direct-acting relief valve   direktgesteuertes (nichtvorgesteuertes) Druckbegrenzungsventil *n*
direct-acting valve   direktgesteuertes (nichtvorgesteuertes, einstufiges) Ventil *n*
direct actuation   *s.* direct control
direct control   *VALV OP* direkte Betätigung (Steuerung) *f*
direct current solenoid   *VALV OP* Gleichstrommagnet *m*
direct impact modulator   *LOG EL* Gegenstrahlelement *n* mit axialer Steuerdüse, direkter Impaktmodulator *m*
direct-mounted valve   Ventil *n* für Rohrleitungseinbau, Leitungsventil *n*, Rohrventil *n*
direct operated pump   direktgesteuerte Pumpe *f*
direct operation   *VALV OP* direkte Betätigung (Steuerung) *f*
direct valve   *VALV OP* direktgesteuertes (nichtvorgesteuertes, einstufiges) Ventil *n*
direction of delivery   *PU/MOT* Förderrichtung *f*
direction of flow   Strömungsrichtung *f*
direction of output flow   *s.* direction of delivery
direction of rotation   *PU/MOT* Drehrichtung *f*
directional control [circuit]   Wegsteuerkreislauf *m*, Wegsteuersystem *n*, Wegsteuerung *f*; Wegesteuereinrich-

tung *f*, Wegesteuerung *f*, Richtungssteuerung *f*
directional control cover *of a logic valve DCV* Richtungssteuerdeckel *m (eines 2-Wege-Einbauventils)*
directional control system s. directional control
directional control valve Wegeventil *n*
directional valve s. directional control valve
dirt *FLUIDS* Schmutz *m*, Verschmutzung *f*
dirt capacity *FILTERS* Schmutztragevermögen *n*, Schmutzaufnahmekapazität *f*
dirt carrying capacity *FLUIDS* Schmutztragevermögen *n*, Schmutztransportvermögen *n*
dirt characteristic *FLUIDS* Schmutzkennwert *m*, Verschmutzungskennwert *m*
dirt class *FLUIDS* Schmutzklasse *f*, Verschmutzungsklasse *f*
dirt concentration *FLUIDS* Schmutzkonzentration *f*, Verschmutzungskonzentration *f*
dirt content *FLUIDS* Schmutzgehalt *m*
dirt dam *RESERVOIRS* Filterwehr *n*
dirt face *FILTERS* Schmutzseite *f*
dirt holding capacity *FILTERS* Schmutztragevermögen *n*, Schmutzaufnahmevermögen *n*
dirt indicator *FILTERS* Verstopfungsanzeige *f*
dirt ingress *FLUIDS* Eindringen *n* von Schmutz
dirt level *FLUIDS* Verschmutzungsgrad *m*, Schmutzniveau *n*

dirt particle *FLUIDS* Schmutzteilchen *n*, Schmutzpartikel *n*, Verschmutzungsteilchen *n*
dirt-sensitive *FLUIDS* schmutzempfindlich, verschmutzungsempfindlich
dirt sensitivity *FLUIDS* Schmutzempfindlichkeit *f*, Verschmutzungsempfindlichkeit *f*
dirt-sensitivity class (grade) *of a component* Schmutzempfindlichkeitsklasse *f eines Geräts*
dirt wiper *CYL* Schmutzabstreifring *m*, Abstreifring *m*, Abstreifer *m*
dirty *FLUIDS* schmutzhaltig, verschmutzt
disc s. disk
discharge *ACCUM* Entladung *f*, Leerung *f*; *PU/MOT* Förderung *f*; Förderstrom *m*
discharge air *COMPR* geförderte Luft *f*, Förderluft *f*
discharge chamber Förderraum *m*, Druckraum *m*, Auslaßkammer *f*, Austrittskammer *f*, Ausgangsraum *m*
discharge channel Förderkanal *m*, Druckkanal *m*, Auslaßkanal *m*, Austrittskanal *m*, Abflußkanal *m*
discharge characteristics *PU/MOT* Förderkennlinie *f*, Förderverhalten *n*
discharge coefficient *THEOR* Widerstandsbeiwert *m*, Durchflußbeiwert *m*, Verlustbeiwert *m*
discharge duct Förderkanal *m*, Druckkanal *m*, Auslaßkanal *m*, Austrittskanal *m*, Abflußkanal *m*
discharge flow *PU/MOT* Förderstrom *m*
discharge line Förderleitung *f*, Druckleitung *f*, Auslaßleitung *f*, Austrittsleitung *f*, Ausgangsleitung *f*, Ab-

flußleitung *f*, Ablaufleitung *f*, Abführleitung *f*, Ableitung *f*
discharge per cycle   PU/MOT Fördervolumen *n* je Umdrehung
discharge port   Förderanschluß *m*, Druckanschluß *m*, Auslaßanschluß *m*, Austrittsanschluß *m*, Abführanschluß *m*
discharge pressure   Förderdruck *m*, Auslaßdruck *m*, Austrittsdruck *m*, Ausgangsdruck *m*, Ablaufdruck *m*
discharge range   PU/MOT Förderstrombereich *m*, Volumenstrombereich *m*
discharge rate   PU/MOT Förderstrom *m*; COMPR Liefermenge *f*
discharge rate control   PU/MOT Förderstromverstellung *f*
discharge rating   Nennförderstrom *m*
discharge side   Förderseite *f*, Druckseite *f*, Auslaßseite *f*, Austrittsseite *f*, Ausgangsseite *f*, Ablaufseite *f*
discharge stroke   Förderhub *m*
discharge valve   ACCUM Flüssigkeitsventil *n*
discharging direction   PU/MOT Förderrichtung *f*
disconnect *v*   *eg hose coupling* entkuppeln, trennen, lösen *z. B. Schlauchkupplung*
disk   FILTERS Filterscheibe *f*, Filterlamelle *f*, Filterplatte *f*
disk filter   Scheibenfilter *m,n*, Lamellenfilter *m,n*, Plattenfilter *m,n*
disk pack   FILTERS Scheibenpaket *n*, Lamellenpaket *n*, Plattenpaket *n*
dispersant   *s.* dispersing agent
dispersing agent   FLUIDS Dispergierungsmittel *n*, Dispergens *n*

displace *v*   verdrängen, ausstoßen, ausschieben; schlucken   Motor
displacement   Verschiebung *f*; Verdrängung *f*; Verdrängungsvolumen *n*; *beim Motor* Motorverdrängungsvolumen *n*, Schluckvolumen *n*; *bei der Pumpe* Pumpenverdrängungsvolumen *n*, Fördervolumen *n*
displacement chamber   PU/MOT Verdrängungsraum *m*
displacement control   *s.* control
displacement element   PU/MOT Verdrängerelement *n*
displacement motor   Verdrängermotor *m*
displacement pump   Verdrängerpumpe *f*
displacement pump *or* **motor**   Verdrängermaschine *f*
displacement sensor   CYL Positionsgeber *m*, Stellungsgeber *m*, Lagegeber *m*, Zylindersensor *m*
displacement volume   PU/MOT Verdrängungsvolumen *n*
displacing chamber   PU/MOT Verdrängungsraum *m*
displacing member   PU/MOT Verdrängerelement *n*
disposable [filter element]   Wegwerffilterelement *n*
dissipating muffler   dissipativer Schalldämpfer *m*
dissolved air   FLUIDS gelöste Luft *f*
dissolved water   FLUIDS gelöstes Wasser *n*
distance thermometer   Fernthermometer *n*
distributor   COMPR Leitrad *n*, Leitapparat *m*
dither *v*   mit geringer Amplitude oszillieren

dither

dither drive *to eliminate breakaway friction* VALV OP Ditherantrieb *m*, Oszillierantrieb *m* (zur Überwindung der Startreibung
dither frequency VALV Ditherfrequenz *f*
divert *v* im Nebenschluß ableiten, umleiten
DOM tubing DOM-Rohr *n*, stangengezogenes Rohr *n*
dope *v* *hydraulic fluids* legieren, additivieren *Hydraulikflüssigkeiten*
doped oil legiertes Öl *n*, additiviertes Öl *n*, HL *n*
double-acting cylinder doppeltwirkender Zylinder *m*
double-acting hand pump doppeltwirkende Handpumpe *f*
double-acting intensifier doppeltwirkender Druckübersetzer *m*
double-acting solenoid VALV OP Umkehrmagnet *m*, Doppelhubmagnet *m*
double-barreled cylinder Doppelmantelzylinder *m*
double-base fluid Zweistoff-Flüssigkeit *f*, zweibasische Flüssigkeit *f*
double check valve wechselseitig entsperrbares Rückschlagventil *n*, Doppelrückschlagventil *n*, Zwillingsrückschlagventil *n*
double-diaphragm element LOG EL Doppelmembranelement *n*, Doppelmembranrelais *n*
double-end poppet CHECKS Doppelkegel *m*
double-end rod CYL beid[er]seitige Kolbenstange *f*
double-flapper valve zweidüsiges Prallplattenventil *n*

double-inlet screw pump zweiflutige (doppelflutige) Schraubenpumpe *f*
double-lip seal Nutring *m*, Doppellippenring *m*, Zweilippenring *m*
double-lip wiper SEALS Doppellippenabstreifer *m*, Zweilippenabstreifer *m*
double nozzle-flapper valve zweidüsiges Prallplattenventil *n*
double piston-row radial pump doppelreihige Radialkolbenpumpe *f*
double-plane swivel joint FITT Zweiebenen-Rohrgelenk *n*
double-poppet coupling FITT Zweistrangsitzkupplung *f*
double-pressure intensifier Eilgangdruckübersetzer *m*
double pump Zweistrompumpe *f*, Zweikreispumpe *f*, Zwillingspumpe *f*, Doppelpumpe *f*
double-rack rotary actuator Doppelzahnstangenschwenkmotor *m*, Drehwinkelmotor *m* mit zwei Zahnstangen
double-rod [end] cylinder Zylinder *m* mit doppelseitiger (beidseitiger) Kolbenstange, Doppelstangenzylinder *m*
double-screw pump Zweispindelschraubenpumpe *f*
double-seat valve Doppelsitzventil *n*
double shut-off *quick disconnects* Zweistrangabsperrung *f*, Zweiwegeabsperrung *f* *Schnellkupplung*
double shut-off coupling Zweistrangabsperrkupplung *f*, Zweiwegeabsperrkupplung *f*
double-solenoid valve Doppelmagnetventil *n*
double-suction screw pump zwei-

flutige (doppelflutige) Schraubenpumpe *f*
double-valved coupling Zweistrangabsperrkupplung *f*, Zweiwegeabsperrkupplung *f*
double-vane rotary actuator Zweiflügelschwenkmotor *m*, Doppelflügelschwenkmotor *m*
double wire braid hose Schlauch *m* mit zwei Drahtgeflechteinlagen
downstream line Leitung *f* hinter (nach) *einem Element*, Leitung *f* unterhalb *eines Elements*, Leitung *f* in Stromrichtung, *einem Element* nachgeschaltete Leitung *f*
downstroke *CYL* Abwärtshub *m*
drag flow *THEOR* Schleppströmung *f*, Mitschleppströmung *f*, Scherströmung *f*
drain *v eg a case* mit einer Leckflüssigkeitsableitung versehen *z. B. ein Gehäuse*; *eg tank* ablassen, entleeren *z. B. Behälter*
drain Kondensatablaß *m*, Kondensatableiter *m*; Wasserablaß *m*; Ablaßöffnung *f*, Entleerungsöffnung *f*
drain leg Kondensatleitung *f*, Entwässerungsleitung *f*, Ablaßleitung *f*
drain line Leckleitung *f*, Leckageleitung *f*
drain line pressure Leckdruck *m*
drain [opening] Ablaßöffnung *f*, Entleerungsöffnung *f*
drain passage Leckflüssigkeitskanal *m*
drain plug *RESERVOIRS* Ablaßschraube *f*, Entleerungsschraube *f*
drain port Leckflüssigkeitsanschluß *m*
drain return Leckflüssigkeitsrücklauf *m*

drain trap *LINES* Wasserfang *m*, Wasserablaß *m*, Entwässerung *f*
drain valve *RESERVOIRS* Ablaßventil *n*, Entleerungsventil *n*
drainage Leckflüssigkeitsabführung *f*, Leckflüssigkeitsableitung *f*; *RESERVOIRS* Entleerung *f*, Ablaß *m*
drainage port Leckflüssigkeitsanschluß *m*
drainback Leckflüssigkeitsrücklauf *m*
draw *v fluid from reservoir into inlet line* ansaugen, saugen *(Flüssigkeit aus dem Behälter in die Saugleitung*
draw-back cylinder Rückzugszylinder *m*
draw-back stroke *CYL* Rückzug *m*, Rücklauf *m*, Rückhub *m*, Einfahrhub *m*
draw-back speed *CYL* Einfahrgeschwindigkeit *f*, Rückzuggeschwindigkeit *f*, Rücklaufgeschwindigkeit *f*
draw in *v through leaks* einsaugen *(durch Leckstellen*drier *s.* dryerdrilled plate maschinengebundene (kreislaufgebundene) Unterplatte *f*, Batterieplatte *f*, Sammelanschlußplatte *f*
drip rate Ölzuführstrom *m*, Ölförderstrom *m*
drive hydraulic circuit *APPL* Fahrhydraulik *f*
drive rotor *of a screw pump* Treibspindel *f einer Schraubenpumpe*
drive shaft *PU/MOT* Antriebswelle *f*
driven screw *of an screw pump* Laufspindel *f einer Schraubenpumpe*
driving horsepower Antriebsleistung *f*, Eingangsleistung *f*
driving rpm Antriebsdrehzahl *f*
driving screw *s.* drive rotor
driving speed Antriebsdrehzahl *f*

driving torque Antriebsdrehmoment *n*
droop Abfall *m*, Abfallen *n* einer im Idealfall horizontalen Kennlinie
drop *v* *eg pressure* abfallen, sinken z. B. Druck
dropping body Fallkörper *m*
dropping-body viscometer Fallviskosimeter *n*
dropping sphere Fallkugel *f*
dropping sphere viscometer Kugelfallviskosimeter *n*
dry ölfrei, öllos, unbeölt, trocken
dry air Trockenluft *f*
dry pneumatics Trockenluftpneumatik *f*
dryer *PREPAR* Trockner *m*
drying compound *PREPAR* Trockenmittel *n*
Dryseal pipe thread *s*. NPTF thread
dual check valve wechselseitig entsperrbares Rückschlagventil *n*, Doppelrückschlagventil *n*, Zwillingsrückschlagventil *n*
dual-diameter accumulator Differentialkolbenspeicher *m*
dual-fluid intensifier Druckübersetzer *m* für verschiedene Fluide, Druckmittelwandler *m*
dual-material seal kombinierte Dichtung *f*, Zweistoffdichtung *f*
dual-operation solenoid *VALV OP* Umkehrmagnet *m*, Doppelhubmagnet *m*
dual-path transmission Hydrogetriebe *n* mit zwei Motoren
dual-pressure circuit Zweidruckkreislauf *m*
dual-pressure pump Schaltpumpe *f*
dual pump Zweistrompumpe *f*, Zweikreispumpe *f*, Zwillingspumpe *f*, Doppelpumpe *f*
dual-vane pump Flügelzellenpumpe *f* mit geteilten Flügeln
duct Strömungsweg *m*, Kanal *m*
ducting system *DCV* Kanalsystem *n*, Kammersystem *n*
dump *v* mit dem Behälter verbinden, entlasten
dump valve Ablaßventil *n*, Bodenventil *n*, Fußventil *n*
duplex filter Doppelfilter *m,n*, Zwillingsfilter *m,n*
dust cap Staubkappe *f*, Schmutzkappe *f*, Schutzkappe *f*
duty cycle Arbeitszyklus *m*, Arbeitsspiel *n*, Betriebszyklus *m*, Betriebsspiel *n*
dwell time *RESERVOIRS* Verweilzeit *f*, Aufenthaltszeit *f*
dynamic fluid sampling *MEASUR* dynamische Flüssigkeitsprobenahme *f*
dynamic pressure *THEOR* dynamischer Druck *m*, Staudruck *m*
dynamic seal Bewegungsdichtung *f*, dynamische Dichtung *f*

E

E-ring *SEALS* E-Ring *m*
ear mount *CYL* Schwenkaugenbefestigung *f*, Augenbefestigung *f*
eccentric cam ring-type vane pump einfachwirkende Flügelzellenpumpe *f*, Exzenterring-Flügelzellenpumpe *f*
edge filter Spaltfilter *m,n*, Kantenfilter *m,n*
edge filtration Spaltfilterung *f*, Kantenfilterung *f*

edgetone amplifier *LOG EL* Schneidentonverstärker *m*
effective area *CYL* wirksame Fläche *f*
efflux adge *DCV* Abströmkante *f*
efflux viscometer Auslaufviskosimeter *n*, Kapillarviskosimeter *n*, Ausflußviskosimeter *n*
EHPM *electrohydraulic pulse motor* elektrohydraulischer Schrittmotor *m*
elastic seal elastische (nachgiebige) Dichtung *f*
elastomeric seal Elastomerdichtung *f*
elbow Winkelverschraubung *f*
45° elbow 45°-Winkelverschraubung *f*
elbow connector (end connection, end fitting) *of a hose line* Winkelendstück *n*, Krümmerverbindung *f* *einer Schlauchleitung*
elbow fitting Winkelverschraubung *f*
elbow port fitting Winkelanschlußverschraubung *f*
ELC *electrostatic liquid cleaner PREPAR* elektrostatischer Flüssigkeitsreiniger *m*
electric actuator *VALV OP* elektrische Stelleinheit (Betätigungseinrichtung) *f*
electric control *s.* electric actuator; *PU/MOT* elektrische Stelleinheit *f*
electric operator *s.* electric actuator
electric stroker *PU/MOT* *s.* electric control
electrical pressure transducer elektrischer Druckwandler *m*
electrical-to-fluid transducer elektrisch-fluidischer Wandler *m*
electro-rheological fluid *with electrostatically controlled viscosity FLUIDS* elektrorheologische Flüssigkeit *f*, ER-Flüssigkeit *f* *mit elektrisch beeinflußbarer Viskosität*

electrohydraulic elektrohydraulisch
electrohydraulic axis *of a robot APPL* elektrohydraulische Achse *f* *eines Roboters*
electrohydraulic control *PU/MOT* elektrohydraulische Servo-Stelleinheit *f*
electrohydraulic pulse motor elektrohydraulischer Schrittmotor *m*
electrohydraulic servovalve elektrohydraulisches Servoventil *n*
electrohydraulic stroker *PU/MOT* elektrohydraulische Servo-Stelleinheit *f*
electrohydraulic transducer elektrisch-hydraulischer (elektrohydraulischer) Wandler *m*, EY-Wandler *m*
electrohydraulics Elektrohydraulik *f*
electromagnetic flowmeter induktiver (elektromagnetischer) Durchflußmesser *m*
electromagnetic pressure transducer induktiver (elektromagnetischer) Druckwandler *m*
electromechanical actuator (control, operator) *VALV OP* elektromechanische Stelleinheit *f* (Betätigungseinrichtung *f*)
electromotor-operated valve elektromotorbetätigtes Ventil *n*
electronic control (stroker) *PU/MOT* elektronische Stelleinheit *f*
electropneumatic elektropneumatisch
electropneumatic transducer elektrisch-pneumatischer (elektropneumatischer) Wandler *m*, EP-Wandler *m*
electropneumatics Elektropneumatik *f*
electrostatic filtration elektrostatische Filterung *f* (Filtration *f*)

element 38

electrostatic liquid cleaner, ELC *PREPAR* elektrostatischer Flüssigkeitsreiniger *m*
element *FILTERS* Filterelement *n*; *s.* hydraulic element; *s.* pneumatic element
ell Winkelverschraubung *f*
elliptical cam ring-type vane pump Ovalring- Flügelzellenpumpe *f*, doppeltwirkende Flügelzellenpumpe *f*
elliptical ring *SEALS* Ovalring *m*
emergency control *VALV OP* Notbetätigung *f*
emergency pump Notpumpe *f*
emergency valve Notventil *n*
emulsibility *FLUIDS* Emulgierbarkeit *f*
emulsible *FLUIDS* emulgierbar
emulsifiability *s.* emulsibility
emulsifiable *s.* emulsible
emulsification *FLUIDS* Emulgierung *f*, Emulsionsbildung *f*
emulsification resistance *FLUIDS* Emulgierwiderstand *m*, Beständigkeit *f* gegen Emulgieren
emulsifier *FLUIDS* Emulgator *m*, Emulsionsbildner *m*
emulsify *v* *FLUIDS* emulgieren
emulsifying agent *s.* emulsifier
emulsion *FLUIDS* Emulsion *f*
  emulsion stability *FLUIDS* Emulsionsbeständigkeit *f*
enclosed hydraulic panel *APPL* Hydraulikschrank *m*
end *s.* cylinder end
  end cap *s.* cylinder end cap
  end closure *s.* cylinder end closure
  end connection Endstück *n*
  end coupling *s.* hose end coupling
  end cover *s.* cylinder end cover; *of a valve stack* Sperrplatte *f*, Abschlußplatte *f*, Dichtungsplatte *f* *einer Steuersäule*
end fitting *s.* end connection
end load test *FILTERS* Längsfestigkeitsprüfung *f*
end of stroke *CYL* Hubende *n*
end-of-stroke indicator *CYL* Hubendeanzeiger *m*
end-of-stroke locking *CYL* Endlagenverriegelung *f*
end-of-stroke position *CYL* Hubendeposition *f*
end-of-stroke sensor *CYL* Hubendeanzeiger *m*
end plate *of a valve stack* Sperrplatte *f*, Abschlußplatte *f*, Dichtungsplatte *f*
end position *CYL* Endposition *f*, Endstellung *f*, Endlage *f*
energize *v* *a solenoid* erregen *einen Elektromagneten*; *a seal* vorspannen, spannen *eine Dichtung, s z. B.* rubber-energized seal
Engler's viscometer *efflux-type viscometer* Engler-Viskosimeter *n* *ein Auslaufviskosimeter*
enlargement loss *THEOR* Erweiterungsverlust *m*
entrain *v* *eg air* mitschleppen, mitreißen *z. B. Luft*
entrained air *FLUIDS* freie (ungelöste) Luft *f*
entrained-flow shock absorber *Stoßdämpfer, bei dem eine kompressible Flüssigkeit eingesaugt und zusammengedrückt wird*
entrainment compressor Strahlverdichter *m*
entrance length *required to attain fully developed flow mode* *THEOR* Anlaufstrecke *f* *bis zur voll-*

ständigen Ausbildung der Strömungsform
entrance loss THEOR Eintrittsverlust *m*, Einlaufverlust *m*
entrapped oil PU/MOT Quetschöl *n*
envelope s. valve envelope
environmental contamination FLUIDS Umgebungsverschmutzung *f*
EP additive extreme-pressure additive FLUIDS Höchstdruckadditiv *n*, Superhochdruckadditiv *n*, EP-Additiv *n*
equalizer s. equalizing valve
equalizing valve FCV Gleichlaufventil *n*, Synchronisierventil *n*
equal-stroking speed cylinder Gleichlaufzylinder *m*, Gleichgangzylinder *m*
equivalent orifice area THEOR äquivalenter Drosselquerschnitt *m*
equivalent pipeline length THEOR äquivalente Rohrlänge *f*
equivalent restrictive area THEOR äquivalenter Drosselquerschnitt *m*
ER fluid electro-rheological fluid, with electrostatically controlled viscosity elektrorheologische Flüssigkeit *f*, ER-Flüssigkeit *f* mit elektrisch beeinflußbarer Viskosität
ester-base fluid FLUIDS esterbasische Flüssigkeit *f*
Euler's equation of fluid motion THEOR Eulersche Gleichung *f* Bewegung von Fluiden
evacuate *v* evakuieren, auspumpen
evaporation deposit FLUIDS Verdampfungsrückstand *m*
evaporation resistance below boiling temperature FLUIDS Verdunstungswiderstand *m* unterhalb des Siedepunkts

excess flow Stromüberschuß *m*, Überstrom *m*
excess-flow loss PCV Überströmverlust *m*
excess-flow valve flow-limiting valve FCV Stromsicherheitsventil *n*, Strombegrenzungsventil *n*
exchanger heat exchanger Wärmeübertrager, Wärme[aus]tauscher *m*
exclusion seal Schutzdichtung *f*, Dichtung *f* gegen Einströmen (Eindringen)
exhaust Auslaß *m*, Austritt *m*, Ausgang *m*, Abfluß *m*, Ablauf *m*, Ausfluß *m*, Abführung *f*
exhaust *v* mit dem Behälter verbinden, entlasten
exhaust air Abluft *f*
exhaust air line Abluftleitung *f*
exhaust air metering PU/MOT Abluftdrosselung *f*, Abluftsteuerung *f*
exhaust chamber Auslaßkammer *f*, Austrittskammer *f*, Ausgangsraum *m*, in einer Pumpe auch Förderraum *m*, Druckraum *m*
exhaust channel Auslaßkanal *m*, Austrittskanal *m*, Abflußkanal *m*, in einer Pumpe auch Förderkanal *m*, Druckkanal *m*
exhaust duct s. exhaust channel
exhaust fluid Rückflüssigkeit *f*, abfließende Flüssigkeit *f*
exhaust line Auslaßleitung *f*, Austrittsleitung *f*, Ausgangsleitung *f*, Abflußleitung *f*, Ablaufleitung *f*, Abführleitung *f*, Ableitung *f*, bei einer Pumpe auch Förderleitung *f*, Druckleitung *f*; Entlüftungsleitung *f*, Abluftleitung *f*
exhaust oil Rücköl *n*, Alböl *n*

exhaust pressure Auslaßdruck *m*, Austrittsdruck *m*, Ausgangsdruck *m*, Ablaufdruck *m*, *bei einer Pumpe auch* Förderdruck *m*; *in der PN auch* Druck *m* in der Entlüftungsleitung
exhaust side Auslaßseite *f*, Austrittsseite *f*, Ausgangsseite *f*, Ablaufseite *f*, *bei einer Pumpe auch* Förderseite *f*, Druckseite *f*
exhaust-speed controller *s.* exhaust throttle
exhaust throttle Entlüftungsdrossel *f*, Abluftdrossel *f*
exhaust water Rückwasser *n*, Abwasser *n*
exit flow [rate] Ausgangs[volumen]strom *m*, Austrittsstrom *m*, Abführstrom *m*, Ablaufstrom *m*
exit loss *THEOR* Austrittsverlust *m*, Auslaufverlust *m*
exit port Auslaßanschluß *m*, Austrittsanschluß *m*, Abführanschluß *m*, *bei einer Pumpe auch* Förderanschluß *m*, Druckanschluß *m*
expanding hose *used to smoothen pressure peaks* Dehnschlauch *m* *für Druckspitzenabbau*
expansion fitting Aufdornringverschraubung *f*, ADR-Verschraubung *f*; *telescopic line* Teleskoprohr *n*
expansion thermometer Ausdehnungsthermometer *n*, Expansionsthermometer *n*
expansion wave *THEOR* Expansionswelle *f*
expansion work *THEOR* Expansionsarbeit *f*, Entspannungsarbeit *f*
expansional viscosity *FLUIDS* Expansionsviskosität *f*, Ausdehnungsviskosität *f*

expose *v* to a pressure mit Druck *m* beaufschlagen, Druck *m* zuführen, *als Adj. auch zusammengesetzt* druckbeaufschlagt
exposed area *CYL* wirksame Fläche *f*
extend *vi and vt* *cylinder* ausfahren *Zylinder*
extend stroke *CYL* Ausfahrhub *m*, Vorhub *m*, Vorlauf *m*
extended length *CYL* Ausfahrlänge *f*
extended position *CYL* Ausfahrstellung *f*
extension *s.* extension stroke length
extension speed *CYL* Ausfahrgeschwindigkeit *f*
extension *CYL* Ausfahrhub *m*, Vorhub *m*, Vorlauf *m*
extension stroke length *CYL* Ausfahrhublänge *f*, Ausfahrhub *m*
exterior admission *PU/MOT* Außenbeaufschlagung *f*
external drainage *DCV* Ablaufdruckentlastung *f*, äußere Flüssigkeitsrückführung *f*
external drain line *reservoir drain* äußere Leckölabführung *f*
external filtration Außenfilterung *f*, Spülung *f*
external gear motor Außenzahnradmotor *m*
external gear pump Außenzahnradpumpe *f*
external gear pump *or* motor Außenzahnradmaschine *f*, Außenzahnradeinheit *f*, Außenzahnradgerät *n*
external leakage äußerer Leckverlust *m*
external piloting *VALV OP* Fremdsteuerung *f*, aktive (externe) Vorsteuerung *f*

external pilot pressure *VALV OP* Fremdsteuerdruck *m*
external pressure Außendruck *m*
external seal Schutzdichtung *f*, Dichtung *f* gegen Einströmen (Eindringen)
external yoke *of a magnetic rodless cylinder* Außenläufer *m* *eines Magnetkolbenzylinders*
externally drained *DCV* ablaufdruckentlastet
externally drained relief valve ablaufdruckentlastetes Druckbegrenzungsventil *n*, Druckbegrenzungsventil *n* mit äußerer Lecköltrückführung
externally-guided radial pump Radialkolbenpumpe *f* mit äußerer Kolbenabstützung (außenliegender Hubkurve)
externally-guided weight-loaded accumulator gewichtsbelasteter Speicher *m* mit Außenführung
externally piloted valve *VALV OP* fremdgesteuertes (extern vorgesteuertes) Ventil *n*
extra-fine fog lubricator *PREPAR* Mikronebelöler *m*
extreme-pressure additive *FLUIDS* Höchstdruckadditiv *n*, Superhochdruckadditiv *n*, EP-Additiv *n*
extreme-pressure hose Höchstdruckschlauch *m*, Superhochdruckschlauch *m*
extrusion *SEALS* Extrusion *f*
extrusion resistance *SEALS* Extrusionsfestigkeit *f*
extrusion-resistant *SEALS* extrusionsfest
eye mount *CYL* Schwenkbefestigung *f*, Augenbefestigung *f*

F

fabric braid hose Schlauch *m* mit Textilgeflechteinlage[n], textilgeflechtarmierter Schlauch *m*
fabric cover *LINES* Textilumflechtung *f*, Textilmantel *m*
fabric filter Gewebefilter *m,n*, Textilfilter *m,n*
fabric-reinforced seal gewebeverstärkte (textilbewehrte, textilarmierte) Dichtung *f*
fabrication integrity test *FILTERS* Integritätsprüfung *f* *Nachweis der einwandfreien Fertigungsqualität*
face mounting *cap or front end CYL* Stirnflächenbefestigung *f* *boden- oder ausfahrseitig*
face seal axiale Dichtung *f*
face-seal fitting O-Ring-Verschraubung *f*
factory-attached hose fitting herstellerverpreßte (werkverpreßte) Schlauchverbindung *f*
failsafe-retract circuit *CYL* Notrückzugsschaltung *f*
falling ball *MEASUR* Fallkugel *f*
falling-ball viscometer Kugelfallviskosimeter *n*
falling body *MEASUR* Fallkörper *m*
falling-body viscometer Fallviskosimeter *n*
fan-in *LOG EL* Eingangszahl *f*, Fan-in *m* *Zahl der möglichen Steuereingänge*
fan-out *LOG EL* Belastungszahl *f*, Fan-out *m*
fast-acting valve Schnellschaltventil *n*
fast-closure valve schnellschließendes Ventil *n*

fast feed Eilvorschub *m*
fast return Eilrücklauf *m*, Eilrückgang *m*, Eilrückzug *m*
fast return stroke *CYL* Eilrückhub *m*
fast stroke Eilhub *m*
fast traverse piston Eilgangkolben *m*
feather *v* *mit handbetätigtem Ventil lastangepaßt steuern*
feed circuit *APPL* Vorschubkreislauf *m*
feed cylinder *APPL* Vorschubzylinder *m*
feed line Speiseleitung *f*, Einlaßleitung *f*, Eintrittsleitung *f*, Eingangsleitung *f*, Zulaufleitung *f*, Zuführleitung *f*, Zuflußleitung *f*, Zuleitung *f*
feed pressure Speisedruck *m*, Einlaßdruck *m*, Eintrittsdruck *m*, Eingangsdruck *m*, Zulaufdruck *m*, Zuführdruck *m*
feed rate *s.* oil feed rate
feedback duct *VALV OP* Rückführkanal *m*
felt filter Filzfilter *m,n*
female adaptor *of a V-ring assembly* Druckring *m*, Sattelring *m* *eines Dichtungssatzes*
female branch tee T-Verschraubung *f* mit Aufschraubabzweig
female connector Aufschraubverschraubung *f*
female coupling half *FITT* Kupplungsdose *f*
female elbow Aufschraubwinkelverschraubung *f*
female end fitting Aufschraubverschraubung *f*
female rotor *COMPR* Steuerläufer *m*, Nebenläufer *m*

female run tee T-Verschraubung *f* mit Aufschraubkappe im durchgehenden Teil
female side tee T-Verschraubung *f* mit Aufschraubabzweig
female support ring *of a V-ring assembly* Druckring *m*, Sattelring *m* *eines Dichtungssatzes*
female threaded union Übergangsstück *n* mit Innengewinde, Gewindemuffe *f*
ferrule fitting Schneidringverschraubung *f*
fiber filter Faser[stoff]filter *m,n*
fiberglass filter Glasfiberfilter *m,n*, Glasfaserfilter *m,n*
filler *RESERVOIRS* Füllöffnung *f*, Einfüllöffnung *f*
filler-breather [assembly] *RESERVOIRS* Einfüll- Belüftungs-Kombination *f*
filler opening *RESERVOIRS* Füllöffnung *f*, Einfüllöffnung *f*
filler-strainer *RESERVOIRS* Einfüllfilter *m,n*
filling *ACCUM* Aufladung *f*, Ladung *f*, Füllung *f*
filling line Fülleitung *f*, Vorfüllleitung *f*, Zuförderleitung *f*
filter Filter *m,n* *als Oberbegriff sowie als Tiefenfilter und für kleine Schmutzteilchen; vgl* strainer
filter *v* to *x* µm filtern auf *x* µm
filter assembly Filterbatterie *f*
filter body Filtergehäuse *n*, Filterkörper *m*
filter capacity Schmutztragevermögen *n*, Schmutzaufnahmekapazität *f*
filter cart Ölwechselgerät *n*, Ölaufbereitungsgerät *n*, Ölreinigungs-

gerät *n*, Öl-Service-Aggregat *n*, Ölfiltrationswagen *m*
filter cartridge Filtereinsatz *m*, Filterpatrone *f*
filter disk Filterscheibe *f*, Filterlamelle *f*, Filterplatte *f*
filter efficiency Filterwirkungsgrad *m*, Gesamtabscheidegrad *m*
filter element Filterelement *n*
filter fineness Filterfeinheit *f*
filter housing Filtergehäuse *n*, Filterkörper *m*
filter life Filterstandzeit *f*
filter location *in the system* Filteranordnung *f* *im Kreislauf*
filter loss Filter[druck]verlust *m*
filter maintenance Filterwartung *f*
filter material (media, medium) Filtermittel *n*
filter plate Filterscheibe *f*, Filterlamelle *f*, Filterplatte *f*
filter pressure drop Filterdruckabfall *m*
filter pressure loss Filter[druck]verlust *m*
filter rating Filterfeinheit *f*
filter-regulator *PCV* Filter-Druckminderventil *n*, Filter-Reduzierventil *n*
filter-regulator-lubricator Druckluft-Wartungseinheit *f*, Luftaufbereitungseinheit *f*, Luftaufbereiter *m*
filter surface Filterfläche *f*
filter washer Filterscheibe *f*, Filterlamelle *f*, Filterplatte *f*
filter with filtered bypass Filterbypassfilter *m,n*, Filter *m,n* mit gefilterter Umgehung
filterable filterbar, filterfähig, filtrationsfähig
filtering area Filterfläche *f*
filtering gap Filterspalt *m*

filtrable *s.* filterable
filtration Filterung *f*, Filtration *f*
filtration area Filterfläche *f*
filtration fineness (rating) Filterfeinheit *f*
filtration ratio Filter[ungs]grad *m*, *vgl* beta ratio
final position *CYL* Endposition *f*, Endstellung *f*, Endlage *f*
fine filter Feinfilter *m,n*
fine filtration Feinfilterung *f*
fine-meshed feinmaschig
fine test dust *MEASUR* Prüffeinstaub *m*
fineness *s.* filtration fineness
fingertip actuated *VALV OP* drucktastenbetätigt
finned tube Rippenrohr *n*
fintube *s.* finned tube
fire point *FLUIDS* Brennpunkt *m*
fire resistance *FLUIDS* Schwerentflammbarkeit *f*
fire resistance test *FLUIDS* Prüfung *f* der Schwerentflammbarkeit
fire resistant fluid schwerentflammbare Flüssigkeit *f*
fitting *s.* hose fitting; *s.* port fitting; *s.* tube fitting
fitting pressure *of a seal* Vorspanndruck *m* *einer Dichtung*
five-port valve Ventil *n* mit fünf Anschlüssen
five-way valve Fünfwegeventil *n*, 5-Wege-Ventil *n*
fixed axial clearance gear pump Zahnradpumpe *f* mit festem Seitenspiel
fixed-block radial pump Radialkolbenpumpe *f* mit feststehendem Kolbenträger

fixed-displacement motor nicht [ver]stellbarer Motor *m*, Konstantmotor *m*
fixed-displacement pump nicht [ver]stellbare Pumpe *f*, Konstantpumpe *f*
fixed-displacement transmission Konstantgetriebe *f*, hydraulische Welle *f*
fixed mount *CYL* nichtnachgiebige (starre) Befestigung *f*
fixed pressure reduction valve Druckdifferenzventil *n*, *für Differenz zwischen Ein- und Ausgangsdruck auch* Druckgefälleventil *n*
fixed restriction Konstantdrossel *f*, Festdrossel *f*, nichtverstellbare Drossel *f*
flammability *FLUIDS* Entflammbarkeit *f*, Entzündbarkeit *f*
flammability test *FLUIDS* Entflammbarkeitsprüfung *f*, Entzündungsprüfung *f*
flammable *FLUIDS* entflammbar, entzündbar
flame-proof *FLUIDS* nicht entflammbar, unentflammbar, unentzündbar, nicht brennbar, flammbeständig
flameproofness *FLUIDS* Nichtentflammbarkeit *f*, Unentflammbarkeit *f*, Unentzündbarkeit *f*, Unbrennbarkeit *f*, Flammbeständigkeit *f*
flange connection Flanschverbindung *f*
flange half Flanschhälfte *f*
flange mount Flanschbefestigung *f*
flange-mount gear pump Anflanschzahnradpumpe *f*, Anbauzahnradpumpe *f*

flange-mount pump Anflanschpumpe *f*, Flanschpumpe *f*
flange port Flanschanschluß *m*
flanged cartridge valve Ventil *n* für Bohrungseinbau mit Flanschbefestigung, Flansch-Einbauventil *n*, Einbauflanschventil *n*
flanged gear pump Anflanschzahnradpumpe *f*, Anbauzahnradpumpe *f*
flanged mount Flanschbefestigung *f*
flanged port Flanschanschluß *m*
flanged pump Anflanschpumpe *f*, Flanschpumpe *f*
flanged seal Hutmanschette *f*, Innenlippenring *m*
flapper *of the flapper-and-nozzle valve VALV OP* Prallplatte *f des Düse-Prallplatte-Ventils*
flapper[-and-nozzle] valve *VALV OP* Düse-Prallplatte- Ventil *n*
flare *v FITT* bördeln
flare angle *FITT* Bördelwinkel *m*
flared fitting Bördelverschraubung *f*
flareless fitting bördellose Verbindung *f*
flash point *FLUIDS* Flammpunkt *m*
flat-diaphragm cylinder Glattmembranzylinder *m*
flat-faced fitting O-Ring-Verschraubung *f*
flat-plate clearance *THEOR* Plattenspalt *m*
flat slide valve *DCV* Flachschieberventil *n*, Plattenschieberventil *n*
flat-valve axial piston pump wegegesteuerte (ventillose) Axialkolbenpumpe *f*
flexible hose Schlauch *m*, Schlauchleitung *f*
flexible line flexible (biegsame) Leitung *f*; Schlauchleitung *f*

flexible liner accumulator Schlauchspeicher *m*
flexible metallic hose Metallschlauch *m*
flexible seal elastische (nachgiebige) Dichtung *f*
flexible-tube pump Schlauchpumpe *f*
flexible-vane pump Pumpe *f* mit biegsamen Flügeln
flexible-wall cylinder Balgzylinder *m*
flexing-diaphragm element *LOG EL* Folienelement *n*
float-actuated (-controlled) *VALV OP* schwimmerbetätigt
float drain *FILTERS* schwimmerbetätigter Ablaß *m*
float level switch *RESERVOIRS* Schwimmerschalter *m*
float-operated *VALV OP* schwimmerbetätigt
float position *DCV* Freigangstellung *f*
floating piston *ACCUM* Freikolben *m*, fliegender Kolben *m*, Flugkolben *m*
flocculant *s.* flocculating agent
flocculating agent *FLUIDS* Ausflockungsmittel *n*, Flockungsmittel *n*, Flocker *m*
flood *v* *of a rotary screw compressor* *s.* lubricate *v*
flooded suction *PU/MOT* Ansaugen *n* unter Vorfülldruck
flow Förderstrom *m*; *flow rate* Volumenstrom *m*, Durchflußstrom *m*, *oft nur* Durchfluß *m*; *fluid motion* Fluidströmung *f*, Flüssigkeitsströmung *f*, Strömung *f*; *stream, current* Fluidstrom *m*, Flüssigkeitsstrom *m*, Strom *m*
flow *v* *eg through a restriction* strömen, fließen *z. B. durch eine Öffnung*, durchströmen, durchfließen *z. B. eine Öffnung*
flow amplification *LOG EL* Stromverstärkung *f*
flow amplifier *LOG EL* Stromverstärker *m*
flow board *s.* flowboard
flow capacity *PU/MOT* Durchflußkapazität *f*, Durchflußleistung *f*
flow cartridge Stromventil *n* für Bohrungseinbau, Einbaustromventil *n*
flow characteristics *of a restriction* *THEOR* Drosselcharakteristik *f*, Drosselverhalten *n*, Öffnungsverhalten *n*
flow coefficient *THEOR* Widerstandsbeiwert *m*, Durchflußbeiwert *m*, Verlustbeiwert *m*
flow-combiner valve Stromvereinigungsventil *n*
flow-compensated pump förderstromgeregelte Pumpe *f*
flow compensation *FCV* Stromkompensation *f*
flow compensator *pump control* Förderstromregler *m* Pumpenregeleinrichtung
flow control *FCV* Volumenstromsteuerung *f*, Volumenstromsteuereinrichtung *f*; *s.* flow-control circuit; *s.* flow-control valve
flow-control circuit Volumenstromsteuerkreislauf *m*, Volumenstromsteuersystem *n*, Volumenstromsteuerung *f*
flow-control fitting Stromsteuerverschraubung *f*, Dosierverschraubung *f*
flow-control servovalve *VALV OP* stromregelndes Servoventil *n*

flow-control system *s.* flow-control circuit
flow-control valve Stromventil *n*
flow curve *of a pressure regulator* Kennlinie *f*, Regelkurve *f* *eines Druckminderventils*
flow development *THEOR* Strömungsausbildung *f*
flow direction Strömungsrichtung *f*
flow-displacement profile plot *CIRC* Volumenstrom-Weg-Diagramm *n*, Volumenstrom-Zyklusprofil *n*, Volumenstromdiagramm *n*
flow-divider valve Stromteilventil *n*
flow equalizer *FCV* Gleichlaufventil *n*, Synchronisierventil *n*
flow equation *THEOR* Strömungsgleichung *f*
flow factor *s.* flow coefficient
flow fatigue strength *FILTERS* Stromschwankungsdauerfestigkeit *f*
flow fatigue test *FILTERS* Stromschwankungsdauerfestigkeitsprüfung *f*, Dauerermüdungsprüfung *f*
flow force *THEOR* Strömungskraft *f*
flow fuse *FCV* Leitungsbruchventil *n*, Rohrbruchventil *n*
flow gain *LOG EL* Stromverstärkung *f*, Stromübertragungsfaktor *m*
flow indicator Flüssigkeitsstromanzeiger *m*, Strömungsanzeiger *m*, Strömungswächter *m*
flow integrator Stromvereinigungsventil *n*
flow-limiting valve Stromsicherheitsventil *n*, Strombegrenzungsventil *n*
flow line Leitung *f*; Stromlinie *f*
flow measurement Volumenstrommessung *f*, Fluidstrommessung *f*, Durchflußmessung *f*, Strommessung *f*
flow meter *s.* flowmeter
flow mode *THEOR* Strömungsform *f*
flow nozzle *MEASUR* Meßdüse *f*, Düsendurchflußmesser *m*
flow path Strömungsweg *m*, Kanal *m*
flow pattern *THEOR* Strömungsform *f*
flow pattern plot *CIRC* Volumenstrom-Weg-Diagramm *n*, Volumenstrom-Zyklusprofil *n*, Volumenstromdiagramm *n*
flow peak *THEOR* Stromspitze *f*
flow plot *s.* flow pattern plot
flow-produced force *THEOR* Strömungskraft *f*
flow range Volumenstrombereich *m*, Durchflußstrombereich *m*
flow rate Volumenstrom *m*, Durchflußstrom *m*, *oft nur* Durchfluß *m*; *COMPR* Liefermenge *f*; *PU/MOT* Förderstrom *m*
flow rate measurement Volumenstrommessung *f*, Fluidstrommessung *f*, Flüssigkeitsstrommessung *f*, Durchflußmessung *f*, Strommessung *f*
flow rating Nennförderstrom *m*
flow recorder *MEASUR* Volumenstromschreiber *m*, Flüssigkeitsstromschreiber *m*, Durchflußschreiber *m*, Stromschreiber *m*
flow regulator Stromregelventil *n*, Stromregler *m*, Strombegrenzungsventil *n*
flow requirements Förderstrombedarf *m*
flow resistance *THEOR* Strömungs-

widerstand *m*, fluidischer Widerstand *m*, Fluidwiderstand *m*
flow resistance value *THEOR* Widerstandsbeiwert *m*, Durchflußbeiwert *m*, Verlustbeiwert *m*
flow sensor Flüssigkeitsstromsensor *m*, Volumenstromsensor *m*, Durchflußstromsensor *m*, Stromsensor *m*
flow separation *THEOR* Strömungsablösung *f*, Strömungsabriß *m*
flow stream Fluidstrom *m*, Flüssigkeitsstrom *m*, Strom *m*; *fluid motion* Fluidströmung *f*, Strömung *f*
flow surge *THEOR* Volumenstromspitze *f*, Stromspitze *f*
flow switch *MEASUR* Strömungsschalter *m*
flow temperature *FLUIDS* Fließpunkt *m*, Pourpoint *m*, Stockpunkt *m   als Pourpoint angegeben*
flow transducer *MEASUR* Volumenstromwandler *m*, Stromwandler *m*
flow velocity *THEOR* Strömungsgeschwindigkeit *f*
flowboard *CIRC* Kreislaufsimulator *m*, Schaltungssimulator *m*, Modellunterplatte *f*
flowmeter Volumenstrommesser *m*, Fluidstrommesser *m*, Durchflußmesser *m*, Strommesser *m*
fluctuate *v   pressure, flow* pulsieren, schwanken *Druck, Volumenstrom*
fluctuation *s.* pressure fluctuation; *s.* output fluctuation
flueric *LOG EL* Strahlelement *n*, Strömungselement *n*, Fluidic *n*; fluidisch, Strahl ... , Strömungs ... , Fluidik ...
flueric amplifier *LOG EL* Strahlverstärker *m*, Strömungsverstärker *m*, Fluidikverstärker *m*
fluid *FLUIDS* Flüssigkeit *f*, Fluid *n* (*pl* -e *oder* -s), *Zusammensetzungen s. auch unter* fluid power, hydraulic, hydro-, oil; *s auch* hydraulic fluid; *s. auch* oil; *s. auch* working fluid
fluid absorber Flüssigkeitsdämpfer *m*, viskoser Dämpfer *m*
fluid admittance   *flow rate/pressure drop THEOR* fluidische Admittanz *f*, Fluidadmittanz *f   Volumenstrom/ Druckabfall*
fluid amplifier *s.* fluidic amplifier
fluid analog device *LOG EL* analoges Fluidikelement *n*
fluid breadboard *CIRC* Kreislaufsimulator *m*, Schaltungssimulator *m*, Modellunterplatte *f*
fluid capacitance *THEOR* fluidische Kapazität *f*, Fluidkapazität *f*
fluid clutch hydrodynamische Kupplung *f*, Strömungskupplung *f*, Flüssigkeitskupplung *f*
fluid compatibility *SEALS* Flüssigkeitsverträglichkeit *f*
fluid component Hydraulikgerät *n*, Hydraulikbauteil *n*, Hydraulikelement *n*, Hydraulikkomponente *f*, Hydrogerät *n*, fluidtechnisches Bauelement *n*
fluid conditioning *PREPAR* Flüssigkeitsaufbereitung *f*, Flüssigkeitswartung *f*, Flüssigkeitspflege *f*
fluid conductance *THEOR* fluidischer Leitwert *m*, Fluidleitwert *m   reeller Leitwert*, Fluidkonduktanz *f*
fluid consumption Flüssigkeitsverbrauch *m*

fluid 48

fluid control hydraulische Steuerung *f*, Hydrauliksteuerung *f*, Hydrosteuerung *f*
fluid coupling hydrodynamische Kupplung *f*, Strömungskupplung *f*, Flüssigkeitskupplung *f*
fluid cylinder Hydraulikzylinder *m*, Hydrozylinder *m*
fluid drive *APPL* hydraulischer Antrieb *m*, Hydraulikantrieb *m*, Hydroantrieb *m*, fluidtechnischer Antrieb *m*
fluid dynamics *THEOR* Fluiddynamik *f*, Dynamik *f*, flüssiger und gasförmiger Körper, Strömungslehre *f*
fluid equipment manufacturer Fluidtechnikhersteller *m*
fluid flow Flüssigkeitsströmung *f*, Strömung *f*
fluid friction *THEOR* Flüssigkeitsreibung *f*; viskose (innere, flüssige) Reibung *f*
fluid gauge transducer hydraulische Kraftmeßdose *f*
fluid hammer *THEOR* hydraulischer Stoß *m*, Druckstoß *m*, Druckschlag *m*
fluid impedance *pressure drop/flow THEOR* fluidische Impedanz *f*, Fluidimpedanz *f* *Druckabfall/ Volumenstrom*
fluid inductivity *THEOR* fluidische Induktivität *f*, Fluidinduktivität *f*
fluid level *RESERVOIRS* Flüssigkeitsstand *m*
fluid life Flüssigkeitsgebrauchsdauer *f*, Flüssigkeitsgrenznutzungsdauer *f*
fluid logic *LOG EL* Fluidik *f*, Fluidlogik *f*, fluidische Digitaltechnik *f*
fluid logic device Logikelement *n*

fluid maintenance *PREPAR* Flüssigkeitsaufbereitung *f*, Flüssigkeitswartung *f*, Flüssigkeitspflege *f*
fluid make-up *in a closed circuit* Leckölergänzung *f*, Kreislaufspülung *f* *in einem geschlossenen Kreislauf*
fluid mechanics *THEOR* Fluidmechanik *f*, Mechanik *f* flüssiger und gasförmiger Körper
fluid motion Flüssigkeitsströmung *f*, Strömung *f*
fluid motor Hydraulikmotor *m*, Hydromotor *m*
fluid needs Förderstrombedarf *m*
fluid-operated hydraulisch angetrieben (getrieben, betrieben)
fluid port Hydraulikanschluß *m*
fluid power *fluid power technology* Fluidtechnik *f*, *s auch* hydraulic, *s auch* pneumatic, *Zusammensetzungen s auch unter* fluid, hydraulic, hydro-, oil; *THEOR* hydraulische Leistung *f*; pneumatische Leistung *f*
fluid power-actuated *VALV OP* fluidtechnisch (fluidisch) betätigt, fluidbetätigt
fluid power circuit Hydraulikkreislauf *m*, Hydrauliksystem *n*, Hydraulikschaltung *f*, Hydrokreislauf *m*, Hydrosystem *n*, fluidtechnischer Kreislauf *m*, fluidtechnisches System *n*
fluid power circuit diagram Hydraulikschaltplan *m*
fluid power component *s.* fluid component
fluid power connection Hydraulikanschluß *m*
fluid power-controlled *VALV OP*

fluidtechnisch (fluidisch) betätigt, fluidbetätigt
fluid power cylinder Hydraulikzylinder *m*, Hydrozylinder *m*
fluid power designer Fluidtechniker *m*, Fluidingenieur *m*
fluid-powered hydraulisch angetrieben (getrieben, betrieben)
fluid power engineer Fluidtechniker *m*, Fluidingenieur *m*
fluid power engineering Fluidtechnik *f*, *s auch* hydraulic, *s auch* pneumatic, *Zusammensetzungen s auch unter* fluid, hydraulic, hydro-, oil
fluid power maintainer Fluidtechnik-Instandhaltungsingenieur *m*
fluid power medium Arbeitsflüssigkeit *f*, Arbeitsmedium *n*, Arbeitsmittel *n*, Betriebsflüssigkeit *f*, Druck[übertragungs]mittel *n*, Energieübertragungsmittel *n*, Druckflüssigkeit *f*, *vgl* hydraulic fluid
fluid power-operated *VALV OP* fluidtechnisch (fluidisch) betätigt, fluidbetätigt
fluid power system Hydrauliksystem *n*, Hydraulikanlage *f*, Hydraulik *f*, Hydrosystem *n*, Hydroanlage *f*, fluidtechnisches System *n*, fluidtechnische Anlage *f*; *s.* fluid power circuit
fluid power technology *hydraulics* Hydraulik *f*, Ölhydraulik *f*; *s.* fluid power engineering
fluid replenishment *in a closed circuit* Leckölergänzung *f*, Kreislaufspülung *f* *in einem geschlossenen Kreislauf*
fluid reservoir Flüssigkeitsbehälter *m*, Hydraulikbehälter *m*, Ölbehälter *m*, *umgangssprachlich noch üblich:* Tank *m*
fluid resistance *THEOR* ohmscher *Widerstand, Eigenschaft* fluidischer Widerstand *m*, Fluidwiderstand *m*; Strömungswiderstand *m*, fluidischer Widerstand *m*, Fluidwiderstand *m*
fluid sample *FLUIDS* Flüssigkeitsprobe *f*
fluid sensor Flüssigkeitsstromsensor *m*, Volumenstromsensor *m*, Durchflußsensor *m*, Stromsensor *m*
fluid shock *THEOR* hydraulischer Stoß *m*, Druckstoß *m*, Druckschlag *m*
fluid statics *THEOR* Fluidstatik *f*, Statik *f* flüssiger und gasförmiger Körper
fluid tank Flüssigkeitsbehälter *m*, Hydraulikbehälter *m*, Ölbehälter *m*, *umgangssprachlich noch üblich:* Tank *m*
fluid-tight flüssigkeitsdicht, dicht
fluid-to-electrical transducer fluidisch-elektrischer Wandler *m*
fluid-to-fluid extrusion *from superpressure to somewhat lower pressure APPL* Gegendruckfließpressen *n* *ein Höchstdruckverfahren*
fluid-to-fuel heat exchanger kraftstoffgekühlter Wärmeübertrager *m*
fluid transfer cart *PREPAR* Ölwechselgerät *n*, Ölaufbereitungsgerät *n*, Ölreinigungsgerät *n*, Öl-Service- Aggregat *n*, Ölfiltrationswagen *m*
fluid trap Flüssigkeitstasche *f*, Totflüssigkeitsraum *m*
fluid under pressure Druckflüssigkeit *f*, Flüssigkeit *f* unter Druck
fluid valve Hydraulikventil *n*, Hydroventil *n*

fluid wear Flüssigkeitsabnutzung *f*, Flüssigkeitsverschleiß *m*
fluidborn aus der Flüssigkeit stammend, in der Flüssigkeit enthalten, Flüssigkeits ...
fluidic *LOG EL* Strahlelement *n*, Strömungselement *n*, Fluidic *n*; fluidisch, Strahl ... , Strömungs ... , Fluidik ...
fluidic-actuated *VALV OP* strahlelementbetätigt
fluidic amplifier *LOG EL* Strahlverstärker *m*, Strömungsverstärker *m*, Fluidikverstärker *m*
fluidic-controlled *VALV OP* strahlelementbetätigt
fluidic ear *LOG EL* pneumatischakustischer Sensor *m*
fluidic-operated *LOG EL* strahlelementbetätigt
fluidic oscillator *LOG EL* fluidischer Oszillator *m*
fluidic timer *LOG EL* fluidischer Zeitgeber *m*
fluidics Fluidik *f*, Fluidlogik *f*, fluidische Digitaltechnik *f*; *oft nur im engeren Sinn: Einsatz von Logikelementen ohne bewegte Teile*
fluidity *reciprocal of viscosity FLUIDS* Fluidität *f*, Flüssigkeit *f*, Fließvermögen *n* *Kehrwert der Viskosität*
flush *v* spülen
flushing fluid Spülflüssigkeit *f*
flushing pump Spülpumpe *f*
flutter *v* *eg of valves* rattern, schnarren, flattern *z. B. von Ventilen*
foam *FLUIDS* Schaum *m*
foam *v* *FLUIDS* schäumen
foam depressant *s.* foam inhibitor

foam formation *FLUIDS* Schaumbildung *f*, Schäumen *n*
foam inhibitor *FLUIDS* Anti-Schaum-Additiv *n*, Entschäumadditiv *n*, Schaumhemmer *m*
foaming resistance *FLUIDS* Widerstand *m* gegen Schaumbildung, Schäumwiderstand *m*
foaming tendency *FLUIDS* Schaumneigung *f*, Schäumneigung *f*
focused-jet amplifier *LOG EL* Ringstrahlverstärker *m*, Bündelstrahlverstärker *m*
fog *oil fog* Ölnebel *m*
fog lubricator *PREPAR* Nebelöler *m*, Ölnebelgerät *n*
follower control *APPL* Folgesteuerung *f*, Nachlaufsteuerung *f*
foot-actuated valve *%Drehpunkt am Pedalende* pedalbetätigtes (fußbetätigtes) Ventil *n*, Pedalventil *n*, Fußventil *n* *vgl aber* foot valve; *treadle-actuated valve* pedalbetätigtes (fußbetätigtes) Ventil *n*, Pedalventil *n*, Fußwippenventil *n*
foot mount *CYL* Fußbefestigung *f*
foot-mounted *VALV* unterflächenmontiert, *s auch* surface-mounted
foot-operated valve *s.* foot-actuated valve
foot valve Ablaßventil *n*, Bodenventil *n*, Fußventil *n*; *s.* counterbalance valve
force motor *VALV OP* Steuermotor *m* mit linearer Bewegung, Proportionalmagnet *m*
force rating Nennkraft *f*
forward stroke *CYL* Hinhub *m*
four ball wear test *FLUIDS* Vierkugelapparat-Test *m*, VKA-Test *m*

four-bolt flange *FITT* Vierlochflansch *m*, 4-Loch-Flansch *m*
four-lobed ring *SEALS* X-Ring *m*
four-port valve Ventil *n* mit vier Anschlüssen
four-position valve Vierstellungsventil *n*, 4-Stellungsventil *n*
four-way valve Vierwegeventil *n*, 4-Wege-Ventil *n*
FPDA = Fluid Power Distributors Association *USA*
FPEF = Fluid Power Educational Fund *USA*
FPS = Fluid Power Society *USA*
free air atmosphärische Luft *f*, Luft *f* im Ansaugzustand; *entrained (undissolved) air* *FLUIDS* freie (ungelöste) Luft *f*
free-float flowmeter Schwebekörper-Durchflußmesser *m*
free-flow coupling *FITT* Durchgangskupplung *f*
free-flow return Rücklauf *m* mit freiem Durchfluß, freier Rücklauf *m*
free jet *LOG EL* Freistrahl *m*
free of contamination *FLUIDS* verschmutzungsfrei, schmutzfrei
free-passage coupling *FITT* Durchgangskupplung *f*
free piston *ACCUM* Freikolben *m*, fliegender Kolben *m*, Flugkolben *m*
free return *s.* free-flow return
free speed *PU/MOT* Nullastdrehzahl *f*, Durchgangsdrehzahl *f*
free-surface accumulator Speicher *m* ohne Trennwand
free water *in the oil* freies Wasser *n*, *im Öl*
FR fluid *fire resistant fluid* schwerentflammbare Flüssigkeit *f*
fresh oil Neuöl *n*, Frischöl *n*

friction damper Reibungsdämpfer *m*
friction factor *s.* flow coefficient
friction loss *THEOR* Reibungsverlust *m*
friction torque Reib[ungs]moment *n*
frictional force Reib[ungs]kraft *f*
frictionless flow *THEOR* reibungsfreie Strömung *f*
FRL *filter-regulator-lubricator* Druckluft-Wartungseinheit *f*, Luftaufbereitungseinheit *f*, Luftaufbereiter *m*
front cover *s.* cylinder front cover
front end *CYL* Kolbenstangenraum *m*, Kolbenringraum *m*, Ringraum *m*; Kolbenstangenseite *f*, Ringraumseite *f*, Ausfahrseite *f*; kolbenstangenseitiges (ausfahrseitiges, ringraumseitiges) Zylinderende *n*
front-end chamber *CYL* Kolbenstangenraum *m*, Kolbenringraum *m*, Ringraum *m*
front-end pressure *CYL* kolbenstangenseitiger (ausfahrseitiger, ringraumseitiger) Druck *m*
front mont *CYL* ausfahrseitige Befestigung *f*
froth *FLUIDS* Schaum *m*
froth *v* *FLUIDS* schäumen
froth formation *FLUIDS* Schaumbildung *f*, Schäumen *n*
full displacement *PU/MOT* Vollverdrängungsvolumen *n*, Vollförderung *n*
full-displacement position *PU/MOT* Vollförderstellung *f*
full-flow filter Vollstromfilter *m,n*, Gesamtstromfilter *m,n*, Zwanglauffilter *m,n*, Hauptstromfilter *m,n*
full-flow filtration Vollstromfilterung *f*, Gesamtstromfilterung *f*,

function 52

Zwanglauffilterung *f*, Hauptstromfilterung *f*
full flow pressure *PCV* Druck *m* bei maximalem Strom
full-stroke position *CYL* Ausfahrstellung *f*
full symbol vollständiges Symbol (Schaltzeichen) *n*
function fitting Funktionsverschraubung *f mit Steuerfunktionen*
functional circuit Funktionsablaufplan *m*, Ablaufdiagramm *n*
functional symbol Funktionssymbol *n*, Funktionsschaltzeichen *n*
fuse *FCV* Leitungsbruchventil *n*, Rohrbruchventil *n*; Membransicherheitsventil *n*

G

gage *USA* s. gauge
gaiter *SEALS* Faltenbalg *m*
gallery Strömungsweg *m*, Kanal *m*
gallons per minute Einheit des Volumenstroms: $75{,}8 \cdot 10^{-6}$ m^3/s = $4{,}546$ L/min
gang *v valves* verketten *Ventile*
gang mounting Modulverkettung *f*, Batterieverkettung *f*
gang valve Ventil *n* für Modulverkettung (Batterieverkettung), Modulventil *n*
ganged *VALV OP* verkettungsfähig, verkettbar
ganged subbase (subplate) Verkettungsunterplatte *f*, verkettbare Unterplatte *f*
ganged valve Ventilbatterie *f*, *wenn höhenverkettet auch* Steuersäule *f*

gap *leakage gap* Leckspalt *m*; *filtering gap* Filterspalt *m*
gap formula *THEOR* Spaltformel *f*
gap height *THEOR* Spalthöhe *f*, Spaltweite *f*
gap length *THEOR* Spaltlänge *f*
gap loss *THEOR* Spaltverlust *m*
gap-type restriction *THEOR* Spaltdrossel *f*
gas bottle *ACCUM* Gasflasche *f*
gas chamber *ACCUM* Gasraum *m*
gas charge *ACCUM* Gasfüllung *f*
gas-charged accumulator gasbelasteter (hydropneumatischer) Speicher *m*, Gasdruckspeicher *m*
gas charging valve *ACCUM* Gas[füll]ventil *n*
gas-dynamic aerodynamisch, gasdynamisch
gas-loaded accumulator s. gas-charged accumulator
gas loading *ACCUM* s. gas charge
gas loading valve *ACCUM* s. gas charging valve
gas-oil accumulator s. gas-charged accumulator
gas precharge valve s. gas charging valve
gas release *FLUIDS* Gasausscheidevermögen *n*, Gasabscheidevermögen *n*
gas side *ACCUM* Gasseite *f*
gas solubility *FLUIDS* Gaslösungsvermögen *n*
gas-tight gasdicht
gas under pressure Druckgas *n*, Gas *n* unter Druck
gasdynamics Aerodynamik *f*, Gasdynamik *f*
gasket Flachdichtung *f*

gasket-mounted *of valves* flächenmontiert *von Ventilen*
gasmechanics Aeromechanik *f*, Gasmechanik *f*
gate valve Querschieberventil *n*
gauge Manometer *n*
gauge cock Manometerabsperrventil *n*
gauge glass *RESERVOIRS* Füllstandsglas *n*, Füllstandsauge *n*
gauge isolating valve Manometerabsperrventil *n*
gauge pressure Druck *m*, Überdruck *m*
gauge pulsation damper *s.* gauge pulsation snubber
gauge pulsation snubber Manometerdämpfer *m*
gauge tapping point Druckmeßstelle *f*, Druckmeßabzweig *m*
gear motor Zahnradmotor *m*
gear-on-gear motor Außenzahnradmotor *m*
gear-on-gear pump Außenzahnradpumpe *f*
gear-on-gear pump *or* motor Außenzahnradmaschine *f*, Außenzahnradeinheit *f*, Außenzahnradgerät *n*
gear pump Zahnradpumpe *f*
gear pump *or* motor Zahnradmaschine *f*, Zahnradeinheit *f*, Zahnradgerät *n*
gear pump with pressure-dependent axial clearance (with pressurized sideplates) Zahnradpumpe *f* mit axialem Spielausgleich (mit druckabhängigem Axialspalt)
gear-tip clearance *in a gear pump or motor* Zahnkopfspiel *n in einer Zahnradmaschine*

generated contamination *FLUIDS* Betriebsverschmutzung *f*, erzeugte Verschmutzung *f*
geometric displacement [volume] *PU/MOT* geometrisches Verdrängungsvolumen *n* (Fördervolumen *n*)
gerotor motor *progressing-tooth gear motor with fixed internal gear* Gerotormotor *m* trochoidenverzahnter Innenzahnradmotor mit feststehendem Zahnring
gerotor pump *vgl* gerotor motor
glass-fiber filter Glasfaserfilter *m,n*, Glasfiberfilter *m,n*
globe valve Plansitzventil *n*, Plattensitzventil *n*, Tellerventil *n*
governor *s.* speed governor
gpm *gallons per minute* Einheit des Volumenstroms: $75{,}8 \cdot 10^{-6}\ m^3/s = 4{,}546\ L/min$
gradual contraction *THEOR* allmähliche Verengung *f*
gradual enlargement *THEOR* allmähliche Erweiterung *f*
graphical diagram Schaltplan *m*
gravitational head *THEOR* Schweredruck *m*
gravity drain line Leckflüssigkeitsgefälleleitung *f*, Gefälleleckleitung *f*
gravity feeding *PU/MOT* Schwerkraftvorfüllung *f*
gravity-feed oiler Tropföler *m*
gravity flooding *s.* gravity feeding
gravity head *THEOR* Schweredruck *m*
gravity-held check valve gewichtsbelastetes Rückschlagventil *n*
gravity return *CYL* Schwerkraftrückzug *m*
gravity-return cylinder Zylinder *m* mit Schwerkraftrückzug

gravity-return ram schwerkraftrückgezogener Tauchkolben *m*
grip fitting Klemmringverschraubung *f*, Keilringverschraubung *f*, Druckringverschraubung *f*
groove *s.* circumferential groove; *s.* metering groove
guided cable cylinder Zylinder *m* mit geführtem Kabel
guided poppet geführter Ventilkegel *m*
gum *v* *eg valve spool* verklemmen, verkleben *z. B. Ventilkolben*
gum content *FLUIDS* Harzgehalt *m*
gum formation *FLUIDS* Harzbildung *f*, Verharzung *f*
gumming *of a valve spool* hydraulisches Verklemmen *n*, Klemmen *n*, Verkleben *n*, Kleben *n* *des Ventilkolbens*

H

halogenated fluid Flüssigkeit *f* auf Basis halogenierter Kohlenwasserstoffe
hand-actuated *VALV OP* handbetätigt
hand control *PU/MOT* Handstelleinheit *f*, manuelle Stelleinheit *f*
hand-controlled *s.* hand-actuated
hand-operated *s.* hand-actuated
hand pump Handpumpe *f*
hand stroker *PU/MOT* *s.* hand control
handwheel-actuated (controlled, operated) *VALV OP* handradbetätigt
handwheel stroker *PU/MOT* Handradstelleinheit *f*

hat seal Hutmanschette *f*, Innenlippenring *m*
head *THEOR* Druckhöhe *f*, Druck *m* *nutzbarer Druck*; *CYL* Zylinderkopf *m*; Kolbenkopf *m*
head cap (cover) *CYL* Zylinderkopfdeckel *m*, Zylinderkopf *m*, Dichtungsgehäuse *n*
head end [chamber] *CYL* Kolbenstangenraum *m*, Kolbenringraum *m*, Ringraum *m*
head end *CYL* Kolbenstangenseite *f*, Ringraumseite *f*, Ausfahrseite *f*; kolbenstangenseitiges (ringraumseitiges, ausfahrseitiges) Zylinderende *n*
head end pressure *CYL* kolbenstangenseitiger (ringraumseitiger, ausfahrseitiger) Druck *m*
head loss *THEOR* Druckverlust *m*
head mount *CYL* ausfahrseitige Befestigung *f*
headed piston *CYL* Kolben *m* mit Kopf; Scheibenkolben *m*
header Mehrfachverteiler *m*, Rohrverteiler *m*, Leitungsverteiler *m*, Verteilerverschraubung *f*
header [line] Verteilerleitung *f*, Sammelleitung *f*
heat balance Wärmebilanz *f*
heat conductivity *FLUIDS* Wärmeleitfähigkeit *f*
heat exchanger Wärmeübertrager *m*, Wärme[aus]tauscher *m*
heat regeneration *of a desiccant* Warmregenerierung *f* *von Trockenmittel*
heater *PREPAR* Heizer *m*, Vorwärmer *m*
heavy-walled *LINES* dickwandig

heel *of a lip ring* Manschettenrücken *m*
helical Bourdon tube Schraubenrohrfeder *f*
helical compressor Schraubenverdichter *m*
helical gear pump außenschrägverzahnte Zahnradpumpe *f*
helical spline rotary actuator Steilgewindeschwenkmotor *m*, Schraubkolbenschwenkmotor *m*, Schwenkmotor *m* mit Drehkeilwelle
helix rotary actuator *s.* helical spline rotary actuator
herringbone-gear pump pfeilverzahnte Zahnradpumpe *f*
HFI *s.* Hydraulic Fluid Index
HIC *hydraulic integrated circuit VALV* integrierter Hydraulikschaltkreis *m*, Hydraulik-IC *m*
high-flow coupling *FITT* Durchgangskupplung *f*
high-flow pump Pumpe *f* mit großem Förderstrom, großvolumige Pumpe *f*
high-grade pressure gauge Feinmeßmanometer *n*
high-low circuit Zweidruckkreislauf *m*
high-low pump Schaltpumpe *f*
high pressure *in HY from about 1500 up to about psi; in PN from about 25 up to about 150 psi* Hochdruck *m*, HD *in der HY von etwa 100 bis etwa 600 bar; in der PN von etwa 2 bis etwa 10 bar*
high-pressure circuit Hochdruckkreislauf *m* *auch: der Kreislauf mit dem höheren Druck*, HD-Kreislauf *m*

high-pressure filter Druckfilter *m,n*, Hochdruckfilter *m,n*
high-pressure hose Hochdruckschlauch *m*
high-pressure line Hochdruckleitung *f*
high-pressure pump Hochdruckpumpe *f*
high-pressure side Hochdruckseite *f*, HD-Seite *f* *die Seite mit dem höheren Druck*
high-pressure spray test *of fire resistance FLUIDS* Hochdruck-Sprühprüfung *f* *der Schwerentflammbarkeit*
high-speed [low-torque] motor Schnelläufermotor *m*
high-swell fluid stark quellende Flüssigkeit *f*
high-temperature fluid Hochtemperaturflüssigkeit *f*
high-torque [low-speed] motor Langsamläufermotor *m*, Langsamläufer *m*, Hochmomentmotor *m*
high viscosity index oil Mehrbereichsöl *n*
high viscous *FLUIDS* hochviskos, dickflüssig
high water base (content) fluid hochwasserhaltige Flüssigkeit *f*, "dickes Wasser" *n*
hi-lo ... *s.* high-low ...
HIP *hot isostatic pressing APPL* isostatisches Heißpressen *n*
hold position *DCV* Haltstellung *f*
holding circuit Haltekreislauf *m*
holding current *of a solenoid VALV OP* Haltestrom *m*, Dauerstrom *m* *eines Elektromagneten*
holding valve Halteventil *n*

hole configuration *VALV* Anschluß[loch]bild *n*, Bohrbild *n*
hollow O-ring *SEALS* Hohlrundring *m*
hollow spool *DCV* Hohlschieber *m*
homogeneous seal unverstärkte (unbewehrte, nicht armierte) Dichtung *f*
horizontal valve stacking Horizontalverkettung *f*, Längsverkettung *f*
horsepower limiter control *of a pump* Leistungsregelung *f einer Pumpe*
horsepower output Abtriebsleistung *f*, Abgabeleistung *f*, abgegebene Leistung *f*, Ausgangsleistung *f*
horsepower rating Nennleistung *f abgegebene Leistung*
horseshoe-shaped Bourdon tube Hufeisenrohrfeder *f*
hose Schlauch *m*, Schlauchleitung *f*
hose assembly Schlauchleitung *f einbaufertig montiert*; Schlaucheinbindung *f*
hose-break valve Schlauchbruchventil *n*
hose bundle Schlauchbündel *n*, Multischlauch *m*
hose carrier Schlauchstütze *f*, Schlauchsteg *m*, Schlauchbrücke *f*
hose clamp Schlauchschelle *f*, Schlauchklemme *f*
hose coupling *s.* hose fitting
hose cover Schlauchmantel *m*, Schlauchumhüllung *f*, Schlauchdecke *m*
hose end coupling *s.* hose fitting
hose fitting Schlauchverbindung *f*, Schlauchverschraubung *f*, *vgl* quick-disconnect coupling
hose guard Schlauchschutzhülle *f*
hose inner tube Schlauchseele *f*
hose kinking Schlauchverschlingung *f*
hose line Schlauchleitung *f*
hose protector Schlauchschutzhülle *f*
hose reel Schlauchaufroller *m*
hose reinforcement (support) Schlaucharmierung *f*, Schlauchverstärkung *f*
hot isostatic pressing *APPL* isostatisches Heißpressen *n*
housing *s.* filter housing; *s.* pump housing
H. P. *high pressure; in HY from about 1500 up to about psi; in PN from about 25 up to about 150 psi* Hochdruck *m*, HD *in der HY von etwa 100 bis etwa 600 bar; in der PN von etwa 2 bis etwa 10 bar*; *Zusammensetzungen s unter* high pressure
HPU *hydraulic power unit* Hydraulikaggregat *n*, Pumpenaggregat *n*, Motor-Pumpen-Aggregat *n*
HST *hydrostatic transmission* hydrostatisches Getriebe *n*
HTLS motor *high-torque low-speed motor* Langsamläufermotor *m*, Langsamläufer *m*, Hochmomentmotor *m*
HWBF *s.* HWCF
HWCF *high water base (content) fluid* hochwasserhaltige Flüssigkeit *f*, "dickes Wasser" *n*
hydraulic hydraulisch, *Zusammensetzungen s auch unter* hydro ... , fluid power, fluid, oil
hydraulic accessories *pl* Hydraulikzubehör *n*
hydraulic accumulator Druckflüssigkeitsspeicher *m*, Flüssigkeitsspeicher *m*, Hydraulikspeicher *m*, Hydrospeicher *m*

hydraulic actuator *VALV OP*
hydraulische Stelleinheit *f* (Betätigungseinrichtung *f*)
hydraulic amplifier *VALV OP*
hydraulischer Verstärker *m*
hydraulic axis *for positioning APPL* hydraulische Achse *f* *für Positionieraufgaben*
hydraulic brake *APPL* hydraulische Bremse *f*, Öldruckbremse *f*, Flüssigkeitsbremse *f*
hydraulic cabinet Hydraulikschrank *m*
hydraulic centering *VALV OP* hydraulische Zentrierung *f*, Druckzentrierung *f*
hydraulic circuit Hydraulikkreislauf *m*, Hydrauliksystem *n*, Hydraulikschaltung *f*, Hydrokreislauf *m*, Hydrosystem *n*, fluidtechnischer Kreislauf *m*, fluidisches System *n*
hydraulic circuit diagram Hydraulikschaltplan *m*
hydraulic clutch *APPL* hydrodynamische Kupplung *f*, Strömungskupplung *f*, Flüssigkeitskupplung *f*
hydraulic compliance *THEOR* hydraulische Nachgiebigkeit *f*
hydraulic component Hydraulikgerät *n*, Hydraulikbauteil *n*, Hydraulikelement *n*, Hydraulikkomponente *f*, Hydrogerät *n*, fluidtechnisches Bauteil *n*
hydraulic connection Hydraulikanschluß *m*
hydraulic control *APPL* hydraulische Steuerung *f*, Hydrauliksteuerung *f*, Hydrosteuerung *f*; *VALV OP* hydraulische Stelleinheit *f* (Betätigungseinrichtung *f*)
hydraulic coupling *s.* hydraulic clutch

hydraulic cylinder Hydraulikzylinder *m*, Hydrozylinder *m*
hydraulic diameter *THEOR* hydraulischer Durchmesser *m*
hydraulic drive *APPL* hydraulischer Antrieb *m*, Hydraulikantrieb *m*, Hydroantrieb *m*, fluidtechnischer Antrieb *m*
hydraulic element Hydraulikgerät *n*, Hydraulikbauteil *n*, Hydraulikelement *n*, Hydraulikkomponente *f*, Hydrogerät *n*, fluidtechnisches Bauteil *n*
hydraulic enclosure Hydraulikschrank *m*
hydraulic engineer Hydraulikingenieur *m*, Hydrauliker *m*
hydraulic excavator *APPL* Hydraulikbagger *m*
hydraulic failure *APPL* Ausfall *m* (Versagen *n*) der Hydraulik
hydraulic filter Hydraulikfilter *m,n*, Hydrofilter *m,n*, Flüssigkeitsfilter *m,n*
hydraulic fitting Hydraulikverschraubung *f*, Hydraulikverbindung *f*, *umgangssprachlich auch* Hydroarmatur *f*
hydraulic fluid Hydraulikflüssigkeit *f*, Hydraulikmedium *n*, *vgl* working fluid
hydraulic fluid chamber *ACCUM* Flüssigkeitsraum *m*
hydraulic fluid connection *ACCUM* Flüssigkeitsanschluß *m*
hydraulic fluid discharge valve *ACCUM* Flüssigkeitsventil *n*
hydraulic fluid index *SEALS Verhältnis der je Zeiteinheit verbrauchten Flüssigkeitsmenge zur in allen Anlagen enthaltenen Flüssigkeitsmenge = Maß für die Gesamtleckverluste*

hydraulic 58

z. B. in einem Betrieb; eingeführt von Mobil Oil Corp., New York, N. Y.
hydraulic fluid port *ACCUM* Flüssigkeitsanschluß *m*
hydraulic fluid side *ACCUM* Flüssigkeitsseite *f*
hydraulic fluid valve *ACCUM* Flüssigkeitsventil *n*
hydraulic-follower amplifier *VALV OP* Folgekolbenverstärker *m*, Steuerkolbenverstärker *m*
hydraulic-follower feedback *VALV OP* Folgekolbenrückführung *f*
hydraulic fuse *PCV* Membransicherheitsventil *n*
hydraulic horsepower hydraulische Leistung *f*
hydraulic hose Hydraulikschlauch *m*
hydraulic integrated circuit *VALV* integrierter Hydraulikschaltkreis *m*, Hydraulik-IC *m*
hydraulic jack *APPL* hydraulischer Hebel *m*, *s auch* jack
hydraulic liquid *s.* hydraulic fluid
hydraulic lock *of a valve spool* hydraulisches Verklemmen *n*, Klemmen *n*, Verkleben *n*, Kleben *n des Ventilkolbens*
hydraulic logic element (valve) *DCV* Logikventil *n*, Zweiwege-Einbauventil *n*, Wegesitzventil *n*, Cartridge *n*, Hydrologikventil *n*
hydraulic medium Hydraulikflüssigkeit *f*, Hydraulikmedium *n*, *vgl* working fluid
hydraulic motor Hydraulikmotor *m*, Hydromotor *m*
hydraulic nut runner *APPL* Hydraulikschrauber *m*
hydraulic oil Hydrauliköl *n*
hydraulic operator *VALV OP* hydraulische Stelleinheit (Betätigungseinrichtung) *f*
hydraulic output *of a pump* Förderleistung *f*
hydraulic port Hydraulikanschluß *m*
hydraulic power hydraulische Leistung *f*
hydraulic power pack (unit) *APPL* Hydraulikaggregat *n*, Pumpenaggregat *n*, Motor-Pumpen-Aggregat *n*
hydraulic pump Hydraulikpumpe *f*, Hydropumpe *f*
hydraulic radius *THEOR* hydraulischer Radius *m*
hydraulic reservoir Flüssigkeitsbehälter *m*, Hydraulikbehälter *m*, Ölbehälter *m*, *umgangssprachlich noch üblich:* Tank *m*
hydraulic resistor *THEOR* hydraulischer Widerstand *m*
hydraulic shock *THEOR* hydraulischer Stoß *m*, Druckstoß *m*, Druckschlag *m*
hydraulic stiffness *THEOR* hydraulische Steife *f*
hydraulic supply *APPL* Hydroversorgungseinheit *f*, *s auch* hydraulic power pack
hydraulic symbol Hydrauliksymbol *n*, Hydraulikschaltzeichen *n*
hydraulic system Hydraulikanlage *f*, Hydrauliksystem *n*, Hydraulik *f*, Hydrosystem *n*, Hydroanlage *f*, fluidtechnisches System *n*, fluidtechnische Anlage *f*; Hydraulikkreislauf *m*, Hydrauliksystem *n*, Hydraulikschaltung *f*, Hydrokreislauf *m*, Hydrosystem *n*, fluidtechnischer Kreislauf *m*, fluidisches System *n*
hydraulic tank Flüssigkeitsbehälter *m*, Hydraulikbehälter *m*, Ölbehäl-

ter *m*, *umgangssprachlich noch üblich:* Tank *m*
hydraulic trainer APPL Hydraulik-Lehrversuchsstand *m*
hydraulic valve Hydraulikventil *n*, Hydroventil *n*
hydraulic wrench APPL Hydraulikschrauber *m*
hydraulically detented VALV OP hydraulisch verriegelt
hydraulically piloted VALV OP hydraulisch vorgesteuert
hydraulically powered hydraulisch angetrieben (getrieben, betrieben)
hydraulics Hydraulik *f*, Ölhydraulik *f*
hydraulics industry APPL Hydraulikindustrie *f*
hydraulics manufacturer APPL Hydraulikhersteller *m*
hydraulics user APPL Hydraulikanwender *m*
hydro ... : *Zusammensetzungen s auch unter* hydraulic, fluid power, fluid, oil
hydrocarbon-base fluid kohlenwasserstoffbasische Flüssigkeit *f*
hydrodifferential transmission Getriebe *n* mit äußerer Leistungsverzweigung
hydrodynamic hydrodynamisch
hydrodynamic bearing APPL hydrodynamisches Lager *n*
hydrodynamic coupling APPL hydrodynamische Kupplung *f*, Strömungskupplung *f*, Flüssigkeitskupplung *f*
hydrodynamic lubrication hydrodynamische Schmierung *f*
hydrodynamic machine hydrodynamische Maschine *f*, Strömungsmaschine *f*
hydrodynamics Hydrodynamik *f*

hydroforming *hydrostatic forming of sheet metal* APPL Hydroformen *n* hydrostatisches Ziehen oder Tiefziehen
hydrokinetic machine s. hydrodynamic machine
hydrolytic stability FLUIDS Hydrolysebeständigkeit *f*
hydromechanic hydromechanisch, flüssigkeitsmechanisch
hydromechanical control PU/MOT mechanisch-hydraulische Stelleinheit *f*
hydromechanical transmission leistungsverzweigtes Getriebe *n*
hydromechanics Hydromechanik *f*, Flüssigkeitsmechanik *f*
hydropneumatic hydropneumatisch, pneumohydraulisch
hydropneumatic accumulator luftbelasteter Speicher *m*, hydropneumatischer Speicher *m*, Luftdruckspeicher *m*; gasbelasteter (hydropneumatischer) Speicher *m*, Gasdruckspeicher *m*
hydropneumatic pump hydropneumatische Pumpe *f*, Drucklufthydraulikpumpe *f*
hydropneumatics Hydropneumatik *f*, Pneumohydraulik *f*
hydrostat *pressure compensator in a compensated flow-control valve* Regeldrossel *f*, Aktivdrossel *f*, Druckwaage *f*, Differenzdruckregler *m im Stromregelventil*
hydrostatic hydrostatisch
hydrostatic bearing APPL hydrostatisches Lager *n*
hydrostatic drive hydrostatischer Antrieb *m*
hydrostatic extrusion APPL hydrostatisches Fließpressen *n*

hydrostatic lubrication *APPL* hydrostatische Schmierung *f*
hydrostatic transaxle *APPL* hydrostatische (hydraulische) Antriebsachse *f*, Hydroachse *f*
hydrostatic transmission hydrostatisches Getriebe *n*
hydrostatics Hydrostatik *f*

I

ice scraper *SEALS* Eisabstreifer *m*
ID *inside diameter* Innendurchmesser *m*, ID *m*; *s auch* cylinder inner diameter
ideal delivery (discharge, flow) rate *PU/MOT* idealer (ideeller) Förderstrom *m*
ideal fluid *THEOR* ideales (vollkommenes) Fluid *n*
ideal input flow rate *PU/MOT* idealer (ideeller) Schluckstrom *m*
idle stroke *CYL* Leerhub *m*
idler rotor *of a screw pump* Laufspindel *f* *einer Schraubenpumpe*
idling pressure *PU/MOT* Leerlaufdruck *m*
IFPA = International Fluid Power Exhibition *in Chicago, USA*
ignition test *FLUIDS* Entflammbarkeitsprüfung *f*, Entzündungsprüfung *f*
immersed getaucht, Tauch ... , Unteröl ...
immersed solenoid *VALV OP* öldruckdichter (nasser) Magnet *m*
immersed torque motor *VALV OP* öldruckdichter (nasser) Torquemotor *m*
immersion filter Behälterfilter *m,n*, Sumpffilter *m,n*
immersion thermometer Tauchthermometer *n*, Eintauchthermometer *n*
impact absorber Stoßdämpfer *m*, Stoßfänger *m*, Prallfänger *m*
impact loss *THEOR* Stoßverlust *m*
impact modulator *LOG EL* Gegenstrahlelement *n*, Impaktmodulator *m*
impedance *pressure drop/flow* *THEOR* Impedanz *f* Druckabfall/ Volumenstrom
impeller *COMPR* Laufrad *n*
impressed flow rate *CIRC* eingeprägter Volumenstrom *m*
impressed pressure *CIRC* eingeprägter Druck *m*
improver *FLUIDS* Verbesserer *m*, verbesserndes Additiv *n*
impulse resistance *LINES* Impulsfestigkeit *f*
impulse-type turbine motor Gleichdruckturbinenmotor *m*, Aktionsturbinenmotor *m*
in-depth filter Tiefenfilter *m,n*
in-depth filtration Tiefenfilterung *f*
in-line mounted valve Ventil *n* für Rohrleitungseinbau, Leitungsventil *n*, Rohrventil *n*
in-line axial piston pump Axialkolbenpumpe *f* mit antriebsachsparallelem Kolbenträger
in-line filter Filter *m,n* für Rohrleitungseinbau, Leitungsfilter *m,n*
in-line mounted valve Ventil *n* für Rohrleitungseinbau, Leitungsventil *n*, Rohrventil *n*
in-line mounting Rohrleitungseinbau *m*, Leitungseinbau *m*
in-line piston motor Reihenkolbenmotor *m*

in-line piston pump Reihenkolbenpumpe *f*
in-line valve *s.* in-line mounted valve
in-reservoir filter Behälterfilter *m,n*, Sumpffilter *m,n*
in stroke *CYL* Einfahrhub *m*, Rückhub *m*
in-tank filter *s.* in-reservoir filter
inching speed *PU/MOT* Kriechdrehzahl *f*
inclined-piston pump Schrägkolbenpumpe *f*
inclusion seal Dichtung *f* gegen Ausströmen
incompressible inkompressibel, nicht zusammendrückbar
index of performance *LOG EL* Leistungsverhältnis *n*
indicator *FILTERS* *s.* clogging indicator
inductance *THEOR* Induktivität *f*, induktiver Widerstand *m*
induction (inductive) flowmeter induktiver (elektromagnetischer) Durchflußmesser *m*
induction (inductive) pressure transducer induktiver (elektromagnetischer) Druckwandler *m*
industrial hydraulics *APPL* Industriehydraulik *f*, Stationärhydraulik *f*
industrial pneumatics *APPL* Industriepneumatik *f*, Stationärpneumatik *f*
industrial pressure gauge Betriebsmanometer *n*, Standardmanometer *n*
industrial-type fluid *FLUIDS* Industrieflüssigkeit *f*
inertial load Trägheitslast *f*
infinite-position valve *DCV* Ventil *n* mit Zwischenstellungen, Stetigventil *n*

inflammability *FLUIDS* Entflammbarkeit *f*, Entzündbarkeit *f*
inflammability test *FLUIDS* Entflammbarkeitsprüfung *f*, Entzündungsprüfung *f*
inflammable *FLUIDS* entflammbar, entzündbar
inflation pressure *ACCUM* Fülldruck *m*, Aufladedruck *m*, Vorspanndruck *m*
inflow Einlaß *m*, Eintritt *m*, Eingang *m*, Zulauf *m*, Zufluß *m*, Zutritt *m*, Zuführung *f*
ingest *v* einsaugen *durch Leckstellen*
ingress *v* *eg particles* eindringen lassen *z. B.* Schmutzteilchen
ingressed contamination *FLUIDS* eingedrungene Verschmutzung *f*
inhibited oil legiertes (additiviertes) Öl *n*, HL *n*
inhibiter *s.* inhibitor
inhibitor *FLUIDS* Inhibitor *m*, Verzögerungsmittel *n*, Hemmstoff *m*
initial contamination *FLUIDS* Anfangsverschmutzung *f*, Grundverschmutzung *f*, Urverschmutzung *f*
injected compressor einspritzgekühlter Verdichter *m*
injection lubrication Einspritzschmierung *f*
injection pump Einspritzpumpe *f*
injector *s.* oil injector; *s.* injection pump
inlet Einlaß *m*, Eintritt *m*, Eingang *m*, Zulauf *m*, Zufluß *m*, Zutritt *m*, Zuführung *f*
inlet chamber Einlaßkammer *f*, Eintrittskammer *f*, Zulaufkammer *f*, Zuflußkammer *f*, *in einer Pumpe auch* Saugraum *m*, Ansaugraum *m*, in

inlet

einem Verbraucher auch Druckkammer *f*, Druckraum *m*
inlet channel (duct) Einlaßkanal *m*, Eintrittskanal *m*, Zulaufkanal *m*, Zuflußkanal *m*, *in einer Pumpe auch* Saugkanal *m*, Ansaugkanal *m*, *in einem Verbraucher auch* Druckkanal *m*
inlet filter Saugfilter *m,n*
inlet flow [rate] Einlaß[volumen]strom *m*, Eintrittsstrom *m*, Zulaufstrom *m*, *bei einer Pumpe auch* Saugstrom *m*, Ansaugstrom *m*, *bei einem Motor auch* Schluckstrom *m*
inlet kidney *PU/MOT* Saugniere *f*
inlet line Einlaßleitung *f*, Eintrittsleitung *f*, Eingangsleitung *f*, Zulaufleitung *f*, Zuführleitung *f*, Zuflußleitung *f*, Zuleitung *f*, *an einer Pumpe auch* Saugleitung *f*, Ansaugleitung *f*, *an einem Verbraucher auch* Druckleitung *f*
inlet port Einlaßanschluß *m*, Eintrittsanschluß *m*, Zulaufanschluß *m*, Zuführanschluß *m*, Zuflußanschluß *m*, *bei einer Pumpe auch* Sauganschluß *m*, Ansauganschluß *m*, *bei einem Verbraucher auch* Druckanschluß *m*
inlet pressure Einlaßdruck *m*, Eintrittsdruck *m*, Eingangsdruck *m*, Zulaufdruck *m*, Zuführdruck *m*, *an einer Pumpe auch* Saugdruck *m*, Ansaugdruck *m*, *an einem Verbraucher auch* Speisedruck *m*
inlet section *of ganged valves* Eintrittsdeckel *m*, Zulaufdeckel *m einer Ventilbatterie*
inlet side Einlaßseite *f*, Eintrittsseite *f*, Eingangsseite *f*, Zulaufseite *f*, Zuflußseite *f*, *bei einer Pumpe auch* Saugseite *f*, *bei einem Verbraucher auch* Druckseite *f*
inlet throttling *COMPR* Ansaugdrosselung *f*, Saugdrosselung *f*
inlet valve *PU/MOT* Saugventil *n*, Einlaßventil *n*
inner diameter Innendurchmesser *m*, ID *m*; *s auch* cylinder inner diameter
inner tube *s.* hose inner tube
inner wall *s.* cylinder inner wall
input Einlaß *m*, Eintritt *m*, Eingang *m*, Zulauf *m*, Zufluß *m*, Zutritt *m*, Zuführung *f*
input air Zuluft *f*
input chamber Einlaßkammer *f*, Eintrittskammer *f*, Zulaufkammer *f*, Zuflußkammer *f*, *in einer Pumpe auch* Saugraum *m*, Ansaugraum *m*, *in einem Verbraucher auch* Druckkammer *f*, Druckraum *m*
input channel (duct) Einlaßkanal *m*, Eintrittskanal *m*, Zulaufkanal *m*, Zuflußkanal *m*, *in einer Pumpe auch* Saugkanal *m*, Ansaugkanal *m*, *in einem Verbraucher auch* Druckkanal *m*
input flow [rate] Einlaß[volumen]strom *m*, Eintrittsstrom *m*, Zulaufstrom *m*, *bei einer Pumpe auch* Saugstrom *m*, Ansaugstrom *m*, *bei einem Motor auch* Schluckstrom *m*
input horsepower Antriebsleistung *f*, Eingangsleistung *f*
input impeller *of a hydrodynamic coupling* Pumpenrad *n einer Strömungskupplung*
input jet *LOG EL* Eingangsstrahl *m*
input line Einlaßleitung *f*, Eintrittsleitung *f*, Eingangsleitung *f*, Zulaufleitung *f*, Zuführleitung *f*, Zuflußleitung *f*, Zuleitung *f*, *an einer*

Pumpe auch Saugleitung *f*, Ansaugleitung *f*, *an einem Verbraucher auch* Druckleitung *f*
input nozzle *of the jet-pipe valve* Strahldüse *f* *des Strahlrohrventils*
input port Einlaßanschluß *m*, Eintrittsanschluß *m*, Zulaufanschluß *m*, Zuführanschluß *m*, Zuflußanschluß *m*, *bei einer Pumpe auch* Sauganschluß *m*, Ansauganschluß *m*, *bei einem Verbraucher auch* Druckanschluß *m*
input power *s.* input horsepower
input pressure Einlaßdruck *m*, Eintrittsdruck *m*, Eingangsdruck *m*, Zulaufdruck *m*, Zuführdruck *m*, *an einer Pumpe auch* Saugdruck *m*, Ansaugdruck *m*, *an einem Verbraucher auch* Speisedruck *m*
input rpm Antriebsdrehzahl *f*
input shaft Antriebswelle *f*
input side Einlaßseite *f*, Eintrittsseite *f*, Eingangsseite *f*, Zulaufseite *f*, Zuflußseite *f*, *bei einer Pumpe auch* Saugseite *f*, *bei einem Verbraucher auuch* Druckseite *f*
input speed Antriebsdrehzahl *f*
input torque Antriebs[dreh]moment *n*
input tube *of the jet-pipe valve* Strahlrohr *n* *des Strahlrohrventils*
inrush current *VALV OP* Einschaltstrom *m*
insert *of a hose fitting* Tülle *f*, Nippel *m*
insert fitting Stecknippelverbindung *f*
inside diameter Innendurchmesser *m*, ID *m*; *s auch* cylinder inside diameter
inside nominal diameter Nennweite *f*, NW *f*, Nenndurchmesser *m*, ND *m* Innendurchmesser
inside-out flow filter element von innen nach außen durchströmtes Filterelement *n*
inside support ring *of a V-ring assembly* Stützring *m* *eines Dichtungssatzes*
install *v* **in parallel** parallelschalten
install *v* **in series** hintereinanderschalten, in Serie (Reihe) *f* schalten
installation-actuated seal vorgespannte Dichtung *f*
instrument air *for measurement and control PREPAR* Steuerluft *f* *für Meß- und Regelzwecke*
intake Einlaß *m*, Eintritt *m*, Eingang *m*, Zulauf *m*, Zufluß *m*, Zutritt *m*, Zuführung *f*
intake air Zuluft *f*; *COMPR* Ansaugluft *f*
intake air metering *PU/MOT* Zuluftdrosselung *f*, Zuluftsteuerung *f*
intake capacity *PU/MOT* Saugvermögen *n*, Ansaugvermögen *n*, Saugfähigkeit *f*
intake chamber Einlaßkammer *f*, Eintrittskammer *f*, Zulaufkammer *f*, Zuflußkammer *f*, *in einer Pumpe auch* Saugraum *m*, Ansaugraum *m*, *in einem Verbraucher auch* Druckkammer *f*, Druckraum *m*
intake channel (duct) Einlaßkanal *m*, Eintrittskanal *m*, Zulaufkanal *m*, Zuflußkanal *m*, *in einer Pumpe auch* Saugkanal *m*, Ansaugkanal *m*, *in einem Verbraucher auch* Druckkanal *m*
intake characteristics *PU/MOT* Saugverhalten *n*, Ansaugverhalten *n*
intake filter Saugfilter *m,n*

intake flow [rate] Einlaß[volumen]strom *m*, Eintrittsstrom *m*, Zulaufstrom *m*, *bei einer Pumpe auch* Saugstrom *m*, Ansaugstrom *m*, *bei einem Motor auch* Schluckstrom *m*
intake line Einlaßleitung *f*, Eintrittsleitung *f*, Eingangsleitung *f*, Zulaufleitung *f*, Zuführleitung *f*, Zuflußleitung *f*, Zuleitung *f*, *an einer Pumpe auch* Saugleitung *f*, Ansaugleitung *f*, *an einem Verbraucher auch* Druckleitung *f*
intake port Einlaßanschluß *m*, Eintrittsanschluß *m*, Zulaufanschluß *m*, Zuführanschluß *m*, Zuflußanschluß *m*, *bei einer Pumpe auch* Sauganschluß *m*, Ansauganschluß *m*, *bei einem Verbraucher auch* Druckanschluß *m*
intake section *of ganged valves* Eintrittsdeckel *m*, Zulaufplatte *f*
intake side Einlaßseite *f*, Eintrittsseite *f*, Eingangsseite *f*, Zulaufseite *f*, Zuflußseite *f*, *bei einer Pumpe auch* Saugseite *f*, *bei einem Verbraucher auch* Druckseite *f*
intake starvation *of a pump* unvollständige Füllung *f*, Abschnappen *n* *einer Pumpe*
intake throttling *COMPR* Ansaugdrosselung *f*, Saugdrosselung *f*
integral bypass filter Bypassfilter *m,n*, Filter *m,n* mit Umgehung
integral check valve eingebautes Rückschlagventil *n*
integral reservoir (tank) strukturintegrierter Behälter *m*
integral transmission, packaged transmission Kompaktgetriebe *n*

integrated circuit *s.* hydraulic integrated circuit
intensifier Druckübersetzer *m*
intensifier-operated druckübersetzergespeist
intensify *v* *pressure* verstärken Druck
interaction chamber *LOG EL* Wirkraum *m*, Wirkkammer *f*
interconnecting plate *in a valve stack* Zwischenplatte *f* *in einer Steuersäule*
intercooler Zwischenkühler *m*
intercooling Zwischenkühlung *f*
interference pressure *of a seal* Vorspanndruck *m einer Dichtung*
interior admission *PU/MOT* Innenbeaufschlagung *f*
intermediate position *DCV* Zwischenstellung *f*
intermediate trunnion mount *CYL* Mittenzapfenbefestigung *f*
internal backflow *PU/MOT* innerer Leckstrom *m*, Schlupfstrom *m*
internal drainage *VALV OP* innere Flüssigkeitsrückführung *f* *ohne Ablaufdruckentlastung*
internal gear motor Innenzahnradmotor *m*
internal gear pump Innenzahnradpumpe *f*
internal gear pump *or* **motor** Innenzahnradmaschine *f*, Innenzahnradeinheit *f*, Innenzahnradgerät *n*
internal leakage innerer Leckverlust *m*
internal locking device *CYL* innere Verriegelung *f*
internal piloting *VALV OP* Eigensteuerung *f*, Selbststeuerung *f*, passive Vorsteuerung *f*
internal pilot pressure *VALV OP*

Eigensteuerdruck *m*, Selbststeuerdruck *m*
internal pressure Innendruck *m*
internal seal Dichtung *f* gegen Ausströmen
internally drained *DCV* nicht ablaufdruckentlastet
interruptible jet sensor *LOG EL* Strahlunterbrechersensor *m*
interruptible sound beam sensor *LOG EL* akustischer Oszillationssensor *m*
interstage cooler Zwischenkühler *m*
interstage cooling Zwischenkühlung *f*
interval drain *of condensate* zeitgesteuerter Ablaß *m* *von Kondensat*
intra vane *innerer Flügel der Flügelzellenpumpe mit Doppelflügeln*
intrinsic viscosity *FLUIDS* Grenzviskosität *f*, Grundviskosität *f*
invert emulsion *FLUIDS* Wasser-in-Öl-Emulsion *f*
inverted flare fitting Rückbördelverbindung *f*
inviscid fluid *THEOR* nichtviskose (reibungsfreie) Flüssigkeit *f*
involute damper *to smoothen pressure pulsation* Zentrifugaldämpfer *m* *gegen Druckpulsation*
inward stroke *CYL* Einfahrhub *m*, Rückhub *m*
iodine value *FLUIDS* Iodzahl *f*, IZ *f*
IOP = Internationale Fachmesse für Ölhydraulik und Pneumatik *in Zürich/CH*
IP *index of performance LOG EL* Leistungsverhältnis *n*
I/P-transducer *MEASUR* Stromstärke-Druck-Wandler *m*
irrotational flow *THEOR* nichtrotierende (drehungsfreie) Strömung *f*

isolate *v* *eg flow, line* [ab]sperren, blockieren *z. B. Strom, Leitung*
isolating valve Absperrventil *n*
isolator *s.* shock absorber
isostatic pressing *APPL* isostatisches Pressen *n*
I. V. *s.* iodine value

J

jack *APPL* Hubzylinder *m*
jam *v* *eg valve spool* verklemmen, verkleben *z. B. Ventilkolben*
jet Strahl *m*
jet-action valve Strahlrohrventil *n*
jet angle Strahlwinkel *m*
jet compressor Strahlverdichter *m*
jet-controlled *LOG EL* strahlgesteuert
jet deflection element *LOG EL* Strahlablenkelement *n*
jet impact force *THEOR* Strahlkraft *f*
jet impact pressure *THEOR* Strahldruck *m*
jet pipe *of the jet-pipe valve VALV OP* Strahlrohr *n* *des Strahlrohrventils*
jet-pipe valve *VALV OP* Strahlrohrventil *n*
jet separation *THEOR* Strahlablösung *f*
jet thrust *s.* jet impact force
JIC = Joint Industrial Council *USA*
JIC cylinder *s.* tie-rod cylinder
Joukowski impact *sudden flow change THEOR* Joukowski-Stoß *m* *plötzliche Stromänderung*
joystick actuator (control, operator) *VALV OP* Joysticksteuerung *f*, Joy-

junction

stickbetätigung *f*, Steuerknüppelbetätigung *f*, Knüppelstelleinheit *f*
junction   Verbindungsleitung *f*, Verbindung *f*

K

K-ring   *SEALS* K-Ring *m*
kicker cylinder   *to overcome ram friction in presses APPL* Drückzylinder *m*   überwindet Pressenkolbenreibung
kidney   *PU/MOT* Steuerniere *f*
kidney machine   *s.* oil conditioner
kinematic viscosity   *FLUIDS* kinematische Viskosität *f*
kinetic head   *THEOR* Geschwindigkeitshöhe *f*   Druck
knife   *FILTERS* Spalträumer *m*, Abstreifer *m*, Kratzer *m*

L

L-ported filter   Filter *m,n* in L-Gehäuseausführung, L-Gehäuse-Filter *m,n*
L-ring   *SEALS* L-Ring *m*
L-type filter   *s.* L-ported filter
labyrinth seal   Labyrinthdichtung *f*
laminar   *THEOR* laminar, *s auch DE* Laminarität *f*
laminar flow   *THEOR* laminare Strömung *f*, Laminarströmung *f*
laminar-to-turbulent transition   *THEOR* Laminar-Turbulenz-Umschlag *m*
laminar-type restriction   *THEOR* Laminarwiderstand *m*, Drossel *f*
land   *DCV* Steuerschieberkolben *m*,

Steuerschieberbund *m*, Steuerschiebersteg *m*
land edge   *DCV* Schieberkante *f*, Steuerkolbenkante *f*
***n*-land spool**   *DCV n*kolbenschieber *m*
lap   *DCV* [positive] Überdeckung *f*
large-angle pump   *type of bent-axis axial piston pump* Großwinkelpumpe *f*   *Ausführung der Schrägachsen-Axkopumpe*
large-bore cylinder   Zylinder *m* mit großem Innendurchmesser
last-chance filter   Geräteschutzfilter *m,n*
lathe-cut ring   [form]gedrehter Dichtring *m*
lay line   *on hose outer wall* Kennzeichnungszeile *f*   *auf dem Schlauchmantel*
leak   Leckstelle *f*
leak *v*   lecken, leck sein;   *joint leaks oil*   durch eine Leckstelle Öl austreten lassen;   *air leaks into a chamber* eindringen, hineingelangen   *Luft durch eine Leckstelle in einen Raum*
leak-free   leckagefrei
leakage   Leckage *f*, Leckverlust *m*; Leckflüssigkeit *f*
leakage air   Leckluft *f*
leakage area   Leckquerschnitt *m*
leakage clearance   Leckspalt *m*
leakage compensation   Leckageausgleich *m*, Leckflüssigkeitsausgleich *m*
leakage current   *s.* leakage flow
leakage exhaust pressure   Leckdruck *m*
leakage flow   Leckflüssigkeitsstrom *m*, Leckverluststrom *m*, Leckagestrom *m*, Leckstrom *m*

leakage flow [rate]   Leck[volumen]-strom *m*, Verluststrom *m*
leakage fluid   Leckflüssigkeit *f*
leakage gap   Leckspalt *m*
leakage line   Leck[age]leitung *f*
leakage oil   Lecköl *n*
leakage passage   Leckflüssigkeitskanal *m*
leakage path   Leckweg *m*, Sickerweg *m*
leakage rate   Leck[volumen]strom *m*, Verluststrom *m*
leakage water   Leckwasser *m*
leakproof   *s.* leaktight
leaktight   [flüssigkeits]dicht
leaky   leck, leckend, undicht
leather packing   Lederdichtung *f*
leg   *LOG EL* Kanal *m*;   *drain leg* Kondensatleitung *f*, Entwässerungsleitung *f*, Ablaßleitung *f*
length of cushion   *CYL* Bremsweg *m*
length of hose   Schlauchstück *n*
length of stroke   *CYL* Hublänge *f*, Hub *m*
lens-shaped port (valve) plate   *PU/MOT* Linsensteuerplatte *f*
letter rings   Dichtringe mit Querschnitt in Form eines Buchstabens
level   *s.* fluid level
 level-control switch   *RESERVOIRS* Flüssigkeitsstandsschalter *m*, Füllstandsschalter *m*
 level-controlled drain   niveaugesteuerter Ablaß *m*
 level gauge   *s.* level indicator
 level indicator   *RESERVOIRS* Flüssigkeitsstandanzeiger *m*, Füllstandsanzeiger *m*
 level of contamination   *FLUIDS* Verschmutzungsgrad *m*, Schmutzniveau *n*
 level sight glass   Füllstandsglas *n*, Füllstandsauge *n*
 level switch   *s.* level control switch
lever-actuated valve   hebelbetätigtes Ventil *n*, Hebelventil *n*
lever control   *PU/MOT* Hebelstelleinheit *f*, Gestängestelleinheit *f*
lever feedback   *VALV OP* Hebelrückführung *f*
lever-operated valve   *s.* lever-actuated valve
lever stroker   *s.* lever control
life   *s.* filter life;   *s.* fluid life;   *s.* oil life
lift   *of valve poppet* Hub *m* im Sitzventil
lift   *vt und vi*   *poppet off (from) a seat* abheben   Ventilelement von einem Sitz
lift circuit   *APPL* Hubkreislauf *m*, Hubschaltung *f*
lifting cylinder   *APPL* Hubzylinder *m*
lightweight cylinder   Leicht[bau]zylinder *m*
limit valve   Ventil-Endschalter *m*, Endlagenschaltventil *n*, Grenzschaltventil *n*
limited rotation actuator   *s.* rotary actuator
line   Leitung *f*
 line area   *s.* line cross-sectional area
 line branching   Rohrverzweigung *f*, Leitungsverzweigung *f*
 line cross-sectional area   Leitungsquerschnittsfläche *f*, Leitungsquerschnitt *m*
 line filter   Filter *m,n* für Rohrleitungseinbau, Leitungsfilter *m,n*
 line loss   *THEOR* Leitungsverlust *m*   *häufig im pl gebraucht*
 line mounted valve   Ventil *n* für

Rohrleitungseinbau, Leitungsventil *n*, Rohrventil *n*
line mounting Rohrleitungseinbau *m*, Leitungseinbau *m*
line pressure Leitungsdruck *m*
line seal Liniendichtung *f*
line shock *s*. hydraulic shock
line size *s*. line cross-sectional area
line-to-line lap *DCV* Nullüberdeckung *f*
line trap Wasserfang *m*, Wasserablaß *m*, Entwässerung *f*
line under pressure unter Druck stehende (druckführende) Leitung *f*, Druckleitung *f*
line valve Ventil *n* für Rohrleitungseinbau, Leitungsventil *n*, Rohrventil *n*
linear actuator Schubkolbentrieb *m*, Zylinderantrieb *m*
linear force motor *VALV OP* Steuermotor *m* mit linearer Bewegung, Proportionalmagnet *m*
liner *s*. valve liner
lip seal Lippendichtung *f*, Lippenring *m*
liquid *s*. hydraulic fluid; *s*. fluid
liquid contamination *FLUIDS* Flüssigverschmutzung *f*
liquid current Flüssigkeitsstrom *m*, *s auch* flow
liquid-expansion thermometer Flüssigkeitsthermometer *n*
liquid flow *s*. liquid current
liquid metal *FLUIDS* Flüssigmetall *n*
liquid-piston (-ring) compressor Flüssigkeitsringverdichter *m*
liquid sealant Flüssigdichtstoff *m*
liquid spring Flüssigkeitsfeder *f*
liquid stream *s*. liquid current

load *v* with fluid, gas *ACCUM* aufladen, laden, füllen *mit Flüssigkeit, Gas*
load and flow-sensing pump *s*. load-compensated pump
load-carrying capacity *FLUIDS* Lasttragfähigkeit *f*
load cell hydraulische Kraftmeßdose *f*
load-compensated pump Load-sensing-Pumpe *f*, lastdruckgesteuerte Pumpe *f*
load-displacement cycle plot *CIRC* Last-Weg-Zyklusdiagramm *n*, Lastdiagramm *n*
load-flow characteristics Last-Volumenstrom-Verhalten *n*
load-holding circuit Haltekreislauf *m*
load-holding valve *DCV* Halteventil *n*
load-lowering valve Absenkventil *n*, Senk[brems]ventil *n*
load-magnitude cycle plot *CIRC* Lastgrößen-Zyklusdiagramm *n*, Lastdiagramm *n*
load plot *s*. load-displacement cycle plot; *s*. load-magnitude cycle plot
load pressure Lastdruck *m*
load-pressure feedback *VALV OP* Lastdruckrückführung *f*
load return *CYL* Lastrückzug *m*, Lastrückführung *f*
load-sensing control of a pump Load-sensing-Steuerung *f*, Lastdrucksteuerung *f*
load-sensing controller Load-sensing-Steuereinheit *f*
load-sensing directional control valve Load-sensing-Wegeventil *n*, LS-Ventil *n*

load-sensing flow compensator
s. load-sensing controller
load-sensing/pressure-limiting control Load-sensing-Steuerung (Lastdrucksteuerung) *f* mit Druckabschneidung *f*
load-sensing pump Load-sensing-Pumpe *f*, lastdruckgesteuerte Pumpe *f*
load sensitivity Lastempfindlichkeit *f*, Lastabhängigkeit *f*
load torque Lastmoment *n*
load variation Lastschwankung *f*
...-loaded accumulator ... belasteter Speicher *m*
loaded seal vorgespannte Dichtung *f*
loading *ACCUM* Aufladung *f*, Ladung *f*, Füllung *f*
loading line Fülleitung *f*, Vorfüllleitung *f*, Zuförderleitung *f*
lobed-element motor Drehkolbenmotor *m*, Rootsmotor *m*
lobed-element pump Drehkolbenpumpe *f*, Rootspumpe *f*
lobed ring *SEALS* X-Ring *m*
lobed-rotor compressor Kreiskolbenverdichter *m*
local pressure loss *THEOR* örtlicher Druckverlust *m*, *s auch DE* Formverlust *m*
lock *of a valve spool* hydraulisches Verklemmen *n*, Klemmen *n*, Verkleben *n*, Kleben *n* des *Ventilkolbens*
lock *v eg valve spool* verklemmen, verkleben *z. B. Ventilkolben*
lock-out valve Verriegelungsventil *n*
lock valve *DCV* Halteventil *n*
locking cylinder Verriegelungszylinder *m*
locking device *at any midstroke*

position CYL Feststelleinheit *f* *Feststellung in beliebiger Hubposition*
locking mechanism *VALV OP* Verstellsicherung *f*
logic cartridge valve Logikventil *n*, Zweiwege-Einbauventil *n*, Wegesitzventil *n*, Cartridge *n*, Hydrologikventil *n*
logic element s. logic cartridge valve
long hole restriction *THEOR* rohrförmiger Drosselwiderstand *m*
long-stroke cylinder Langhubzylinder *m*, Zylinder *m* mit überlangem Hub
long-stroke type liquid spring Langhubflüssigkeitsfeder *f*
loop flushing valve *CHECKS* Nachsaugeventil *n*, Spülventil *n*
loss Verlust *m* *wenn unbestimmt, meist im pl:* Verluste
loss coefficient Widerstandsbeiwert *m*, Durchflußbeiwert *m*, Verlustbeiwert *m*
loss pressure *THEOR* Verlustdruck *m*
lossless verlustlos, verlustfrei
low-flow efficiency Wirkungsgrad *m* bei niedrigem Volumenstrom *m*
low-flow pump Pumpe *f* mit niedrigem Förderstrom
low-friction cylinder reibungsarmer Zylinder *m*
low pressure *in HY up to about 1500 psi; in PN up to about 1.5 psi* Niederdruck *m*, ND *in der HY unter etwa 100 bar; in der PN unter etwa 0,1 bar*
low-pressure air Niederdruckluft *f*
low-pressure circuit Niederdruck-

kreislauf *m*   auch: der Kreislauf mit dem niederen Druck, ND-Kreislauf *m*
low-pressure filter   Niederdruckfilter *m,n*
low-pressure hose   Niederdruckschlauch *m*
low-pressure line   Niederdruckleitung *f*
low-pressure pump   Niederdruckpumpe *f*
low-pressure relief valve   *PU/MOT* Niederdruckbegrenzungsventil *n*
low-pressure side   Niederdruckseite *f*, ND-Seite *f*   die Seite mit dem niederen Druck
low-speed [high-torque] motor Langsamläufermotor *m*, Langsamläufer *m*, Hochmomentmotor *m*
low-temperature characteristics *FLUIDS* Tieftemperaturverhalten *n*, Kälteverhalten *n*
low-temperature fluidity   *FLUIDS* Kältefließfähigkeit *f*, Kältefließvermögen *n*
low-temperature oil   Kälteöl *n*
low-temperature viscosity   *FLUIDS* Kälteviskosität *f*, Tieftemperaturviskosität *f*
low viscous   *FLUIDS* niedrigviskos, dünnflüssig
low-volume pump   Pumpe *f* mit niedrigem Förderstrom
lowering valve   *DCV* Absenkventil *n*, Senk[brems]ventil *n*
L. P.   **low pressure**   **in HY up to about 1500 psi; in PN up to about 1.5 psi** Niederdruck *m*, ND in der HY unter etwa 100 bar; in der PN unter etwa 0,1 bar; Zusammensetzungen *s* unter low pressure

LSHT motor   *s.* low-speed high-torque motor
lubricate *v*   eg compressed air beölen, schmieren   *z. B.* Druckluft; *v* water fetten   Druckwasser
lubricated compressor   geschmierter (nicht ölloser) Verdichter *m*
lubrication   Schmierung *f*   *z. B.* von Luft, Fettung *f*   von Druckwasser
lubrication rate   Ölzuführstrom *m*, Ölförderstrom *m*
lubricator   *PREPAR* Ölgerät *n*, Öler *m*
lubricity   *FLUIDS* Schmierfähigkeit *f*
lubricity additive   *FLUIDS* Schmierfähigkeitsverbesserer *m*
lug mount   *CYL* Fußbefestigung *f*

M

Mach number   ratio of local flow to sound velocity *THEOR* Machzahl *f*, Ma   Verhältnis zwischen lokaler Strömungs- und Schallgeschwindigkeit
magnet piston   *s.* magnetic piston
magnet rodless cylinder   *s.* magnetic piston cylinder
magnetic cartridge   *FILTERS* Magnetpatrone *f*, Magnetsäule *f*
magnetic element   *FILTERS* Magnetelement *n*
magnetic filter   Magnetfilter *m,n*
magnetic piston   *CYL* Magnetkolben *m*
magnetic piston cylinder   Magnet[kolben]zylinder *m*
magnetic plug   *FILTERS* Magnetstopfen *m*
magnetic pre-separation   *FILTERS* Magnetvorabscheidung *f*

magnetic separator *FILTERS* Magnetabscheider *m*
main circuit Hauptkreislauf *m*
main flow Hauptstrom *m*
main jet *LOG EL* Hauptstrahl *m*, Leistungsstrahl *m*
main line Arbeitsleitung *f*, Hauptleitung *f*
main pump *of a hydrostatic transmission* Arbeitspumpe *f*, Hauptpumpe *f* *eines hydrostatischen Getriebs*
main spool *of a two-stage valve VALV OP* Hauptsteuerschieber *m*, Hauptkolben *m* *eines zweistufigen Ventils*
main stage *of a servovalve VALV OP* Hauptstufe *f* *eines Servoventils*
main valve *VALV OP* Hauptventil *n*
mains pressure Netzdruck *m*
mains tap Netzanschluß *m*
maintain position valve *LOG EL* Speicherventil *n*, Speicher *m*
make up *v* *for leakage* ausgleichen, kompensieren *Leckverluste*
make-up check Nachsaugeventil *n*, Spülventil *n*
make-up fluid Nachfüllflüssigkeit *f*, Ergänzungsflüssigkeit *f*, Frischöl *n*
make-up line Nachfülleitung *f*, Leckölergänzungsleitung *f*
make-up pump *in a closed circuit* Spülpumpe *f*, Hilfspumpe *f*
male adaptor *of a V-ring assembly* Stützring *m* *eines Dichtungssatzes*
male branch tee T-Verschraubung *f* mit Einschraubabzweig
male connector Einschraubverschraubung *f*
male coupling half *FITT* Kupplungsstecker *m*, Kupplungsnippel *m*

male elbow Einschraubwinkelverschraubung *f*
male end fitting Einschraubverschraubung *f*
male rotor *COMPR* Antriebsläufer *m*, Hauptläufer *m*
male run tee T-Verschraubung *f* mit Einschraubzapfen im durchgehenden Teil
male side tee T-Verschraubung *f* mit Einschraubabzweig
male support ring *of a V-ring assembly* Stützring *m* *eines Dichtungssatzes*
male threaded union Übergangsstück *n* mit Außengewinde, Gewindestutzen *m*
manifold *VALV* Unterplatte *f*, Anschlußplatte *f*
manifold *v* mittels Unterplatte[n] verketten, auf Unterplatte montieren
manifold block *VALV* maschinengebundene (kreislaufgebundene) Unterplatte *f*, Batterieplatte *f*, Sammelanschlußplatte *f*
manifold ignition test *FLUIDS* Entflammbarkeitsprüfung *f* an heißer Fläche
manifold mounting *VALV* Blockanbau *m*; *subplate mounting* Unterplattenanbau *m*
manifold valve Ventil *n* für Unterplattenanbau, Unterplattenventil *n*
manual control *PU/MOT* Handstelleinheit *f*, manuelle Stelleinheit *f*
manual drain Handablaß *m*
manual override *VALV OP* Handeingriff *m*
manual stroker *s.* manual control
manually actuated (controlled, operated) *VALV OP* muskelkraft-

betätigt; *hand-actuated (controlled, operated)* handbetätigt
manufacturer of compressed air equipment Pneumatikhersteller *m*
manufacturer of fluid power equipment Fluidtechnikhersteller *m*
manufacturer of hydraulic equipment Hydraulikhersteller *m*
marine hydraulics Schiffshydraulik *f*
mass flow *THEOR* Massetransport *m*, Massestrom *m*
mass flow [rate] Massestromstärke *f*, Massestrom *m*
master cylinder Geberzylinder *m*, Hauptzylinder *m*, Steuerzylinder *m*
material compatibility test *FILTERS* Werkstoffverträglichkeitsprüfung *f*, Materialverträglichkeitsprüfung *f*
maximum flow control valve Leitungsbruchventil *n*, Rohrbruchventil *n*
maximum rotation [angle] *PU/MOT* maximaler Drehwinkel *m* (Rotationswinkel *m*)
MCV *modulating control valve* modulationsgesteuertes Ventil *n*
mean bulk modulus *FLUIDS* Sekantenkompressionsmodul *m*, mittlerer Kompressionsmodul *m*
mean effective pressure Effektivdruck *m*
mechanical actuator (control) *VALV OP* mechanische Stelleinheit (Betätigungseinrichtung) *f*
mechanical control *PU/MOT* mechanische Stelleinheit *f*
mechanical efficiency *PU/MOT* mechanischer Wirkungsgrad *m*
mechanical loss mechanischer Verlust *m*

mechanical operator *s.* mechanical actuator
mechanical servo control *PU/MOT* mechanisch-hydraulische Stelleinheit *f*
mechanical stroker *s.* mechanical control
medium *s.* fluid power medium
medium-pressure hose Mitteldruckschlauch *m*
medium-pressure pump Mitteldruckpumpe *f*
medium-speed motor Mittelläufermotor *m*
medium-viscosity ... *FLUIDS* mittelviskos, mittelzähflüssig
membrane filter *eg of cellulose acetate* Membranfilter *m,n* *z. B. aus Celluloseacetat*
memory valve *LOG EL* Speicherventil *n*, Speicher *m*
MEP *mean effective pressure* Effektivdruck *m*
mesh *FILTERS* Maschenweite *f*
mesh leakage *PU/MOT* Zahneingriffsleckverlust *m*, Eingriffsverlust *m*
mesh size *FILTERS* Maschenweite *f*
metal-cased seal metallgefaßte Dichtung *f*
metal filter Metallfilter *m,n*
metal flexible hose Metallschlauch *m*
metal ribbon filter Drahtbandfilter *m,n*
metal scraper *SEALS* Metallabstreifer *m*
metal screen filter Drahtgewebefilter *m,n*, Drahtsiebfilter *m,n*
meter *v* *eg pressure, flow* steuern, [ein]regeln, [ein]stellen *z. B. Druck, Volumenstrom*
meter-in *v* zumessen, im Zulauf

(Zufluß) steuern, *als Adj. auch zusammengesetzt:* zulaufgesteuert
meter-in circuit Zumeßschaltung *f*, Kreislauf *m* mit zuflußseitiger Stromsteuerung
meter-in flow control zulaufseitige (zuflußseitige) Stromsteuerung *f*, Zulaufsteuerung *f*
meter-in valve Zulaufsteuerventil *n*, Zuflußsteuerventil *n*, Zumeßventil *n*, Zusteuerventil *n*
meter-out *v* im Ablauf (Abfluß) steuern, *als Adj. auch zusammengesetzt:* ablaufgesteuert
meter-out circuit Abmeßkreislauf *m*, Hemmschaltung *f*, Kreislauf *m* mit abflußseitiger Stromsteuerung
meter-out flow control, metering-out ablaufseitige (abflußseitige) Stromsteuerung *f*, Abflußsteuerung *f*
meter-out pump Ablaufdosierpumpe *f*, Hemmpumpe *f*
meter-out valve Abflußsteuerventil *n*, Ablaufsteuerventil *n*, Abmeßventil *n*
metering *s.* meter *v*
metering characteristics *of a restriction* Drosselcharakteristik *f*, Drosselverhalten *n*, Öffnungscharakteristik *f*
metering check valve Drosselrückschlagventil *n*, Einwegdrossel *f*
metering edge *DCV* Steuerkante *f*
metering groove *FCV* Steuerkerbe *f*, Drosselkerbe *f*
metering-in zulaufseitige (zuflußseitige) Stromsteuerung *f*, Zulaufsteuerung *f*; Zuluftdrosselung *f*, Zuluftsteuerung *f*
metering notch *DCV* Steuerkante *f*; *to reduce crossover pressure peaks* *DCV* Feinsteuernut *f* zur Verminderung von Schaltdruckspitzen; *FCV* Steuerkerbe *f*, Drosselkerbe *f*
metering orifice *DCV* Steueröffnung *f*, Steuerspalt *m*; *FCV* Steuerblende *f*, Drosselblende *f*
metering-out ablaufseitige (abflußseitige) Stromsteuerung *f*, Abflußsteuerung *f*; Abluftdrosselung *f*, Abluftsteuerung *f*
metering piston *FCV* Steuerkolben *m*, Drosselkolben *m*
metering pump Dosierpumpe *f*
micro cylinder Miniaturzylinder *m*, Minizylinder *m*, Mikrozylinder *m*, Kleinzylinder *m*
micro-fog (mist) lubricator *PREPAR* Mikronebelöler *m*
micro power pack Mikroaggregat *n*, *s auch* power pack
micro pump Miniaturpumpe *f*, Minipumpe *f*, Kleinpumpe *f*, Pumpe *f* in Kleinstbauweise, Mikropumpe *f*
microemulsion *FLUIDS* Mikroemulsion *f*
microfiberglass filter Mikroglasfaserfilter *m,n*, Mikroglasfiberfilter *m,n*
micronic filter Feinstfilter *m,n*, Mikrofilter *m,n*
micronic filtration Feinstfilterung *f*, Mikrofilterung *f*
micropneumatics Kleinpneumatik *f*, Minipneumatik *f*, Mikropneumatik *f*
microprocessor-controlled valve Digitalventil *n*
mid-position *DCV* *s.* middle position
middle position *DCV* Mittelstellung *f*, *vgl* neutral position
midget cylinder *s.* miniature cylinder

midstroke stopping *CYL Anhalten an beliebiger Stelle des Hubs*
mill cylinder *eine robuste Zylinderbauform*
mineral-oil base fluid mineralölbasische Flüssigkeit *f*
mini ... *s.* miniature ... , micro ...
miniature cylinder Miniaturzylinder *m*, Minizylinder *m*, Mikrozylinder *m*, Kleinzylinder *m*
miniature pneumatics Kleinpneumatik *f*, Minipneumatik *f*, Mikropneumatik *f*
miniature power pack Kleinaggregat *n*, Miniaggregat *n*, *s auch* power pack
miniature pump Miniaturpumpe *f*, Minipumpe *f*, Kleinpumpe *f*, Pumpe *f* in Kleinstbauweise, Mikropumpe *f*
miniature valve Miniaturventil *n*
minimum opening flow *PCV* Ansprech[volumen]strom *m*, Ansprechdurchflußstrom *m*
minimum total pulsation *LINES* Mindestimpulszahl *f*, MIZ *f*
miscibility *FLUIDS* Mischbarkeit *f*
missile fluid Raketenflüssigkeit *f*
mist *oil mist* Ölnebel *m*
mixed-flow compressor Diagonalverdichter *m*
mixed hydrostatic/mechanical transmission leistungsverzweigtes Getriebe *n*
mobile hydraulics Mobilhydraulik *f*, Fahrzeughydraulik *f*
modular *VALV* verkettungsfähig, verkettbar
modular manifold Verkettungsunterplatte *f*, verkettbare Unterplatte *f*
modular mounting (stacking) *VALV* Modulverkettung *f*, Batterieverkettung *f*
modular valve Ventil *n* für Modulverkettung *f* (Batterieverkettung *f*), Modulventil *n*
modulating control valve modulationsgesteuertes Ventil *n*
moisture separator *PREPAR* Wasserabscheider *m*
molecular sieve desiccant *PREPAR* Molekularsieb-Trockenmittel *n*
momentum-exchange fluid amplifier *LOG EL* Impulsverstärker *m*
monoblock valve unit Ventilkombination *f* in Monoblockbauweise
motion-controlled cylinder hubgesteuerter Zylinder *m*
motor Hydraulikmotor *m*, Hydromotor *m*
motor body Motorkörper *m*
motor case (casing) Motorgehäuse *n*
motor characteristics Motorkennlinie *f*, Motorverhalten *n*, Schluckverhalten *n*
motor connection *DCV* Motoranschluß *m*
motor control [circuit, system] *CIRC* Motorsteuerkreislauf *m*, Verbrauchersteuerkreislauf *m*, Motorsteuersystem *n*, Motorsteuerung *f*
motor curves plot Motorkenn[linien]feld *n*
motor cylinder Motorzylinder *m*
motor displacement *s.* motor displacement volume
motor displacement control *of a hydrostatic transmission* Sekundärverstellung *f*, Motorverstellung *f* *eines hydrostatischen Getriebes*
motor displacement volume Motor-

verdrängungsvolumen *n*, Schluckvolumen *n*
motor efficiency Motorwirkungsgrad *m*
motor housing Motorgehäuse *n*
motor mount Motorbefestigung *f*
motor noise Motorgeräusch *n*
motor port Motoranschluß *m*
motor/pump[/tank] unit *s.* hydraulic power pack
motor speed Motordrehzahl *f*
motorised valve elektromotorbetätigtes Ventil *n*
moulded lip *SEALS* vorgeformte Lippe *f*, Formlippe *f*
moulded seal vorgeformte Dichtung *f*, Formdichtung *f*, Profildichtung *f*
mount[ing] Befestigung *f*; Befestigungselement *n*
mount[ing] cavity Ventilaufnahme[bohrung] *f*
mount[ing] configuration Befestigungsart *f*
mount[ing] panel Ventilmontageplatte *f*, Ventilaufnahmeplatte *f*, Montagewand *f*
mount[ing] pattern Befestigungsart *f*
mount[ing] plate *VALV* Unterplatte *f*, Anschlußplatte *f*
mount[ing] style Befestigungsart *f*
move *vt and vi* **poppet off a seat** abheben *Ventilelement von einem Sitz*
moving-ball element *LOG EL* Kugelelement *n*
moving-part element *LOG EL* Element *n* mit bewegten Teilen
moving part logic *LOG EL* Steuertechnik *f* mit bewegten Teilen
moving-sphere element *s.* moving-ball element

MPL *s.* moving part logic
muffler *PREPAR* Schalldämpfer *m*, Geräuschdämpfer *m*
muffler-reclassifier Filter-Schalldämpfer *m*, kombinierter Schalldämpfer *m*, Abluftaufbereitungsgerät *n*
multi-connector *FITT* Rohrgruppenverbindung *f*, Multikupplung *f*, Kupplungsträger *m*
multi-fluid system Mehrfluidsystem *n*
multi-function pressure control valve funktionserweitertes Druckbegrenzungsventil *n*
multi-pass test *FILTERS* Multipass-Prüfung *f*
multi-plane swivel joint Mehrebenen-Rohrgelenk *n*, Universalrohrgelenk *n*
multi-pressure system Mehrdrucksystem *n*
multi-stage compressor mehrstufiger Verdichter *m*
multi-stage valve mehrstufiges Ventil *n*, Mehrstufenventil *n*
multi-station subplate *VALV* Mehrfachunterplatte *f*, Mehrfachanschlußplatte *f*
multi-tube connection *s.* multiconnector
multiple control nozzle element *LOG EL* Element *n* mit mehreren Steuerdüsen
multiple-flow pump Mehrstrompumpe *f*, Mehrkreispumpe *f*, Polyblockpumpe *f*
multiple-gear pump Mehrradpumpe *f*
multiple-hole exhaust *PREPAR* Vielröhrchendüse *f*

multiple-orifice shock absorber
PREPAR Drosselreihenstoßdämpfer *m*
multiple-orifice spear *CYL* Mehrfachdrossel-Bremsansatz *m*
multiple pump *s.* multiple-flow pump
multiple pump section Pumpensektion *f* einer Mehrstrompumpe
multiple-section pump *s.* multiple-flow pump
multiple-stage compressor mehrstufiger Verdichter *m*
multiple subbase (subplate, valve manifold) Mehrfachunterplatte *f*, Mehrfachanschlußplatte *f*
multiple wire braid hose Schlauch *m* mit mehrfacher Drahtgeflechteinlage

N

NAHAD = National Association of Hose and Accessories Distributors *USA*
naphthene-base oil naphthenbasisches (naphthenisches) Öl *n*
National Standard Taper thread NPT-Rohrgewinde *n* außen und innen keglig, mit Dichtstoff in den Gewindegängen
National Taper Pipe thread NPTF-Rohrgewinde *n* außen und innen keglig, selbstdichtend
native oil *FLUIDS* natives Öl *n*, Naturöl *n*
NBS flow factor früher in den USA verwendeter Widerstandsbeiwert für Druckluftbauteile (NBS = National Bureau of Standards)

NC *normally closed DCV* in Ruhestellung *f* geschlossen
NC valve *normally closed valve DCV* Öffnungsventil *n*
NCFP = National Conference on Fluid Power *USA*
neck mount *CYL* Befestigung *f* an ausfahrseitigem Gewindeansatz
needle valve Nadel[drossel]ventil *n*
negative droop Anstieg *m* einer im Idealfall horizontalen Kennlinie
negative overlap *DCV* negative Überdeckung *f*, Unterdeckung *f*
net area *CYL* wirksame Fläche *f*
net positive suction head *PU/MOT* erforderliche Saughöhe *f*
neutralization value *FLUIDS* Neutralisationszahl *f*, NZ *f*, Säurezahl *f*, SZ *f*
neutral position *if centered s also center position DCV* Nullstellung *f*, Ruhestellung *f*, Grundstellung *f*, neutrale Stellung *f*
new oil Neuöl *n*, Frischöl *n*
Newt Einheit der kinematischen Viskosität: 1 Newt = $1\ in^2/s$
newtonian fluid *THEOR* newtonsche Flüssigkeit *f*
Newton's law *of viscosity THEOR* Newtonsches Gesetz *n* der Zähigkeit
NFPA = National Fluid Power Association *USA*
nipple *of a hose fitting* Tülle *f*, Nippel *m*; *of a quick-connect coupling* Kupplungsstecker *m*, Kupplungsnippel *m*
NO *normally open DCV* in Ruhestellung *f* offen
NO valve *normally open valve DCV* Schließventil *n*
no-load flow *PU/MOT* Nullastvolumenstrom *m*

no-load stroke CYL Leerhub *m*
no-skive type hose fitting schällose (schärflose) Schlauchverbindung *f*
no-spill coupling nachtropffreie Kupplung *f*
no-spring check federloses Rückschlagventil *n*
nominal air consumption Nennluftverbrauch *m*
nominal area Nennquerschnittsfläche *f*, Nennquerschnitt *m*
nominal bore Nennweite *f*, NW *f*, Nenndurchmesser *m*, ND *m* Innendurchmesser
nominal capacity Nennleistung *f*, abgegebene Leistung
nominal conditions Nennbedingungen *fpl*
nominal delivery rate Nennförderstrom *m*
nominal diameter Nennweite *f*, NW *f*, Nenndurchmesser *m*, ND *m* Innendurchmesser
nominal discharge rate *s.* nominal delivery rate
nominal displacement PU/MOT Nennverdrängungsvolumen *n*
nominal filter fineness (rating) Nennfilterfeinheit *f*
nominal filtration fineness (rating) Nennfilterfeinheit *f*
nominal flow rate Nenn[volumen]strom *m*, Nenndurchfluß[strom] *m*
nominal force Nennkraft *f*
nominal horsepower *s.* nominal output
nominal ID *s.* nominal diameter
nominal output Nennleistung *f*, abgegebene Leistung
nominal output flow Nennförderstrom *m*

nominal pressure Nenndruck *m*, ND *m*
nominal rpm *s.* nominal speed
nominal size Nennquerschnittsfläche *f*, Nennquerschnitt *m*; Nennweite *f*, NW *f*, Nenndurchmesser *m*, ND *m*; Nenngröße *f*, NG *f* *bei Geräten Nennstrom und Nenndruck bzw. Nennweite*
nominal speed Nenndrehzahl *f*
nominal stroke [length] CYL Nennhublänge *f*, Nennhub *m*
nominal temperature Nenntemperatur *f*
nominal torque Nenn[dreh]moment *n*
nominal viscosity FLUIDS Nennviskosität *f*
non-adjustable retrictor FCV Konstantdrossel *f*, Festdrossel *f*, nicht [ver]stellbare Drossel *f*
non-aqueous fluid nicht-wasserhaltige (wasserfreie) Flüssigkeit *f*
non-balanced nicht[druck]entlastet, nicht[druck]ausgeglichen
non-bypass filter Filter *m,n* ohne Umgehungsmöglichkeit
non-centering directional valve Wegeventil *n* ohne selbsttätige Zentrierung
non-compensated *s.* non-balanced
non-corrosive FLUIDS nichtkorrosiv, nichtkorrosionswirksam, nicht korrodierend wirkend
non-cushioned cylinder Zylinder *m* ohne Endlagenbremsung
non-extruding SEALS extrusionsfest
non-flam *s.* non-flammable
non-flammability FLUIDS Nichtentflammbarkeit *f*, Unentflammbar-

normal 78

keit *f*, Unentzündbarkeit *f*, Unbrennbarkeit *f*, Flammbeständigkeit *f*
non-flammable *FLUIDS* nicht entflammbar, unentflammbar, unentzündbar, nicht brennbar, flammbeständig
non-flared fitting bördellose Verbindung *f*
non-inflammability *s*. non-flammability
non-inflammable *s*. non-flammable
non-integral transmission Ferngetriebe *n*, Getriebe *n* in aufgelöster Bauweise
non-lube *PREPAR* ölfrei, öllos, unbeölt, trocken
non-lube compressor Trockenlaufverdichter *m*, ölfreier (ölloser) Verdichter *m*
non-lube pneumatics Trockenluftpneumatik *f*
non-lubricated *s*. non-lube
non-lubricated water ungefettetes Wasser *n*
non-moving part element *LOG EL* reinfluidisches Element *n*, Element *n* ohne bewegte Teile
non-newtonian fluid *THEOR* nichtnewtonsche Flüssigkeit *f*
non-passing: normally non-passing *LOG EL* in Ruhestellung *f* nichtdurchlässig
non-passing position *DCV* gesperrte Stellung *f*, Sperrstellung *f*
non-positive displacement compressor *s*. turbocompressor
non-pressure compensated nicht[druck]entlastet, nicht[druck]ausgeglichen
non-pressurized drucklos, nicht unter Druck *m*, nicht vorgespannt

non-pressurized line drucklose Leitung *f*
non-resinous *FLUIDS* harzfrei
non-return valve Rückschlagventil *n*
non-reversible (-reversing) motor nicht umsteuerbarer Motor *m*
non-rotating rod cylinder verdrehgesicherter Zylinder *m*
non-rotational flow *THEOR* nichtrotierende (drehungsfreie) Strömung *f*
non-separator accumulator Speicher *m* ohne Trennwand
non-viscous fluid nichtviskose (reibungsfreie) Flüssigkeit *f*
non-viscous restriction *THEOR* Turbulenzwiderstand *m*, Blende *f*
non-water fluid *FLUIDS* nichtwasserhaltige (wasserfreie) Flüssigkeit *f*
normal conditions *of temperature and pressure* Normzustand *m*, Normalbedingungen *fpl*, Standardbedingungen *fpl* *von Temperatur und Druck*
normal position *if centered s also center position* *DCV* Nullstellung *f*, Ruhestellung *f*, Grundstellung *f*, neutrale Stellung *f*
normal pressure Normdruck *m*, Standarddruck *m*
normal temperature Normtemperatur *f*, Standardtemperatur *f*
normal volume Normvolumen *n*, Standardvolumen *n*
normally closed *DCV* in Ruhestellung *f* geschlossen
normally closed valve *DCV* Öffnungsventil *n*
normally non-passing *LOG EL* in Ruhestellung *f* nichtdurchlässig

normally open *DCV* in Ruhestellung *f* offen
normally open valve *DCV* Schließventil *n*
normally passing *LOG EL* in Ruhestellung *f* durchlässig
nose mount *CYL* Befestigung *f* an ausfahrseitigem Gewindeansatz
notch *s.* axial notch; *s.* metering notch
nozzle Düse *f*
nozzle-and-flapper valve *VALV OP* Düse-Prallplatte-Ventil *n*
nozzle noise *THEOR* Düsengeräusch *n*, Düsenlärm *m*
nozzle equation *THEOR* Düsengleichung *f*
nozzle throat *THEOR* Düsenverengung *f*, Düseneinschnürung *f*, Düsenhals *m*
NPSH *net positive suction head PU/MOT* erforderliche Saughöhe *f*
NPT port NPT-Anschluß *m*
NPT thread *National Standard Taper thread* NPT-Rohrgewinde *n außen und innen keglig, mit Dichtstoff in den Gewindegängen*
NPTF thread *National Taper Pipe thread* NPTF-Rohrgewinde *n außen und innen keglig, selbstdichtend*
number of drips *air-line lubricator* Tropfenanzahl *f Nebelöler*
number of pistons *PU/MOT* Kolben[an]zahl *f*
number of teeth *PU/MOT* Zähnezahl *f*
number of vanes *PU/MOT* Flügel[an]zahl *f*

nut-and-sleeve flare fitting Bördelverbindung *f* mit Klemmring
nutating-disk flowmeter Treibscheibenzähler *m*, Scheibenzähler *m*
nylon tubing Nylonrohr *n*

O

O-ring *SEALS* Rundring *m*, O-Ring *m*
O-ring plug Rundringstopfen *m*, O-Ringstopfen *m*
obliterate *v eg pores* verstopfen, zusetzen *z. B. Poren*
OD *outside diameter* Außendurchmesser *m*, AD *m*
off-centered spool exzentrisch liegender Kolben *m*
off-line filtration Sonderkreislauffilterung *f*, Zusatzkreisfilterung *f*
off-load period lastloser Zeitabschnitt *m*
offset position *PCV* Betätigungsstellung *f*
oil Öl *n, s auch* fluid, hydraulic fluid, *Zusammensetzungen s auch unter* hydraulic, hydro-, fluid power, fluid
oil remove the oil *from the air* Druckluft entölen
oil-air accumulator luftbelasteter Speicher *m*, hydropneumatischer Speicher *m*, Luftdruckspeicher *m*
oil base *FLUIDS* Grundöl *n*
oil-base fluid *FLUIDS* ölbasische Flüssigkeit *f*
oil brake *APPL* hydraulische Bremse *f*, Ölbremse *f*, Flüssigkeitsbremse *f*
oil breakdown *FLUIDS* Ölversagen *n*

oil carry-over *PREPAR* Öleintrag *m*
oil change Ölwechsel *m*
oil change period Ölwechselfrist *f*
oil circulation Ölumwälzung *f*, Ölumlauf *m*
oil column *THEOR* Ölsäule *f*
oil conditioner Ölwechselgerät *n*, Ölaufbereitungsgerät *n*, Ölreinigungsgerät *n*, Öl-Service-Gerät *n*, Ölfiltrationswagen *m*
oil content Ölgehalt *m*
oil cooler Ölkühler *m*
oil cooling Ölkühlung *f*
oil eye *RESERVOIRS* Füllstandsglas *n*, Füllstandsauge *n*
oil feed rate *PREPAR* Ölzuführstrom *m*, Ölförderstrom *m*
oil filter *PN* Ölfilter *m,n*
oil filtration unit *s.* oil conditioner
oil foam Ölschaum *m*
oil fog *PREPAR* Ölnebel *m*
oil-fog lubrication *PREPAR* Ölnebelschmierung *f*
oil-fog lubricator *PREPAR* Ölnebelgerät *n*, Nebelöler *m*
oil-free *PREPAR* ölfrei, öllos, unbeölt, trocken
oil-free compressor Trockenlaufverdichter *m*, ölfreier (ölloser) Verdichter *m*
oil froth *s.* oil foam
oil-gas accumulator gasbelasteter Speicher *m*, hydropneumatischer Speicher *m*, Gasdruckspeicher *m*
oil grade *SAE FLUIDS* Viskositätsklasse *f*, Viskositätsgrad *m* SAE
oil hydraulic system Ölhydraulikanlage *f*
oil-immersed solenoid *VALV OP* öldruckdichter (nasser) Magnet *m*

oil-immersed torque motor *VALV OP* öldruckdichter (nasser) Torquemotor *m*
oil-in-water emulsion *FLUIDS* Öl-in-Wasser-Emulsion *f*
oil-injected compressor öleingespritzter Kompressor *m*
oil injection Öleinspritzung *f*
oil injector Einspritzöler *m*, Öleinspritzelement *n*
oil level *RESERVOIRS* Ölstand *m*
oil life Ölgebrauchsdauer *f*, Ölgrenznutzungsdauer *f*
oil-lubricated *COMPR* ölgeschmiert
oil mist Ölnebel *m*
oil-mist lubrication *PREPAR* Ölnebelschmierung *f*
oil-mist lubricator *PREPAR* Ölnebelgerät *n*, Nebelöler *m*
oil purification *s.* oil reclamation
oil purifier *s.* oil reclaimer
oil recirculating Ölumwälzung *f*, Ölumlauf *m*
oil reclaimer Ölregenerator *m*, Ölrückgewinnungsgerät *n*, Ölwiederaufbereitungsgerät *n*
oil reclamation Ölregeneration *f*, Ölrückgewinnung *f*, Ölwiederaufbereitung *f*
oil reconditioner Ölwechselgerät *n*, Ölaufbereitungsgerät *n*, Ölreinigungsgerät *n*, Öl-Service-Gerät *n*, Ölfiltrationswagen *m*
oil remover *PREPAR* Ölabscheider *m*, Entöler *m*
oil-resistant ölbeständig, ölfest
oil sample Ölprobe *f*
oil scrubber *PREPAR* Ölabscheider *m*, Entöler *m*
oil-soluble öllöslich
oil sump Ölsumpf *m*

oil-to-air heat exchanger Öl-Luft-Wärmeübertrager *m*
oil-to-air oil cooler luftgekühlter Ölkühler *m*, Luft-Öl- Kühler *m*
oil-to-oil intensifier reinhydraulischer Druckübersetzer *m*
oil-to-water heat exchanger Öl-Wasser-Wärmeübertrager *m*
oil-to-water oil cooler wassergekühlter Ölkühler *m*, Wasser-Öl-Kühler *m*
oil trap Öltasche *f*, Totölraum *m*
oil under pressure Drucköl *n*, Öl *n* unter Druck
oil vapour Öldampf *m*
oil wear Ölabnutzung *f*, Ölverschleiß *m*
oilfree *PREPAR* ölfrei, öllos, unbeölt, trocken
oiliness *FLUIDS* Schmierfähigkeit *f*
oiliness agent *FLUIDS* Schmierfähigkeitsverbesserer *m*
oilless *PREPAR* ölfrei, öllos, unbeölt, trocken
oilless compressor Trockenlaufverdichter *m*, ölfreier (ölloser) Verdichter *m*
oiltight öldicht
on-line/off-line control *COMPR* Leerlaufregelung *f*, Durchlaufregelung *f*
on-off solenoid *VALV OP* Schaltmagnet *m* im Gegensatz zum Proportionalmagneten
one-piece cylinder Zylinder *m* mit nicht lösbar verbundenen Deckeln
one-stage valve direktgesteuertes (nicht vorgesteuertes, einstufiges) Ventil *n*
one-way seal coupling Einstrangabsperrkupplung *f*, Einwegabsperrkupplung *f*
one-way shut-off *FITT* Einstrangabsperrung *f*, Einwegabsperrung *f*
one-way valve Ventil *n* mit Durchfluß in nur einer Richtung
one wire braid hose Schlauch *m* mit einfacher Drahtgeflechteinlage
open: normally open *DCV* in Ruhestellung *f* offen
open: open to atmosphere *HY* mit dem Behälter *m* verbunden; *PN* mit der Atmosphäre *f* verbunden, nach außen entlüftend
open *v* *VALV* [sich] öffnen, *CHECKS* also entsperren, *PCV* also ansprechen
open bubble point *when bubbles appear over the entire surface of the element FILTERS* Offendruck *m* bei dem auf der gesamten Elementoberfläche Blasen austreten
open-center circuit *CIRC* Kreislauf *m* mit Umlaufstellung
open-center position *DCV* Umlaufstellung *f*
open-center valve *DCV* Ventil *n* mit Umlaufstellung
open circuit offener Kreislauf *m*
open-circuit hydrostatic transmission hydrostatisches Getriebe *n* in offenem Kreislauf
open end *of a conduit THEOR* freies Leitungsende *n*
open-loop circuit offener Kreislauf *m*
open-pore *FILTERS* offenporig
open position *DCV* geöffnete Stellung *f*, Offenstellung *f*
open termination *of a conduit THEOR* freies Leitungsende *n*
opening pressure surge *THEOR* Öffnungsdruckstoß *m*, Öffnungsschlag *m*

opening shock *s.* opening pressure surge
opening stroke *PCV* Öffnungsweg *m*
operate *v* *VALV OP* betätigen, [ver]stellen, steuern
operating cycle Arbeitszyklus *m*, Arbeitsspiel *n*, Betriebszyklus *m*, Betriebsspiel *n*
operating fluid Arbeitsflüssigkeit *f*, Arbeitsmedium *n*, Arbeitsmittel *n*, Betriebsflüssigkeit *f*, Druck[übertragungs]mittel *n*, Druckflüssigkeit *f*, *vgl* hydraulic fluid
operating force Stellkraft *f*, Betätigungskraft *f*, Verstellkraft *f*, Steuerkraft *f*, Schaltkraft *f*
operating line Arbeitsleitung *f*, Hauptleitung *f*
operating piston operating *VALV OP* Stellkolben *m*, Betätigungskolben *m*
operating pressure Betriebsdruck *m*, Arbeitsdruck *m*; *VALV OP* Stelldruck *m*, Betätigungsdruck *m*, Steuerdruck *m*
operating speed Betriebsdrehzahl *f*, Arbeitsdrehzahl *f*
operating stroke *CYL* Arbeitshub *m*, Nutzhub *m*
operating temperature Betriebstemperatur *f*, Arbeitstemperatur *f*
operating torque *VALV OP* Stellmoment *n*, Betätigungsmoment *n*, Verstellmoment *n*, Steuermoment *n*
operating viscosity Betriebsviskosität *f*
operator *s.* valve operator
opposed[-cylinders] compressor Verdichter *m* in Boxerbauart
orbit motor Orbitmotor *m*
orifice Turbulenzwiderstand *m*, Blende *f*; Drossel[stelle] *f*, Strömungswiderstand *m*
orifice *v* *flow* drosseln Volumenstrom
orifice area *THEOR* Drosselquerschnitt *m*, Drosselfläche *f*
orifice check valve Drosselrückschlagventil *n*, Einwegdrossel *f*
orifice equation *THEOR* Drosselgleichung *f*
orifice meter Blendendurchflußmesser *m*, Meßblende *f*
orifice union Drossel-Rohrverschraubung *f*, *vgl ED* union
oscillating motor *s.* rotary actuator
oscillating seal Gleitdichtung *f* *Dichtung für hin- und hergehende Bewegung*
oscillatory motor *s.* rotary actuator
out stroke *CYL* Ausfahrhub *m*, Vorhub *m*, Vorlauf *m*
outer diameter Außendurchmesser *m*, AD *m*
outflow *s.* outlet
outgas *v* *FLUIDS* Luft *f* freigeben (ausscheiden), entgasen, ausgasen
outlet Auslaß *m*, Austritt *m*, Ausgang *m*, Abfluß *m*, Ablauf *m*, Ausfluß *m*, Abführung *f*
outlet chamber Auslaßkammer *f*, Austrittskammer *f*, Ausgangsraum *m*, *in einer Pumpe auch* Förderraum *m*, Druckraum *m*
outlet channel Auslaßkanal *m*, Austrittskanal *m*, Abflußkanal *m*, *in einer Pumpe auch* Förderkanal *m*, Druckkanal *m*
outlet duct *s.* outlet channel
outlet flow [rate] Ausgangs[volumen]strom *m*, Austritts-

strom *m*, Abführstrom *m*, Ablaufstrom *m*
outlet kidney *PU/MOT* Druckniere *f*
outlet line Auslaßleitung *f*, Austrittsleitung *f*, Ausgangsleitung *f*, Abflußleitung *f*, Ablaufleitung *f*, Abführleitung *f*, Ableitung *f*, *bei einer Pumpe auch* Förderleitung *f*, Druckleitung *f*
outlet port Auslaßanschluß *m*, Austrittsanschluß *m*, Abführanschluß *m*, *bei einer Pumpe auch* Förderanschluß *m*, Druckanschluß *m*
outlet pressure Auslaßdruck *m*, Austrittsdruck *m*, Ausgangsdruck *m*, Ablaufdruck *m*, *bei einer Pumpe auch* Förderdruck *m*
outlet rate *s.* outlet flow rate
outlet section *of ganged valves* Austrittsdeckel *m*, Ablaufplatte *f* *einer Ventilbatterie*
outlet side Auslaßseite *f*, Austrittsseite *f*, Ausgangsseite *f*, Ablaufseite *f*, *bei einer Pumpe auch* Förderseite *f*, Druckseite *f*
outlet-to-inlet leakage *PU/MOT* Leckstrom *m* zwischen den Anschlüssen
outlet valve *of a pump* Auslaßventil *n*, Druckventil *n* *einer Pumpe*
output *PU/MOT* Förderung *f*; Förderstrom *m*; Pumpenverdrängungsvolumen *n*, Fördervolumen *n*
  output chamber Förderkammer *f*, Druckkammer *f*, Auslaßkammer *f*, Austrittskammer *f*, Ausgangsraum *m*
  output channel Förderkanal *m*, Druckkanal *m*, Auslaßkanal *m*, Austrittskanal *m*, Abflußkanal *m*
  output characteristics Förderkennlinie *f*, Förderverhalten *n*
  output flow Förderstrom *m*
  output flow control Förderstromverstellung *f*
  output flow direction Förderrichtung *f*
  output flow range Förderstrombereich *m*, Volumenstrombereich *m*
  output flow rate Förderstrom *m*
  output flow rating Nennförderstrom *m*
  output fluctuation Förderstrompulsation *f*, Förderstromfluktuation *f*, Förderungleichförmigkeit *f*, Förderschwankung *f*
  output horsepower *of a pump* Förderleistung *f*; Abtriebsleistung *f*, Abgabeleistung *f*, abgegebene Leistung *f*, Ausgangsleistung *f*
  output impeller *of a hydrodynamic coupling* Turbinenrad *n* *einer Strömungskupplung*
  output line Förderleitung *f*, Druckleitung *f*, Auslaßleitung *f*, Austrittsleitung *f*, Ausgangsleitung *f*, Abflußleitung *f*, Ablaufleitung *f*, Abführleitung *f*, Ableitung *f*
  output port Förderanschluß *m*, Druckanschluß *m*, Auslaßanschluß *m*, Austrittsanschluß *m*, Abführanschluß *m*
  output power *s.* output horsepower
  output pressure Förderdruck *m*, Auslaßdruck *m*, Austrittsdruck *m*, Ausgangsdruck *m*, Ablaufdruck *m*
  output pulsation Förderstrompulsation *f*, Förderstromfluktuation *f*, Förderungleichförmigkeit *f*, Förderschwankung *f*

outside

output rating Nennleistung *f* abgegebene Leistung
output ripple *s.* output pulsation
output shaft Abtriebswelle *f*
output side Förderseite *f*, Druckseite *f*, Auslaßseite *f*, Austrittsseite *f*, Ausgangsseite *f*, Ablaufseite *f*
output speed Abtriebsdrehzahl *f*
output torque Abtriebs[dreh]moment *n*, abgegebenes Moment *n*
output tube *LOG EL* Fangdüse *f*, Fangrohr *n*
output volume Pumpenverdrängungsvolumen *n*, Fördervolumen *n*
output volume per cycle Fördervolumen *n* je Umdrehung
outside diameter Außendurchmesser *m*, AD *m*
outside-in flow filter element von außen nach innen durchströmtes Filterelement *n*
outward stroke *CYL* Ausfahrhub *m*, Vorhub *m*, Vorlauf *m*
oval cylinder Ovalzylinder *m*, Flachzylinder *m*
oval ring *SEALS* Ovalring *m*
over-center control *PU/MOT* Übernullsteuerung *f*
over-center pump Pumpe *f* mit umkehrbarer Förderrichtung (mit Übernullsteuerung), Reversierpumpe *f*
over-center valve *PCV* Lastumkehr-Vorspannventil *n*, Umsteuer-Gegendruckventil *n*
overall efficiency *PU/MOT* Gesamtwirkungsgrad *m*
overflow Stromüberschuß *m*, Überstrom *m*

overflow loss *PCV* Überströmverlust *m*
overflow valve *s.* relief valve; *s.* bleed-off valve
overhead reservoir (tank) Hochbehälter *m*, Fallbehälter *m*
overlap *DCV* [positive] Überdeckung *f*
overload protection Überlastschutz *m*, Überlastsicherung *f*
overpressure unerwünschter Überdruck *m*
overpressure valve *s.* relief valve
override Eingriff *m* in einen automatischen Ablauf; *s* auch pressure override
override control *VALV OP* Korrekturbetätigung *f*, Hilfsbetätigung *f*, Zusatzbetätigung *f*
overrunning load durchgehende Last *f*, ziehende Last *f*
overspeed Überdrehzahl *f*
overspeed *v eg air motor* durchdrehen, durchfallen *z. B.* Druckluftmotor
oxidation inhibitor *FLUIDS* Oxidationsinhibitor *m*, Oxidationshemmer *m*
oxidation stability *FLUIDS* Oxidationsbeständigkeit *f*, Oxidationsstabilität *f*

P

pack *FILTERS s.* disk pack; *APPL s.* powerpack
package *APPL s.* power package
packaged control unit with integral sensing elements *APPL* Ventilinsel *f* Kompakteinheit mit integrierter Elektronik und Sensorik

packaged transmission Kompaktgetriebe *n*
packed spool *DCV* Schieber *m* mit Dichtelementen, gedichteter Schieber *m*
packing [seal] Dichtung *f*, Dichteinrichtung *f*; Dichtungssatz *m*, Dichtsatz *m*, Packung *f*
packingless-spool valve *DCV* schiebergedichtetes Ventil *n*
palm button-actuated (-operated) valve *VALV OP* handtasterbetätigtes Ventil *n*
pancake cylinder *Kurzhubzylinder mit sehr großem Durchmesser-Hub-Verhältnis und extrem kurzem Hub*
panel *s.* valve panel
paper-disk filter Papierscheibenfilter *m,n*
  paper filter Papierfilter *m,n*
  paper-plate filter Papierscheibenfilter *m,n*
  paper-filter Papierbandfilter *m,n*
  paper-filter Papierscheibenfilter *m,n*
parabolic spear *CYL* parabolischer Bremsansatz *m*
paraffin-base oil *FLUIDS* paraffinbasisches (paraffinisches) Öl *n*
parallel *v* parallelschalten
  parallel circuit *CIRC* Parallelschaltung *f*
  parallel [flow] heat exchanger Gleichstromwärmeübertrager *m*
  parallel thread *FITT* gerades Gewinde *n*
  parallel-walled clearance (gap) *THEOR* Parallelspalt *m*
partial circuit *CIRC* Teilkreislauf *m*
  partial delivery (discharge) *PU/MOT* Teilförderung *f*
  partial-filter Teilstromfilter *m,n*

partial-filtration Teilstromfilterung *f*
partial output *PU/MOT* Teilförderung *f*
particle *FLUIDS* Teilchen *n*, Partikel *n*
  particle count *FLUIDS* Teilchenzählung *f*, Partikelzählung *f*
  particle counter *FLUIDS* Teilchenzähler *m*, Partikelzähler *m*
particulate [matter] *FLUIDS* Menge *von* Teilchen *n*, Feststoffteilchen *n*, Schmutzteilchen *n*
partload efficiency *COMPR* Teillastwirkungsgrad *m*
Pascal's law *of hydrostatic pressure transmission THEOR* Pascalsches Gesetz *n* hydrostatische Druckfortpflanzung
pass *v* hindurchströmen (durchfließen, passieren) lassen; *v eg [through] a restriction* strömen, fließen *z. B. durch eine Öffnung*, durchströmen, durchfließen *z. B. eine Öffnung*
passage area *THEOR* Durchflußquerschnitt *m*, Durchlaßquerschnitt *m*
passageway Strömungsweg *m*, Kanal *m*
passing: normally passing *LOG EL* in Ruhestellung *f* durchlässig
passing position *DCV* geöffnete Stellung *f*, Offenstellung *f*
patch test *of oil contamination* Fleckvergleichsprüfung *f* *für den Schmutzgehalt des Öls*
patchboard *CIRC* Kreislaufsimulator *m*, Schaltungssimulator *m*, Modellunterplatte *f*
path line *THEOR* Stromlinie *f*, Strombahn *f*
PDC transducer *pressure to direct*

peak

current   *MEASUR* Gleichspannungsdruckwandler *m*
peak pressure   Spitzendruck *m*
pedal-actuated (-operated) valve   *Drehpunkt am Pedalende* pedalbetätigtes (fußbetätigtes) Ventil *n*, Pedalventil *n*, Fußventil *n*   *vgl aber* foot valve
perfect fluid   *THEOR* ideales (vollkommenes) Fluid *n*
peripheral leakage   *in a gear pump* Umfangsleckverlust *m*, Zahnkopfleckverlust *m*, Kopfverlust *m in einer Zahnradpumpe*
peripherally ported radial-piston pump   außenbeaufschlagte Radialkolbenpumpe *f*, Radialkolbenpumpe *f* mit Außenbeaufschlagung
permanent exhaust   Dauerentlüftung *f*
permanent fitting   nichtwiederverwendbare Verbindung *f*
permanent operating (working) pressure   Dauerbetriebsdruck *m*, Dauerarbeitsdruck *m*
permeability   *of a seal* Durchlässigkeit *f*, Lässigkeit *f*   *einer Dichtung*
petroleum[-based] fluid   *FLUIDS* mineralölbasische Flüssigkeit *f*
phase-shift muffler   Phasenkompensationsschalldämpfer *m*
phosphate ester base fluid   *FLUIDS* phosphatesterbasische Flüssigkeit *f*
phosphate ester fluid   *FLUIDS* Phosphatesterflüssigkeit *f*
piccolo spear   *multiple-orifice spear CYL* Mehrfachdrossel-Bremsansatz *m*
2-piece flare fitting   Bördelverbindung *f* ohne Klemmring
3-piece flare fitting   Bördelverbindung *f* mit Klemmring
piezoelectric pressure transducer   piezoelektrischer Druckwandler *m*, *oft* Quarzdruckwandler *m*
piezoresistive pressure transducer   piezoresistiver Druckwandler *m*
piggyback mounting   *PU/MOT* Huckepackbefestigung *f*
pilot   *s.* pilot valve
pilot *v*   *VALV OP* vorsteuern
pilot-actuate *v*   *s.* pilot *v*
pilot-actuated valve   *VALV OP* vorgesteuertes (indirekt betätigtes) Ventil *n*
pilot actuation   *VALV OP* Vorsteuerung *f*
pilot check [valve]   entsperrbares (gesteuertes) Rückschlagventil *n*
pilot circuit   *VALV OP* Vorsteuerkreislauf *m*, Steuerkreislauf *m*
pilot control   *PU/MOT* hydraulische Servo-Stelleinheit *f*
pilot control line   *VALV OP* Vorsteuerleitung *f*, Steuerleitung *f*
pilot fluid   *VALV OP* Vorsteuerflüssigkeit *f*
pilot line   *VALV OP* Vorsteuerleitung *f*, Steuerleitung *f*
pilot-operate *v*   *VALV OP* vorsteuern
pilot-operated check valve   entsperrbares (gesteuertes) Rückschlagventil *n*
pilot-operated relief valve   vorgesteuertes Druckbegrenzungsventil *n*
pilot-operated valve   *VALV OP* vorgesteuertes (indirekt betätigtes) Ventil *n*
pilot piston   *VALV OP* Vorsteuerkolben *m*, Vorsteuerschieber *m*
pilot port   *VALV OP* Vorsteueranschluß *m*
pilot pressure   *VALV OP* Vorsteuerdruck *m*

pilot pressure line *VALV OP* Vorsteuerleitung *f*, Steuerleitung *f*
pilot spool *s.* pilot piston
pilot stage *VALV OP* Vorsteuerstufe *f*
pilot stroker *PU/MOT* hydraulische Servo-Stelleinheit *f*
pilot valve *VALV OP* Vorsteuerventil *n*
piloted relief valve vorgesteuertes Druckbegrenzungsventil *n*
piloted valve *VALV OP* vorgesteuertes (indirekt betätigtes) Ventil *n*
pin *s.* solenoid pin
pin mounting *CYL* Stiftbefestigung *f*
pintle-ported radial-piston pump Radialkolbenpumpe *f* mit Steuerzapfen, wegegesteuerte (ventillose) Radialkolbenpumpe *f*
pintle valve *PU/MOT* Mittelzapfen *m*, Steuerzapfen *m*
pintle-valve radial-piston pump *s.* pintle-ported radial-piston pump
pipe Rohr *n* *gröberer Toleranz, größeren Innendurchmessers, größerer Wanddicke, Zusammensetzungen s auch unter* tube, line
pipe *v* *eg fluid to actuator* leiten *z. B. Flüssigkeit zum Verbraucher*; mit Rohren ausstatten, verrohren; Rohre verlegen
pipe *v* **in parallel** parallelschalten
pipe *v* **in series** hintereinanderschalten, in Serie (Reihe) *f* schalten
pipe bend Rohrkrümmer *m*, Rohrbogen *m*
pipe bend loss *THEOR* Rohrkrümmerverlust *m*, Rohrbogenverlust *m*, *s auch DE* Knieverlust *m*
pipe bender Rohrbiegevorrichtung *f*

pipe branching *THEOR* Rohrverzweigung *f*, Leitungsverzweigung *f*
pipe-cleaner test *of fire resistance FLUIDS* Pfeifenreinigerprüfung *f* der Schwerentflammbarkeit
pipe cushion Rohrschelle *f* *elastisch und dämpfend*
pipe cutter Rohrschneidegerät *n*, Rohrabschneider *m*
pipe deburrer Rohrentgrater *m*
pipe entrance friction *THEOR* Rohreintrittsreibung *f*, Rohreinlaufreibung *f*
pipe fitting Rohrverschraubung *f*, Rohrverbindung *f*
pipe fracture Rohrbruch *m*
pipe friction *THEOR* Rohrreibung *f*
pipe friction loss *THEOR* Rohrreibungsverlust *m*
pipe plug Rohrverschluß *m*
pipe roughness *THEOR* Rohrrauhigkeit *f*, Rohrrauheit *f*
pipe run Rohrleitungsabschnitt *m*
pipe support Rohrschelle *f*, Rohrhalter *m*
pipe thread *FITT* kegliges (konisches) Gewinde *n*
pipe trap Wasserfang *m*, Wasserablaß *m*, Entwässerung *f*
pipe wall Rohrwand[ung] *f*
pipeline Leitung *f* *allgemein*; Rohrleitung *f*
pipework *s.* piping
piping Gesamtheit von Rohren, Rohrnetz *n*, Rohrleitungssystem *n*, Leitungsnetz *n*, Verrohrung *f*, *s auch* pipe
piston *CYL* Kolben *m*; Scheibenkolben *m*
piston accumulator Kolbenspeicher *m*

piston-and-helix rotary actuator Steilgewindeschwenkmotor *m*, Schraubkolbenschwenkmotor *m*, Schwenkmotor *m* mit Drehkeilwelle
piston area *CYL* Kolbenfläche *f*
piston area ratio *CYL* Kolbenflächenverhältnis *n*
piston bearing *CYL* Kolbenführung *f*
piston-chain rotary actuator Kettenkolbenschwenkmotor *m*
piston-crank rotary actuator Kurbelschwenkmotor *m*
piston damper Kolbendämpfer *m*
piston diameter *CYL* Kolbendurchmesser *m*
piston displacement *CYL* Kolbenweg *m*, Kolbenbewegung *f*, Kolbenvorschub *m*
piston element *LOG EL* Kolbenelement *n*
piston face *CYL* Kolbenfläche *f*, Kolbenstirnseite *f*
piston force *CYL* Kolbenkraft *f*
piston head *CYL* Kolbenkopf *m*
piston head seal *CYL* Kolbenkopfdichtung *f*
piston motor *PU/MOT* Kolbenmotor *m*
piston movement *CYL* Kolbenweg *m*, Kolbenbewegung *f*, Kolbenvorschub *m*
piston number *PU/MOT* Kolben[an]zahl *f*
piston position *CYL* Kolbenposition *f*, Kolbenstellung *f*
piston pressure gauge Kolbenmanometer *n*
piston pressure switch Kolbendruckschalter *m*

piston priming *PU/MOT* Vorfüllung *f* eines Zylinderraums
piston pump *PU/MOT* Kolbenpumpe *f*
piston pump *or* **motor** Kolbenmaschine *f*, Kolbeneinheit *f*, Kolbengerät *n*
piston-rack rotary actuator Zahnstangenschwenkmotor *m*, Kolbenstange-Ritzel-Schwenkmotor *m*
piston ring *SEALS* Kolbenring *m*
piston rod *CYL* Kolbenstange *f*, Zusammensetzungen *s* unter rod
piston seal *CYL* Kolbendichtung *f*
piston sensor *CYL* Positionsgeber *m*, Stellungsgeber *m*, Lagegeber *m*, Zylindersensor *m*
piston separator *ACCUM* Trennkolben *m*
piston shoe *PU/MOT* Gleitschuh *m*
piston speed *CYL* Kolbengeschwindigkeit *f*
piston stroke *CYL* Kolbenhub *m*
piston-to-rod seal *CYL* Dichtung *f* zwischen Kolben und Kolbenstange
piston trajectory *PU/MOT* Kolbenlaufbahn *f*
piston travel *CYL* Kolbenweg *m*, Kolbenbewegung *f*, Kolbenvorschub *m*
piston traverse *s*. piston travel
piston-type cylinder Scheibenkolbenzylinder *m*
piston-type flowmeter Kolbenzähler *m*
piston valve *DCV* Schieberventil *n*, Kolbenventil *n*
piston wear *CYL* Kolbenverschleiß *m*
pitot-static tube for measurement of dynamic pressure Prandtlsches Staurohr *n*, Prandtlrohr *n* zur Messung des Geschwindigkeitsstaudrucks

pitot tube *for pressure measurement of air currents* Pitotrohr *n zur Druckmessung in Luftströmen*
pivot mount *CYL* Schwenkbefestigung *f*
plant air *PREPAR* Betriebsdruckluft *f*
plant air mains Werksdruckluftnetz *n*, Betriebsdruckluftnetz *n*
plastic tubing Plastikrohr *n*, Kunststoffrohr *n als Halbzeug*
plate *of a gear pump* Platte *f einer Zahnradpumpe*; *FILTERS* Filterscheibe *f*, Filterlamelle *f*, Filterplatte *f*
plate filter Scheibenfilter *m,n*, Lamellenfilter *m,n*, Plattenfilter *m,n*
plate heat exchanger Plattenwärmeübertrager *m*
plate pack *FILTERS* Scheibenpaket *n*, Lamellenpaket *n*, Plattenpaket *n*
plate valve *DCV* Flachschieberventil *n*, Plattenschieberventil *n*
pleated filter element Sternfilterelement *n*
pleated paper filter Papiersternfilter *m,n*
pleated screen filter Siebsternfilter *m,n*
plug *s.* coupling plug; *s.* threaded plug; *s.* valve plug
plug *vt a port* verstopfen, verschließen *eine Öffnung*; *vi a small orifice* verstopfen, [sich] zusetzen, zuwachsen *enge Drosselquerschnitte*; *FILTERS* verstopfen, zusetzen, *bei Oberflächenfiltern auch* verlegen
plug end fitting Einschraubverschraubung *f*
plug-in fitting Steckverbindung *f*, Einsteckverbindung *f*

plug-in pump Einbaupumpe *f*, Steckpumpe *f*
plug valve *DCV* Drehschieberventil *n*
plugging *FILTERS* Verstopfung *f*, Zusetzen *n*, *bei Oberflächenfiltern auch* Verlegung *f*
plunger *CYL* Tauchkolben *m*; *VALV* Ventilstößel *m*
plunger-actuated valve stößelbetätigtes Ventil *n*
plunger cylinder Tauchkolbenzylinder *m*
plunger-operated valve stößelbetätigtes Ventil *n*
plunger pump Tauchkolbenpumpe *f*, Kolbenpumpe *f*; *peripherally ported radial-piston pump* außenbeaufschlagte Radialkolbenpumpe *f*
plunger valve *DCV* Schieberventil *n*, Kolbenventil *n*
pneumatic pneumatisch, *Zusammensetzungen s auch unter* pneumo-, compressed air, air
pneumatic accessories *pl* Pneumatikzubehör *n*
pneumatic actuator pneumatische Stelleinheit (Betätigungseinrichtung) *f*, Druckluftstelleinheit *f*
pneumatic axis *for positioning APPL* pneumatische Achse *f für Positionieraufgaben*
pneumatic bearing *APPL* Luftlager *n*
pneumatic circuit Pneumatikschaltung *f*, Pneumatiksystem *n*, Druckluftschaltung *f*, Druckluftsystem *n*, Luftsystem *n*
pneumatic clutch *APPL* pneumatische Kupplung *f*, Druckluftkupplung *f*

pneumatic component Pneumatikgerät *n*, Pneumatikbauteil *n*, Pneumatikelement *n*, Pneumatikkomponente *f*
pneumatic connection Pneumatikanschluß *m*, Druckluftanschluß *m*, Luftanschluß *m*
pneumatic control *APPL* Pneumatiksteuerung *f*, Druckluftsteuerung *f*; *VALV OP* pneumatische Stelleinheit (Betätigungseinrichtung) *f*, Druckluftstelleinheit *f*
pneumatic conveying *APPL* pneumatisches Fördern *n*
pneumatic coupling *s.* one-way seal coupling
pneumatic cylinder Pneumatikzylinder *m*, Druckluftzylinder *m*, Luftzylinder *m*
pneumatic drive *APPL* Pneumatikantrieb *m*, pneumatischer Antrieb *m*, Druckluftantrieb *m*
pneumatic element Pneumatikgerät *n*, Pneumatikbauteil *n*, Pneumatikelement *n*, Pneumatikkomponente *f*
pneumatic engineer Pneumatikingenieur *m*, Pneumatiker *m*
pneumatic filter Druckluftfilter *m,n*, Luftfilter *m,n*, Pneumatikfilter *m,n*
pneumatic hose Druckluftschlauch *m*, Luftschlauch *m*, Pneumatikschlauch *m*
pneumatic line Druckluftleitung *f*, Luftleitung *f*, Pneumatikleitung *f*
pneumatic linear actuator Pneumatikzylinder *m*, Druckluftzylinder *m*, Luftzylinder *m*
pneumatic logic control pneumatische Logik (Digitaltechnik) *f*
pneumatic operator *VALV OP* pneumatische Stelleinheit (Betätigungseinrichtung) *f*, Druckluftstelleinheit *f*
pneumatic power pneumatische Leistung *f*
pneumatic relay pneumatisches Relais *n*
pneumatic symbol Pneumatiksymbol *n*, Pneumatikschaltzeichen *n*
pneumatic system Pneumatikanlage *f*, Pneumatiksystem *n*, Pneumatik *f*, Druckluftanlage *f*; Pneumatikschaltung *f*, Pneumatiksystem *n*, Druckluftschaltung *f*, Luftsystem *n*
pneumatic tool *APPL* Druckluftwerkzeug *n*
pneumatic valve Pneumatikventil *n*, pneumatisches Ventil *n*, Druckluftventil *n*, Luftventil *n*
pneumatically piloted *VALV OP* druckluftvorgesteuert, pneumatisch vorgesteuert
pneumatically powered pneumatisch angetrieben, [druck]luftgetrieben, [druck]luftbetrieben
pneumatics Pneumatik *f*
pneumatics industry Pneumatikindustrie *f*
pneumatics manufacturer Pneumatikhersteller *m*
pneumatics user Pneumatikanwender *m*
pneumo ... : *Zusammensetzungen s auch unter* pneumatic, compressed air, air
pneumonic control *LOG EL s.* pneumatic logic control
pneumonic element *LOG EL* Strahlelement *n*, Fluidic *n*, Strömungselement *n*
pneumostatic pneumostatisch
pocketed oil *PU/MOT* Quetschöl *n*

point-of-use filter Geräteschutzfilter *m,n*
pole height *FLUIDS* Viskositätspolhöhe *f*, Polhöhe *f*, VPh *f*
pole viscosity *FLUIDS* Polviskosität *f*
polyethylene tubing Polyethylenrohr *n*
polyglycol ester fluid polyglycolesterbasische Flüssigkeit *f*
polyglycol solution wäßrige Polymerlösung *f*, Wasser-Glycol-Lösung *f*, glycolbasische Flüssigkeit *f*
polyol ester fluid Polyolesterflüssigkeit *f*
polyurethane tubing Polyurethanrohr *n*
poppet *DCV* Kegelsitzelement *n* *stumpfer Kegel*, Sitzventilkolben *m*, Ventilkegel *m*
poppet check valve Kegelrückschlagventil *n*
poppet valve Kegelsitzventil *n*, Kolbensitzventil *n*
porous filter Porenfilter *m,n*
porous plug *LOG EL* Porenkörper *m*
port Anschluß *m*
port *v* verbinden, anschließen; mit Anschluß *m* versehen; *vt eg fluid to actuator* leiten *z. B. Flüssigkeit zum Verbraucher*
port area Anschlußquerschnitt *m*, Anschlußweite *f*
port configuration *VALV* Anschluß[loch]bild *n*, Bohrbild *n*
port connection Geräteverschraubung *f*, Geräteverbindung *f*, Anschlußverbindung *f*
port cross-section Anschlußquerschnitt *m*, Anschlußweite *f*
port fitting Geräteverschraubung *f*, Geräteverbindung *f*, Anschlußverbindung *f*
port plate *connecting with the intake and outlet pipes PU/MOT* Anschlußplatte *mit den Anschlüssen für Zu- und Ableitung*; *valve plate PU/MOT* Steuerplatte *f*, Steuerscheibe *f*, Ventilplatte *f*
port-plate axial piston pump *s.* port-valve axial piston pump
port-plate controlled pump schlitzgesteuerte (wegegesteuerte, flächengesteuerte) Pumpe *f*
port relief valve Rücklauf-Druckbegrenzungsventil *n*
port size Anschlußquerschnitt *m*, Anschlußweite *f*
***n*-port valve** *DCV* Ventil *n* mit *n* Anschlüssen
port-valve axial piston pump wegegesteuerte (ventillose) Axialkolbenpumpe *f*
portable filter unit Ölwechselgerät *n*, Ölaufbereitungsgerät *n*, Ölreinigungsgerät *n*, Öl-Service- Gerät *n*, Ölfiltrationswagen *m*
porting pintle *PU/MOT* Mittelzapfen *m*, Steuerzapfen *m*
position-indicating cylinder Positionsanzeigezylinder *m*
position sensing *CYL* Positionserfassung *f*
position sensor Positionsgeber *m*, Stellungsgeber *m*, Lagegeber *m*, Zylindersensor *m*
***n*-position valve** *DCV n*-Stellungsventil *n*
positional cylinder Mehrpositionszylinder *m*, Mehrstellungszylinder *m*, Digitalzylinder *m*

positive-displacement

positive-displacement compressor Verdrängungsverdichter *m*
positive-displacement motor Verdrängermotor *m*
positive-displacement pump Verdrängerpumpe *f*
positive-displacement pump or motor Verdrängermaschine *f*, Verdrängereinheit *f*, Verdrängergerät *n*
positive metering *PU/MOT* volumetrische Dosierung (Zuteilung) *f*
potentiometric pressure transducer Potentiometerdruckwandler *m*
pounds per square inch *Einheit des Drucks: 1 psi = 0,06895 bar*
pour point *FLUIDS* Fließpunkt *m*, Pourpoint *m*, Stockpunkt *m als Pourpoint angegeben*
pour-point depressant *FLUIDS* Fließpunkterniedriger *m*, Pourpointerniedriger *m*, Stockpunkterniedriger *m*
power-assisted steering *APPL* Servolenkung *f*
power brake *APPL* Servobremse *f*
power compensator *PU/MOT* Leistungsregler *m*
power cylinder *APPL* Arbeitszylinder *m*, Kraftzylinder *m*, Leistungszylinder *m*
power demand pattern plot *s.* power-displacement plot
power-displacement plot *CIRC* Leistung-Weg-Diagramm *n*, Leistungszyklusprofil *n*, Leistungsdiagramm *n*
power jet *LOG EL* Hauptstrahl *m*, Leistungsstrahl *m*
power loss Leistungsverlust *m*
power-matching control *of a pump* Load-sensing-Steuerung *f*, Lastdrucksteuerung *f einer Pumpe*

power motion *CYL* Leistungsbewegung *f*
power output Abtriebsleistung *f*, Abgabeleistung *f*, abgegebene Leistung *f*, Ausgangsleistung *f*
power package *APPL* Hydraulikaggregat *n*, Pumpenaggregat *n*, Motor-Pumpen-Aggregat *n*
power plot *CIRC s.* power-displacement plot
power pump *eg of a hydraulic press* Preßpumpe *f einer Umformmaschine*
power stage valve *VALV OP* Hauptventil *n*
power steering *APPL* Servolenkung *f*
power stroke *CYL* Arbeitshub *m*, Nutzhub *m*
power transmitting agent Arbeitsflüssigkeit *f*, Arbeitsmedium *n*, Arbeitsmittel *n*, Betriebsflüssigkeit *f*, Druck[übertragungs]mittel *n*, Energieübertragungsmittel *n*, Druckflüssigkeit *f*, *vgl* hydraulic fluid
power unit *APPL* Hydraulikaggregat *n*, Pumpenaggregat *n*, Motor-Pumpen-Aggregat *n*
power valve *VALV OP* Hauptventil *n*
powerpack *APPL* Hydraulikaggregat *n*, Pumpenaggregat *n*, Motor-Pumpen-Aggregat *n*
PQ manifold block *pressure/flow control section VALV* Druck-Strom-Steuerplatte *f*
Prandtl number *heat conduction of fluid flow THEOR* Prandtl-Zahl *f*, Pr *Wärmeleitung in strömenden Medien*
precharge *v with gas ACCUM* [auf]laden, füllen *mit Gas*

precharge pressure *ACCUM* Fülldruck *m*, Aufladedruck *m*, Vorspanndruck *m*
precision choke s. precision restriction
precision pressure gauge Feinmeßmanometer *n*
precision restriction (throttle) *FCV* Feinstelldrossel *f*
precompression Vorverdichtung *f*, Vorkompression *f*
prefill *v* *pump* [vor]füllen *Pumpe*
prefill circuit *presses APPL* Vorfüllkreislauf *m* *Umformmaschinen*
prefill pump Speisepumpe *f*, Füllpumpe *f*, Vorfüllpumpe *f*, Zuförderpumpe *f*, Ladepumpe *f*
prefill tank Füllbehälter *m*, Vorfüllbehälter *m*
prefill valve Füllventil *n*, Vorfüllventil *n*
prefilter Vorfilter *m,n*
prefiltration Vorfilterung *f*
preflaring device *FITT* Bördelmaschine *f*
preformed lip *SEALS* Formlippe *f*, vorgeformte Lippe *f*
preformed seal Formdichtung *f*, vorgeformte Dichtung *f*, Profildichtung *f*
preheat *v* vorwärmen
preheater Heizer *m*, Vorwärmer *m*
preheating motor Heizmotor *m*
preload pressure *ACCUM*
s. precharge pressure
preloaded seal vorgespannte Dichtung *f*
preloading pressure *of a seal* Vorspanndruck *m* einer Dichtung
prepackaged control unit *with integral sensing elements APPL* Ventilinsel *f* Kompakteinheit mit integrierter Elektronik
preset pressure [level] *PCV* Einstelldruck *m*, Druckeinstellwert *m*
press cylinder *APPL* Pressenzylinder *m*, Preßzylinder *m*
press plunger *APPL* Pressenkolben *m*
press ram s. press cylinder;
s. press plunger
press stroke *APPL* Preßhub *m*
press water *APPL* Pressenwasser *n*
pressure *gauge pressure THEOR* Druck *m*, Überdruck *m*; *pressure level* Druck *m*, Druckwert *m*; charge *v* with pressure, expose *v* to a pressure, apply *v* pressure to Druck *m* zuführen, mit Druck *m* beaufschlagen, *als Adj. auch zusammengesetzt*: druckbeaufschlagt; under pressure,
pressurized unter Druck *m*
pressure-activated seal selbstwirkende (druckgespannte) Dichtung *f*
pressure-actuated *VALV OP* druckbetätigt, druckgesteuert
pressure-actuated drain druckgesteuerter Ablaß *m*
pressure-actuated seal selbstwirkende (druckgespannte) Dichtung *f*
pressure-adjustment relief valve einstellbares Druckbegrenzungsventil *n*
pressure air Druckluft *f*, Luft *f* unter Druck
pressure amplification *INTENS* Druckverstärkung *f*
pressure amplifier *VALV OP* Druckverstärker *m*; *INTENS* Druckübersetzer *m*
pressure balance *DCV* Druckentlastung *f*, Druckausgleich *m*

pressure-balance *v* vom Druck *m* entlasten, Druck kompensieren, Druck ausgleichen, *als Adj. auch zusammengesetzt:* druckentlastet, druckkompensiert, druckausgeglichen
pressure-balanced pump druckausgeglichene (ausgeglichene, kompensierte, entlastete) Pumpe *f*
pressure-biased valve druckbelastetes Ventil *n*
pressure boost *INTENS* Druckverstärkung *f*
pressure build-up Druckaufbau *m*
pressure cap *druckdichte* Verschlußkappe *f*
pressure carrier *of a hose* Druckträger *m eines Schlauchs*
pressure-centered valve druckzentriertes Ventil *n*
pressure centering *VALV OP* hydraulische Zentrierung *f*, Druckzentrierung *f*
pressure chamber Druckkammer *f, bei einem Verbraucher auch* Einlaßkammer *f*, Eintrittskammer *f*, Zulaufkammer *f*, Zuflußkammer *f, bei einer Pumpe auch* Förderkammer *f*, Auslaßkammer *f*, Austrittskammer *f*, Ausgangsraum *m*
pressure change Druckänderung *f*
pressure channel Druckkanal *m, bei einem Verbraucher auch* Einlaßkanal *m*, Eintrittskanal *m*, Eingangskanal *m*, Zulaufkanal *m*, Zuflußkanal *m, bei einer Pumpe auch* Förderkanal *m*, Auslaßkanal *m*, Austrittskanal *m*, Abflußkanal *m*
pressure collapse Druckzusammenbruch *m*
pressure-compensate *v* vom Druck *m* entlasten, Druck kompensieren, Druck ausgleichen, *als Adj. auch zusammengesetzt:* druckentlastet, druckkompensiert, druckausgeglichen
pressure-compensated flow-control valve Stromregelventil *n*, Stromregler *m*, Strombegrenzungsventil *n*
pressure-compensated pump druckausgeglichene (ausgeglichene, kompensierte, entlastete) Pumpe *f*; Nullhubpumpe *f*
pressure-compensated vane pump Flügelzellenpumpe *f* mit druckentlastetem Rotor
pressure compensation Druckentlastung *f*, Druckausgleich *m*
pressure compensator *PU/MOT* Nullhubregler *m*; *controlled orifice in a compensated flow-control valve* Regeldrossel *f*, Aktivdrossel *f*, Druckwaage *f*, Differenzdruckregler *m im Stromregelventil*
pressure-compensator control *PU/MOT* Nullhubregelung *f*
pressure control Drucksteuerung *f*, Drucksteuereinrichtung *f*; *s.* pressure control circuit; *VALV OP* Druckbetätigung *f*
pressure control circuit Drucksteuerkreislauf *m*, Drucksteuersystem *n*, Drucksteuerung *f*
pressure control system *s.* pressure control circuit
pressure-control valve Druckventil *n*
pressure-controlled *VALV OP* druckbetätigt, druckgesteuert
pressure-controlled pump Nullhubpumpe *f*
pressure decay Druckabfall *m*, Drucktal *n*

pressure decrease Druckabnahme *f*, Druckverringerung *f*, Drucksenkung *f*
pressure-dependent druckabhängig
pressure dewpoint Drucktaupunkt *m*
pressure difference Differenzdruck *m* *Meßgröße*
pressure differential Druckdifferenz *f* *Wert*
pressure differential switch Differenzdruckschalter *m*
pressure-displacement characteristics *PCV* Druck-Schieberweg-Verhalten *n*
pressure-displacement plot *CIRC* Druck-Weg-Diagramm *n*, Druckzyklusprofil *n*, Druckdiagramm *n*
pressure distribution Druckverteilung *f*, Druckprofil *n*
pressure-dividing valve *PCV* Druckteilventil *n*
pressure drop Druckabfall *m*, Druckgefälle *n*
pressure efficiency Druckwirkungsgrad *m*
pressure-energized seal selbstwirkende (druckgespannte) Dichtung *f*
pressure energy Druckenergie *f*
pressure failure Druckausfall *m*
pressure fatigue test Druckdauerfestigkeitsprüfung *f*
pressure feedback *VALV OP* Druckrückführung *f*
pressure field Druckfeld *n*
pressure filter Druckfilter *m,n*, Hochdruckfilter *m,n*
pressure/flow control section *VALV* Druck-Strom-Steuerplatte *f*
pressure flow generator *PU/MOT* Druckstromerzeuger *m*
pressure fluctuation *PU/MOT* Druckpulsation *f*, Druckungleichförmigkeit *f*

pressure fluid Arbeitsflüssigkeit *f*, Arbeitsmedium *n*, Arbeitsmittel *n*, Betriebsflüssigkeit *f*, Druck[übertragungs]mittel *n*, Energieübertragungsmittel *n*, Druckflüssigkeit *f*, *vgl* hydraulic fluid; *fluid under pressure* Druckflüssigkeit *f*, Flüssigkeit *f* unter Druck
pressure freeze *of a valve spool* hydraulisches Verklemmen *n*, Klemmen *n*, Verkleben *n*, Kleben *n* *des Ventilkolbens*
pressure fuse *PCV* Membransicherheitsventil *n*
pressure gain *INTENS* Druckverstärkung *f*
pressure gas Druckgas *n*, Gas *n* unter Druck
pressure gauge Manometer *n*
pressure generator *PU/MOT* Druckstromerzeuger *m*
pressure gradient Druckgradient *m*, Druckgefälle *n* *Neigung bzw. Steigung einer Druckverlaufskurve*
pressure head Druckhöhe *f* *nutzbarer Druck*
pressure-holding circuit Druckhaltekreislauf *m*
pressure hose Druckschlauch *m*
pressure imbalance Druckungleichgewicht *n*
pressure-independent druckunabhängig
pressure indicator Druckanzeiger *m*, Druckwächter *m*
pressure-insensitive druckunabhängig
pressure intensification (amplification) Druckverstärkung *f*
pressure intensifier Druckübersetzer *m*

pressure kidney *PU/MOT* Druckniere *f*
pressure level Druck *m*, Druckwert *m*
pressure limitation *pump control* Druckabschneidung *f* Pumpenregelung
pressure-limiting valve *s.* relief valve
pressure line Druckleitung *f*, *bei einem Verbraucher auch* Einlaßleitung *f*, Eintrittsleitung *f*, Eingangsleitung *f*, Zulaufleitung *f*, Zuführleitung *f*, Zuflußleitung *f*, Zuleitung *f*, Speiseleitung *f*, *bei einer Pumpe auch* Förderleitung *f*, Auslaßleitung *f*, Austrittsleitung *f*, Ausgangsleitung *f*, Abflußleitung *f*, Ablaufleitung *f*, Abführleitung *f*, Ableitung *f*; *line under pressure* unter Druck stehende (druckführende) Leitung *f*, Druckleitung *f*
pressure-line filter Druckfilter *m,n*, Hochdruckfilter *m,n*
pressure-load *v* Druck *m* zuführen, mit Druck *m* beaufschlagen, *als Adj. auch zusammengesetzt* druckbeaufschlagt
pressure-loaded valve druckbelastetes Ventil *n*
pressure loss *THEOR* Druckverlust *m*
pressure measurement Druckmessung *f*
pressure oil Drucköl *n*, Öl *n* unter Druck
pressure-operated *VALV OP* druckbetätigt, druckgesteuert
pressure oscillation Druckschwingung *f*
pressure override *Differenz zwischen dem Druck bei maximalem Volumenstrom und dem Ansprechdruck beim Druckbegrenzungsventil*
pressure overshoot Drucküberschwingweite *f*
pressure pattern plot Druck-Weg-Diagramm *n*, Druckzyklusprofil *n*, Druckdiagramm *n*
pressure peak Druckspitze *f*
pressure plate *PU/MOT* Seitenplatte *f*
pressure plot *s.* pressure pattern plot
pressure port Druckanschluß *m*, *bei einem Verbraucher auch* Einlaßanschluß *m*, Eintrittsanschluß *m*, Zulaufanschluß *m*, Zuführanschluß *m*, Zuflußanschluß *m*, *bei einer Pumpe auch* Förderanschluß *m*, Auslaßanschluß *m*, Austrittsanschluß *m*, Abführanschluß *m*
pressure profile Druckverteilung *f*, Druckprofil *n*
pressure pulsation *PU/MOT* Druckpulsation *f*, Druckungleichförmigkeit *f*
pressure pulse Druckimpuls *m*
pressure range Druckbereich *m*
pressure rating Nenndruck *m*, ND *m*
pressure ratio Druckverhältnis *n*
pressure recorder *MEASUR* Druckschreiber *m*
pressure recovery Druckrückgewinn *m*
pressure-reducing valve Druckminderventil *n*, Druckreduzierventil *n*, Reduzierventil *n*
pressure regulator *PREPAR* Druckluftregler *m*; *PCV* Druckminderventil *n*, Reduzierventil *n*

pressure relief Druckentlastung *f*, Druckabschaltung *f*, Entspannung *f*
pressure-resistant unter Druck *m* einsetzbar, druckfest
pressure-resistant sensor druckfester Sensor *m*
pressure ripple *PU/MOT* Druckpulsation *f*, Druckungleichförmigkeit *f*
pressure rise Druckanstieg *m*, Druckerhöhung *f*, Drucksteigerung *f*
pressure-sensing line Druckaufnahmeleitung *f*
pressure-sensitive druckabhängig
pressure sensor *MEASUR* Drucksensor *m*
pressure setting *PCV* Einstelldruck *m*, Druckeinstellung *f*, Druckeinstellwert *m*
pressure side Druckseite *f*, *bei einem Verbraucher auch* Einlaßseite *f*, Eintrittsseite *f*, Eingangsseite *f*, Zulaufseite *f*, Zuflußseite *f*, *bei einer Pumpe auch* Förderseite *f*, Auslaßseite *f*, Austrittsseite *f*, Ausgangsseite *f*, Ablaufseite *f*
pressure signal Drucksignal *n*
pressure slip [flow] *s.* pressure slippage
pressure slippage *PU/MOT* druckbedingter Schlupfstrom *m* (innerer Leckstrom *m*)
pressure snubber Druckpulsationsdämpfer *m*, Druckschwingungsdämpfer *m*; Druckstoßdämpfer *m*
pressure spike Druckspitze *f*
pressure supply Druckbeaufschlagung *f*, Druckzuführung *f*, Druckversorgung *f*
pressure surge Druckschlag *m*
pressure-swing regeneration *of a*
desiccant Druckwechselregenerierung *f*, Kaltregenerierung *f*, Heatlessregenerierung *f* *von Trockenmittel*
pressure switch Druckschalter *m*
pressure tap Druckmeßstelle *f*, Druckmeßabzweig *m*
pressure-tight druckdicht
pressure-to-pressure extrusion *from superpressure to somewhat lower pressure APPL* Gegendruckfließpressen *n* ein Höchstdruckverfahren
pressure transient Druckübergangsabschnitt *m*, Druckübergang *m*, Drucktransiente *f*
pressure unloading Druckentlastung *f*, Druckabschaltung *f*, Entspannung *f*
pressure variation Druckschwankung *f*
pressure vibration Druckschwingung *f*
pressure-volume compensation *ACCUM* Kompressibilitätsausgleich *m*
pressure-volume diagram *COMPR* Indikatordiagramm *n*
pressure wave Druckwelle *f*
pressureless drucklos, nicht unter Druck *m*, nicht vorgespannt
pressurization Herstellung *f* eines Überdrucks; *pressure supply* Druckbeaufschlagung *f*, Druckzuführung *f*, Druckversorgung *f*
pressurize *v* Druck *m* zuführen, mit Druck beaufschlagen, *als Adj. auch zusammengesetzt:* druckbeaufschlagt; unter Überdruck *m* setzen, unter Druck *m* setzen; *with gas ACCUM* [auf]laden, füllen *mit Gas*
pressurized unter Druck *m*
pressurized air Druckluft *f*, Luft *f* unter Druck

pressurized fluid Druckflüssigkeit *f*, Flüssigkeit *f* unter Druck
pressurized gas Druckgas *n*, Gas *n* unter Druck
pressurized oil Drucköl *n*, Öl *n* unter Druck
pressurized reservoir (tank) vorgespannter Behälter *m*
pressurized volume Druckraum *m*, unter Druck stehender Raum *m*
pressurizing pump *APPL* Vorspannpumpe *f*
prestressed diaphragm *VALV OP* Spannmembran *f*
primary circuit Hauptkreislauf *m*
primary pump *of a hydrostatic transmission* Arbeitspumpe *f*, Hauptpumpe *f* *eines hydrostatischen Getriebes*
primary spool *of a two-stage valve* Hauptsteuerschieber *m*, Hauptkolben *m* *eines zweistufigen Ventils*
primary stage *of a servovalve* Hauptstufe *f* *eines Servoventils*
primary valve Hauptventil *n*
prime *vt* [vor]füllen; *vi* erstmalig ansaugen, anzusaugen beginnen
prime mover Antriebsmaschine *f*, Antriebsmotor *m*
principle jet *LOG EL* Hauptstrahl *m*, Leistungsstrahl *m*
priority valve *PCV* Folge[schalt]ventil *n*, Zuschaltventil *n*
process air *high purity air* Prozeßluft *f* *für höchste Anforderungen*
progressing-tooth gear pump Innenzahnradpumpe *f* mit Trochoidenverzahnung, trochoidenverzahnte Zahnringpumpe *f*
proof pressure Prüfdruck *m*

propel hydraulic circuit *APPL* Fahrhydraulik *f* *z. B. eines Baggers*
propeller flowmeter Flügelrad-Durchflußmesser *m*, Turbinen- Durchflußmesser *m*, Meßturbine *f*
proportional directional control valve Proportionalwegeventil *n*, Wege-Proportionalventil *n*
proportional flow control valve Proportional-Stromventil *n*, Strom-Proportionalventil *n*
proportional flow lubricator *PREPAR* Proportionalnebelöler *m*, Mehrbereichsnebelöler *m*
proportional pressure control valve Proportional-Druckventil *n*, Druck-Proportionalventil *n*
proportional pressure-reducing valve Druckverhältnisventil *n*
proportional solenoid *VALV OP* Steuermotor *m* mit linearer Bewegung, Proportionalmagnet *m*
proportional valve *DCV* Proportionalventil *n*
proportional valve technology Proportionaltechnik *f*
proportioning pressure regulator Druckverhältnisventil *n*
proportioning pump *s.* metering pump
protective seal Schutzdichtung *f*, Dichtung *f* gegen Einströmen (Eindringen)
protector cap Staubkappe *f*, Schmutzkappe *f*, Schutzkappe *f*
provide *v* fördern, liefern, abgeben *Flüssigkeit in (an) das System*, beaufschlagen *Komponente mit Flüssigkeit*, speisen, beliefern, versorgen *System o. ä. mit Flüssigkeit*; Druckluft *f* zuführen, be-

lüften, mit Druckluft speisen (beaufschlagen)
proximity sensor *LOG EL* Näherungssensor *m*
psi *pounds per square inch Einheit des Drucks: 1 psi = 0,06895 bar*
pull-action cylinder Zugzylinder *m*
pull-break type quick-disconnect coupling Schlauchkupplung *f* mit Abreißsicherung
pull stroke *CYL* Zughub *m*
pull-type solenoid *VALV OP* Zugmagnet *m*
pulsate *v* pulsieren, schwingen
pulsation *of the pump output* Förderstrompulsation *f*, Förderstromfluktuation *f*, Förderungleichförmigkeit *f*, Förderschwankung *f*; Druckpulsation *f*, Druckungleichförmigkeit *f*
pulsation attenuation Pulsationsdämpfung *f*, Pulsationsglättung *f*
pulsation damper Druckpulsationsdämpfer *m*, Druckschwingungsdämpfer *m*
pulsation damping Pulsationsdämpfung *f*, Pulsationsglättung *f*
pulsation-free pulsationsfrei
pulsation noise Pulsationsgeräusch *n*, Pulsationslärm *m*
pulsation resistance *LINES* Impulsfestigkeit *f*
pulse filter akustisches Filter *n*
pulsed-flow hydraulics Impulsstromhydraulik *f*
pump Pumpe *f*
pump and motor displacement control *of a hydrostatic transmission* Primär- und Sekundärverstellung *f* *eines hydrostatischen Getriebes*
pump body Pumpenkörper *m*
pump bracket Pumpenträger *m*

pump case (casing) Pumpengehäuse *n*
pump characteristics Pumpenkennlinie *f*, Pumpenverhalten *n*
pump connection Pumpenanschluß *m*
pump control [circuit, system] Pumpensteuerkreislauf *m*, Pumpensteuersystem *n*, Pumpensteuerung *f*
pump curves plot Pumpenkenn[linien]feld *n*
pump cylinder Pumpenzylinder *m*
pump delivery (discharge) [rate] Pumpenstrom *m*, Förderstrom *m*
pump displacement Pumpenverdrängungsvolumen *n*, Fördervolumen *n*
pump displacement control *of a hydrostatic transmission* Primärverstellung *f*, Pumpenverstellung *f* *eines hydrostatischen Getriebes*
pump efficiency Pumpenwirkungsgrad *m*
pump failure Pumpenausfall *m*, Pumpenversagen *n*
pump flow [rate] Pumpenstrom *m*, Förderstrom *m*
pump frequency *shaft frequency multiplied by number of pumping elements* Pumpenfrequenz *f* *Wellendrehfrequenz mal Anzahl der Pumpelemente*
pump gear Förderrad *n* *einer Zahnradpumpe*
pump horsepower output Förderleistung *f*
pump housing Pumpengehäuse *n*
pump/motor unit *functions either as pump or as rotary motor* Pumpen-Motor-Einheit *f* *arbeitet als Pumpe oder Motor*

pump mount Pumpenbefestigung *f*
pump noise Pumpengeräusch *n*
pump output Pumpenstrom *m*, Förderstrom *m*
pump port Pumpenanschluß *m*
pump pulsation (ripple) Förderstrompulsation *f*, Förderstromfluktuation *f*, Förderungleichförmigkeit *f*, Förderschwankung *f*
pump speed Pumpendrehzahl *f*
pump starvation unvollständige Füllung *f*, Abschnappen *n* der Pumpe
pumping element Pumpelement *n*
pumping stroke Förderhub *m*
pure fluid element *LOG EL* reinfluidisches Element *n*, Element *n* ohne bewegte Teile
purely hydraulic vollhydraulisch, reinhydraulisch
purely pneumatic vollpneumatisch, reinpneumatisch
push-action cylinder Druckzylinder *m*
push-in connection (fitting) Einsteckverbindung *f*
push-pull type solenoid *VALV OP* Umkehrmagnet *m*, Doppelhubmagnet *m*
push stroke *CYL* Druckhub *m*
push-type solenoid *VALV OP* Stoßmagnet *m*
pushbutton-actuated *VALV OP* druckknopfbetätigt
pushpin *of a solenoid VALV OP* Magnetstößel *m*

Q

quad ring *SEALS* X-Ring *m*
quasi-irrotational (-nonrotational) flow *THEOR* quasi-nichtrotierende (-drehungsfreie) Strömung *f*

quick-action coupling Schlauchkupplung *f*, Schnellkupplung *f*
quick-closure valve schnellschließendes Ventil *n*
quick-connect (-disconnect) coupling Schlauchkupplung *f*, Schnellkupplung *f*
quick-exhaust valve Schnellentlüftungsventil *n*, Schnellentlüfter *m*
quick feed *CYL* Eilvorschub *m*
quick-release coupling Schlauchkupplung *f*, Schnellkupplung *f*
quick-release valve Schnellentlüftungsventil *n*, Schnellentlüfter *m*
quick return *CYL* Eilrücklauf *m*, Eilrückgang *m*, Eilrückzug *m*
quick return stroke *CYL* Eilrückhub *m*
quick stroke *CYL* Eilhub *m*
quick traverse piston *CYL* Eilgangkolben *m*

R

R. N. *Reynolds' number THEOR* Reynolds-Zahl *f*, Re- Zahl *f*
R & O inhibitor *rust and oxidation inhibitor FLUIDS* Rost- und Oxidationsinhibitor *m*
rack-and-pinion actuated (operated) valve zahnstangenbetätigtes Ventil *n*
rack-and-pinion rotary actuator Zahnstangenschwenkmotor *m*, Kolbenstange-Ritzel-Schwenkmotor *m*
radial-piston motor Radialkolbenmotor *m*
radial-piston pump Radialkolbenpumpe *f*
radial-piston pump *or* motor Radial-

kolbenmaschine *f*, Radialkolbeneinheit *f*, Radialkolbengerät *n*
radial-piston pump with exterior admission außenbeaufschlagte Radialkolbenpumpe *f*
radial-piston pump with interior admission innenbeaufschlagte Radialkolbenpumpe *f*
radial-piston transmission Radialkolbengetriebe *n*
radial seal radiale Dichtung *f*
ram *CYL* Kolben *m* großer Abmessungen *meist Tauchkolben*; Hubzylinder *m*
ram cylinder Tauchkolbenzylinder *m*
range of flow rate Volumenstrombereich *m*, Durchflußstrombereich *m*
rapid advance (approach) pump Eilgangpumpe *f*
rapid-escape valve *DCV* Schnellentlüftungsventil *n*, Schnellentlüfterventil *n*
rapid feed *CYL* Eilvorschub *m*
rapid return *CYL* Eilrücklauf *m*, Eilrückgang *m*, Eilrückzug *m*
rapid return stroke *CYL* Eilrückhub *m*
rapid stroke *CYL* Eilhub *m*
rapid traverse piston *CYL* Eilgangkolben *m*
rapid traverse pump Eilgangpumpe *f*
rate *delivery rate COMPR* Liefermenge *f*; *PU/MOT* Förderstrom *m*; *flow rate* Volumenstrom *m*, Durchflußstrom *m*
rated air consumption Nennluftverbrauch *m*
rated capacity *PU/MOT* Nennverdrängungsvolumen *n*; *nominal output (horsepower)* Nennleistung *f* *abgegebene Leistung*
rated conditions Nennbedingungen *fpl*
rated cross-section Nennquerschnittsfläche *f*, Nennquerschnitt *m*
rated displacement [volume] *PU/MOT* Nennverdrängungsvolumen *n*
rated flow Nenn[volumen]strom *m*, Nenndurchfluß[strom] *m*
rated force Nennkraft *f*
rated horsepower (output) Nennleistung *f* *abgegebene Leistung*
rated pressure Nenndruck *m*, ND *m*
rated rpm (speed) Nenndrehzahl *f*
rated stroke [length] *CYL* Nennhublänge *f*, Nennhub *m*
rated temperature Nenntemperatur *f*
rated torque Nenn[dreh]moment *n*
rated viscosity *FLUIDS* Nennviskosität *f*
rate of pressure change Druckänderungsgeschwindigkeit *f*
rate of shear *THEOR* Schergeschwindigkeit *f*, Schergefälle *n*
rate of stroking *CYL* Hubgeschwindigkeit *f*
rate-sensitive vortex device *LOG EL* Wirbelelement *n* mit rotierender Wirbelkammer
rating s. rated horsepower; *FILTERS* Filterfeinheit *f*
β ratio *(β for X μm particle size = particles > X μm upstream / particles > X μm downstream)* Beta-Wert *m*, β-Wert *m*, Beta-Verhältnis *n* *ein Maß für den Filterwirkungsgrad*
raw particle count Partikelgrobzählung *f*

raw water Rohwasser *n*, unaufbereitetes Wasser *n*
reaction ring *of a radial piston design* Leitring *m*, Führungsring *m* einer Radialkolbenmaschine
reaction-type turbine motor Überdruckturbinenmotor *m*, Reaktionsturbinenmotor *m*
real fluid *THEOR* reales (wirkliches) Fluid *n*
rear end *CYL* Kolbenseite *f*, Bodenseite *f*; kolbenseitiges (bodenseitiges) Zylinderende *n*
rear end [chamber] *CYL* Kolbenraum *m*
rear end pressure *CYL* kolbenseitiger (bodenseitiger) Druck *m*
receiver Druckluftspeicher *m*, Druckluftbehälter *m*, Windkessel *m*
reciprocating compressor Kolbenverdichter *m*
reciprocating intensifier Dauerstromdruckverstärker *m*
reciprocating seal Gleitdichtung *f* Dichtung für hin- und hergehende Bewegung
recirculated oil volume umgewälzte Ölmenge *f*
reclaimer Ölregenerator *m*, Ölrückgewinnungsgerät *n*, Ölwiederaufbereitungsgerät *n*
reclamation Ölregeneration *f*, Ölrückgewinnung *f*, Ölwiederaufbereitung *f*
rectangular piston *CYL* Rechteckkolben *m*
rectangular section ring *SEALS* Rechteckring *m*
rectifying circuit Gleichrichterschaltung *f*
reduced pressure Unterdruck *m*, *oft auch* Vakuum *n*

reducer [fitting] Reduzierverschraubung *f*, Reduzierung *f*; *s.* reducing valve
reducer union Reduzierverschraubung *f* mit zweiseitigem Rohranschluß
reducing nipple *FITT* Reduzierstutzen *m*
reducing valve *PCV* Druckminderventil *n*, Druckreduzierventil *n*, Reduzierventil *n*
Redwood Second veraltete Einheit der kinematischen Viskosität
reference conditions *pl* Bezugszustand *m*, Bezugsbedingungen *fpl*, Referenzbedingungen *fpl*
reference fluid *FLUIDS* Bezugsflüssigkeit *f*, Referenzflüssigkeit *f*
reference pressure Bezugsdruck *m*
reference temperature Bezugstemperatur *f*
refill *v eg with oil* neubefüllen *z.B.* Öl
refill opening *RESERVOIRS* Füllöffnung *f*, Einfüllöffnung *f*
refrigerant-type (refrigerated-air, refrigeration) dryer Kältetrockner *m*
regenerative circuit Umströmungsschaltung *f*, Differentialschaltung *f*, Rückspeiseschaltung *f*, Eilgangschaltung *f*, Eilvorlaufschaltung *f* mit Differentialzylinder
regenerative desiccant regenerierbares Trockenmittel *n*
regenerative position *DCV* Eilgangstellung *f*
regenerative-type dryer Regenerativtrockner *m*
regulation curve *of a pressure regulator* Kennlinie *f*, Regelkurve *f* eines Druckminderventils

reheater *for compressed air* Zwischenheizer *m* *für Druckluft*
reinforced seal verstärkte (bewehrte, armierte) Dichtung *f*
reinforcement *s.* hose reinforcement
reinforcement layer Verstärkungslage *f*, Bewehrungslage *f*, Armierungslage *f*
release *v* mit dem Behälter *m* verbinden, entlasten; *fluid, energy* abgeben *Flüssigkeit, Energie*
release *v* **air** Luft *f* freigeben (ausscheiden), entgasen, ausgasen
relief *s.* relief valve
relief connection (port) *DCV* Entlastungsanschluß *m*
relief valve Überströmventil *n*; *pressure-limiting valve* Druckbegrenzungsventil *n*; *safety valve* Sicherheitsventil *n*
relieve *v* mit dem Behälter *m* verbinden, entlasten
relieving pressure-reducing valve Druckminderventil *n* mit Überlastsicherung (Sekundärdruckbegrenzung), Dreiwege-Druckminderventil *n*
remote-operated directional valve ferngesteuertes Wegeventil *n*
remote pilot-actuated *VALV OP* fernvorgesteuert
remote pilot actuation *VALV OP* Fernvorsteuerung *f*
remote pilot-control valve *VALV OP* Fern-Vorsteuerventil *n*
remote-piloted *VALV OP* fernvorgesteuert
remote pressure gauge Fernmanometer *n*
remote thermometer Fernthermometer *n*

remote-type coupling *FITT* fernbetätigte Kupplung *f*
removable filter Wechselfilter *m,n*
replacement filter cartridge Austauschfiltereinsatz *m*, Austauschfilterpatrone *f*
replenishing fluid Ergänzungsflüssigkeit *f*, Frischöl *n*
replenishing line Nachfülleitung *f*, Leckölergänzungsleitung *f*
replenishing valve Nachsaugeventil *n*, Spülventil *n*
replenishment flow Ergänzungsstrom *m*
required oil volume *ACCUM* gefordertes Ölvolumen *n*
reseat *v* *of a seat valve* [sich] schließen *von einem Sitzventil*
reseat pressure *PCV* Schließdruck *m*
reservoir Flüssigkeitsbehälter *m*, Hydraulikbehälter *m*, Ölbehälter *m*, *umgangssprachlich noch üblich:* Tank *m*
reservoir: connected to reservoir *HY* mit dem Behälter *m* verbunden
reservoir bottom Behälterboden *m*
reservoir capacity Behältervolumen *n*, Behältergröße *f*
reservoir connection *DCV* Behälteranschluß *m*
reservoir cover Behälterdeckel *m*, Behälterdeckplatte *f*
reservoir drain äußere Leckölabführung *f*
reservoir filter Behälterfilter *m,n*, Sumpffilter *m,n*
reservoir mount filter Behälteraufbaufilter *m,n*, Behälteranbaufilter *m,n*
reservoir port *DCV* Behälteranschluß *m*

reservoir size *s.* reservoir volume
reservoir top mount filter *s.* reservoir mount filter
reservoir top plate Behälterdeckel *m*, Behälterdeckplatte *f*
reservoir volume Behältervolumen *n*, Behältergröße *f*
reservoir wall Behälterwand[ung] *f*
reset pressure Rückstelldruck *m*
residual-pressure exhaust valve Restdruckablaßventil *n*, Restentlüfter *m*
resilient seal elastische (nachgiebige) Dichtung *f*
resilient-seat valve Weichsitzventil *n*
resin content *FLUIDS* Harzgehalt *m*
resinfree *FLUIDS* harzfrei
resinification *FLUIDS* Harzbildung *f*, Verharzung *f*
resinous oil verharztes Öl *n*
resistance thermometer Widerstandsthermometer *n*
resistance to deposit formation *FLUIDS* Schlammbildungswiderstand *m*, Schlammtragvermögen *n*
resistance to emulsification *FLUIDS* Emulgierwiderstand *m*, Beständigkeit *f* gegen Emulgieren
resistance to evaporation *below boiling temperature FLUIDS* Verdunstungswiderstand *m*
resistance to fire *FLUIDS* Schwerentflammbarkeit *f*
resistance to flow *THEOR* Strömungswiderstand *m*, fluidischer Widerstand *m*, Fluidwiderstand *m*
resistance to foaming *FLUIDS* Widerstand *m* gegen Schaumbildung, Schäumwiderstand *m*
resistance to oxidation *FLUIDS* Oxidationsbeständigkeit *f*, Oxidationsstabilität *f*
resistance to sludge formation *FLUIDS* Schlammbildungswiderstand *m*, Schlammtragvermögen *n*
resistant to oil ölbeständig, ölfest
resisting force Widerstandskraft *f*
resistive load *CIRC* Widerstandslast *f*
respond *v* *pressure-control valve* ansprechen, [sich] öffnen *Druckventil*
response characteristics *PCV* Ansprechverhalten *n*
response pressure *PCV* Ansprechdruck *m*, Öffnungsdruck *m*
rest time *TANK* Verweilzeit *f*, Aufenthaltszeit *f*
restrict *v flow* drosseln *Volumenstrom*
restriction *THEOR* Drossel[stelle] *f*, Strömungswiderstand *m*; Drosselung *f*
restriction area *THEOR* Drosselquerschnitt *m*, Drosselfläche *f*
restriction characteristics *THEOR* Drosselcharakteristik *f*, Drosselverhalten *n*, Öffnungsverhalten *n*
restriction loss *THEOR* Drosselverlust *m*
restrictive control *FCV* Drosselsteuerung *f*
restrictive element *FCV* Drosselelement *n*
restrictive length *THEOR* Drossellänge *f*, Drosselstrecke *f*
restrictor *THEOR* Drossel[stelle] *f*, Strömungswiderstand *m*; *s.* restrictor valve
restrictor cartridge Stromventil *n* für Bohrungseinbau, Einbaustromventil *n*
restrictor module *of a valve stack* Drosselplatte *n* *einer Steuersäule*
restrictor valve *FCV* Drosselventil *n*

retention capacity *FILTERS* Schmutztragevermögen *n*, Schmutzaufnahmevermögen *n*
retract *vt the cylinder* zurückziehen, einfahren *den Zylinder*; *vi cylinder* einfahren, [sich] zurückziehen *Zylinder*
   retract cylinder Rückzugszylinder *m*
   retract spring *CYL* Rückzugsfeder *f*
   retract stroke *CYL* Einfahrhub *m*, Rückhub *m*
retracted length *CYL* Einbaulänge *f*, Einfahrlänge *f*
retracted position *CYL* Einfahrstellung *f*
retraction speed *CYL* Einfahrgeschwindigkeit *f*, Rückzugsgeschwindigkeit *f*, Rücklaufgeschwindigkeit *f*
retraction stroke *s.* retract stroke
return *s.* return flow
   return *vi* zurückfließen, zurückströmen; *vt* zurückleiten, zurückführen, zurückfließen (zurückströmen) lassen
   return connection *DCV* Rücklaufanschluß *m*
   return cylinder *CYL* Rückzugszylinder *m*
   return filter Rücklauffilter *m,n*
   return flow Rücklauf *m*, Rückfluß *m*; Rück[volumen]strom *m*
   return fluid Rückflüssigkeit *f*, Abflüssigkeit *f*
   return line Rücklaufleitung *f*, Rückflußleitung *f*, Rückleitung *f*
   return-line filter Rücklauffilter *m,n*
   return-line pressure Rück[lauf]druck *m*
   return oil Rücköl *n*, Aböl *n*
   return-orifice check valve Drosselrückschlagventil *n*, Einwegdrossel *f*
   return plate *of ganged valves* Austrittsdeckel *m*, Ablaufplatte *f einer Ventilbatterie*
   return port *DCV* Rücklaufanschluß *m*
   return pressure Rück[lauf]druck *m*
   return section *of ganged valves* Austrittsdeckel *m*, Ablaufplatte *f einer Ventilbatterie*
   return speed *CYL* Einfahrgeschwindigkeit *f*, Rückzuggeschwindigkeit *f*, Rücklaufgeschwindigkeit *f*
   return spring *f CYL* Rückzugfeder *f*
   return stroke *f CYL* Rückzug *m*, Rücklauf *m*, Rückhub *m*, Einfahrhub *m*
   return water Rückwasser *n*, Abwasser *n*
reusable fitting wiederverwendbare Verbindung *f*
reusable filter element reinigungsfähiges (regenerierbares) Filterelement *n*
reverse *vi and vt cylinder* umsteuern *Zylinder*
reverse stroke *CYL* Gegenhub *m*
reversibility response *PU/MOT* Umsteueransprechempfindlichkeit *f*
reversible motor Umsteuermotor *m*
reversible pump Pumpe *f* für umkehrbare Drehrichtung, Umsteuerpumpe *f*; Pumpe *f* mit umkehrbarer Förderrichtung (mit Übernullsteuerung), Reversierpumpe *f*
reversing motor *s.* reversible motor
reversing time *PU/MOT* Umsteuerzeit *f*
Reynolds' number *THEOR* Reynolds-Zahl *f*, Re-Zahl *f*
rigid line starre Leitung *f*
rigid mount *CYL* nichtnachgiebige (starre) Befestigung *f*

ring-lock

ring-lock coupling Ringverschluß-kupplung *f*
ring main Ringleitung *f*
ring seal Ringdichtung *f*, Dicht[ungs]ring *m*
rinse *v* spülen
ripple *PU/MOT* Druckpulsation *f*, Druckungleichförmigkeit *f*; Förderstrompulsation *f*, Förderstromfluktuation *f*, Förderungleichförmigkeit *f*, Förderschwankung *f*
ripple damper Druckpulsationsdämpfer *m*, Druckschwingungsdämpfer *m*
ripple-free *PU/MOT* pulsationsfrei
ripple noise Pulsationsgeräusch *n*, Pulsationslärm *m*
rod *CYL* Kolbenstange *f*
rod area *CYL* Kolbenstangenfläche *f*, Kolbenstangenquerschnitt *m*
rod bearing *CYL* Führungsbuchse *f* *der Kolbenstange*
rod bellows *CYL* Kolbenstangenschutz *m* *als Faltenbalg*
rod boot *s.* rod bellows
rod buckling *CYL* Kolbenstangenknickung *f*
rod bushing *s.* rod bearing
rod diameter *CYL* Kolbenstangendurchmesser *m*
rod end *CYL* Kolbenstangenseite *f*, Ringraumseite *f*, Ausfahrseite *f*; kolbenstangenseitiges (ausfahrseitiges) Zylinderende *n*; *s.* rod end chamber; *s.* rod end coupler
rod end chamber *CYL* Kolbenstangenraum *m*, Kolbenringraum *m*, Ringraum *m*
rod end coupler *CYL* Kolbenstangenende *n*, Kolbenstangenkopf *m*, Kolbenstangenbefestigung *f*
rod end coupling *s.* rod end coupler
rod-end pressure *CYL* kolbenstangenseitiger (ausfahrseitiger, ringraumseitiger) Druck *m*
rod extension *CYL* Ausfahren *n* der Kolbenstange
rod gaiter *CYL* Kolbenstangenschutz *m* *als Faltenbalg*
rod overextension *CYL* übermäßiges Ausfahren *n* der Kolbenstange
rod seal *CYL* Kolbenstangendichtung *f*
rod wiper [seal] *CYL* Kolbenstangenabstreifer *m*, Kolbenstangenabstreifring *m*
rodless cylinder kolbenstangenloser Zylinder *m*
roller-actuated valve Rollenventil *n*
roller lever-actuated valve Rollenhebelventil *n*
roller-lock coupling *FITT* Rollenverschlußkupplung *f*
roller-operated valve Rollenventil *n*
roller plunger-actuated (-operated) valve Rollenstößelventil *n*
roller valve Rollenventil *n*
roller vane *PU/MOT* Rollflügel *m*
roller-vane motor Rollflügelmotor *m*
roller-vane pump Rollflügelpumpe *f*
rolling diaphragm *CYL* Rollmembran *f*, Stulpmembran *f*
rolling-piston radial pump Radialkolbenpumpe *f* mit rollengeführten Kolben
Roots blower *COMPR* Rootsgebläse *n*
rotameter [flowmeter] Rotameter *n*, Schwebekörper-Durchflußmesser *m*
rotary abutment motor Sperrtrommelmotor *m*

rotary abutment pump Sperrtrommelpumpe *f*
rotary actuator Schwenkmotor *m*, Drehwinkelmotor *m*, *PN auch* Drehantrieb *m*
rotary-block radial pump Radialkolbenpumpe *f* mit rotierendem Kolbenträger
rotary compressor Umlaufkolbenverdichter *m*
rotary connection *FITT* Drehverbindung *f*, Drehverschraubung *f*, Drehübertrager *m*, rotierende Verbindung *f*, Schleifring *m*
rotary disc *DCV* Flachdrehschieber *m*, Kreisschieber *m*
rotary-disc valve *DCV* Flachdrehschieberventil *n*, Kreisschieberventil *n*
rotary face seal Gleitringdichtung *f*
rotary flow divider *FCV* Motorstromteiler *m*
rotary joint *FITT* Drehverbindung *f*, Drehverschraubung *f*, Drehübertrager *m*, rotierende Verbindung *f*, Schleifring *m*
rotary motor Rotationsmotor *m*, Motor *m* mit unbegrenztem Drehwinkel
rotary plate *s.* rotary disc
rotary-plate valve *s.* rotary-disc valve
rotary pump Pumpe *f* mit drehendem Förderteil
rotary screw compressor Schraubenverdichter *m*
rotary seal Rotationsdichtung *f*
rotary shaft seal Wellendichtung *f*
rotary spool *DCV* Kolbendrehschieber *m*
rotary spool valve *DCV* Kolbendrehschieberventil *n*

rotary union *FITT* Drehverbindung *f*, Drehverschraubung *f*, Drehübertrager *m*, rotierende Verbindung *f*, Schleifring *m*
rotary valve *DCV* Dreh[schieber]ventil *n*
rotary valve control *PU/MOT* Drehschiebersteuerung *f*
rotary vane compressor Zellenverdichter *m*, Lamellenverdichter *m*
rotating cylinder viscometer Rotationsviskosimeter *n*
rotating distributor *FITT* Drehverbindung *f*, Drehverschraubung *f*, Drehübertrager *m*, rotierende Verbindung *f*, Schleifring *m*
rotating manifold *FITT* Mehrwege-Drehverbindung *f*
rotating union *FITT* *s.* rotary union
rotating-vane pump Flügelzellenpumpe *f* mit rotierendem Flügelträger
rotational viscometer Rotationsviskosimeter *n*
rotodynamic pump Strömungspumpe *f*, Zentrifugalpumpe *f*, Kreiselpumpe *f*, Turbopumpe *f*
rotor *PU/MOT* Rotor *m*; *of a screw pump* Spindel *f*, Schraube *f* einer Schraubenpumpe
round-edged orifice *THEOR* Blende *f* mit abgerundeter Kante
route *vt eg fluid to actuator* leiten *z. B. Flüssigkeit zum Verbraucher*
rpm rating Nenndrehzahl *f*
rubber-cushion accumulator Gummifederspeicher *m*
rubber-energized seal gummigespannte Dichtung *f*
run *pipe run* Rohrleitungsabschnitt *m*; *s.* tee run

runaway speed *PU/MOT* Nullastdrehzahl *f*, Durchgangsdrehzahl *f*
runner *of a hydrodynamic coupling* Turbinenrad *n* einer Strömungskupplung
running efficiency *PU/MOT* Betriebswirkungsgrad *m*, Wirkungsgrad *m* bei Nennbetriebsbedingungen
running leakage *PU/MOT* Leckverluste *mpl* bei Nenndrehzahl
running torque *PU/MOT* Drehmoment *n* bei Nenndrehzahl
rupture disk *PCV* Berstscheibe *f*, Berstmembran *f*, **Reißscheibe** *f*
rust and oxidation inhibitor *FLUIDS* Rost- und Oxidationsinhibitor *m*
rust inhibitor *FLUIDS* Rostinhibitor *m*, Rostschutzadditiv *n*

S

SAE = Society of Automotive Engineers *USA*
SAE flange *FITT* SAE-Flansch *m* ein geteilter 4-Loch- Flansch
SAE port *FITT* SAE-Anschluß *m* gerades Gewinde, mit O- Ring
SAE grade *FLUIDS* s. viscosity grade
SAE thread *FITT* SAE-Rohrgewinde *n* gerade
safety circuit Sicherheitsschaltung *f* z. B. Zweihandbedienung
safety valve Sicherheitsventil *n*
sampling valve Probenahmeventil *n*
sandwich mounting Modulverkettung *f*, Batterieverkettung *f*; *wenn vertikal* Höhenverkettung *f*, Turmverkettung *f*

sandwich valve Ventil *n* für Modulverkettung (Batterieverkettung), Modulventil *n*
sandwiched gear pump Plattenzahnradpumpe *f*, Dreiplattenpumpe *f*
saponification value *FLUIDS* Verseifungszahl *f*, VZ *f*
saturated vapour pressure *FLUIDS* Sättigungsdruck *m*
Saybolt Second Universal Einheit der kinematischen Viskosität
scavenge *v* *exchange part of oil in closed-circuit hydrostatic transmissions* spülen Teil des Öls in hydrostatischen Getrieben mit geschlossenem Kreislauf austauschen
scavenger circuit Kreislauf *m* (Schaltung *f*) mit Spülung
scavenger pump *in a closed circuit* Spülpumpe *f*, Hilfspumpe *f* in einem geschlossenem Kreislauf
scavenger valve *CHECKS* Nachsaugeventil *n*, Spülventil *n*
scfh standard cubic feet per hour Einheit des Volumenstroms eines Gases unter Standardbedingungen
scfm standard cubic feet per minute Einheit des Volumenstroms eines Gases unter Standardbedingungen
schedule *LINES* Baureihennummer, eine Maßangabe für die Rohrwanddicke
SCL seal *special cut lip seal* [form]gedrehter Dichtring *m*
scotch-yoke rotary actuator Kulissenschwenkmotor *m*
scraper [seal] *SEALS* Abstreifring *m*, Abstreifer *m* gering nachgiebig, vgl wiper; *FILTERS* Spalträumer *m*, Abstreifer *m*, Kratzer *m*

screen *v*   filtern   *mittels Oberflächenfilter*
screen [filter]   Siebfilter *m,n*, Gewebefilter *m,n*
screw   *of a screw pump* Spindel *f*, Schraube *f*   *einer Schraubenpumpe*
screw-down valve   Spindelventil *n*
screw-in cartridge valve   Einschraubventil *n*, Ventil-Einschraubpatrone *f*
screw motor   Schraubenmotor *m*
screw pump   Schraubenpumpe *f*
screw pump *or* **motor**   Schraubenmaschine *f*, Schraubeneinheit *f*, Schraubengerät *n*
screw-together fitting   *for hose end* Aufschraubverschraubung *f Schlauchende*
screwed fitting   Gewindeverschraubung *f*, Gewindefitting *m*
seal   Dichtung *f*, Dichteinrichtung *f*; Dicht[ungs]element *n*, Dichtung *f*
seal *v*   [ab]dichten;   *eg reservoir* verschließen   *z. B. Behälter*
seal assembly   Dicht[ungs]satz *m*, Packung *f*
seal cartridge   Einbaudichtsatz *m*, Einbaudichtung *f*
seal cavity   Dichtungsraum *m*, Einbauraum *m*
seal clearance   Dichtspalt *m*
seal compatibility   *FLUIDS* Dichtungsverträglichkeit *f*
seal compatibility index   *FLUIDS* Dichtungsverträglichkeitsindex *m*, DVI *m*
seal friction   Dichtungsreibung *f*
seal gap   Dichtspalt *m*
seal groove   Dichtungsnut *f*
seal life   Dichtungsstandzeit *f*
seal line   Dichtlinie *f*
seal lip   Dicht[ungs]lippe *f*

seal material   Dichtungs[werk]stoff *m*, Dichtungsmaterial *n*
seal plate   *of a pump or motor* Seitenplatte *f*   *einer Verdrängermaschine*
seal pocket   Dichtungsraum *m*, Einbauraum *m*
seal ring   Ringdichtung *f*, Dicht[ungs]ring *m*
seal washer   Scheibendichtung *f*, Dicht[ungs]scheibe *f*
sealant   Dichtstoff *m*, Dichtmittel *n*, Dichtungskitt *m*
sealed pressurized reservoir (tank)   druckdichter Behälter *m*
sealing band   Dichtlinie *f*
sealing device   Dichtung *f*, Dichteinrichtung *f*
sealing edge   Dichtkante *f*
sealing element   Dicht[ungs]element *n*, Dichtung *f*
sealing force   Dichtkraft *f*
sealing lip   Dicht[ungs]lippe *f*
sealing member   Dicht[ungs]element *n*, Dichtung *f*
sealing profile   Dicht[ungs]schnur *f*
sealing surface   Dicht[ungs]fläche *f*
seam-welded steel pipe   geschweißtes Stahlrohr *n*
seamless tubing   nahtlos gezogenes (nahtloses) Rohr *n*
seat   *of a valve* Ventilsitz *m*
seat valve   Sitzventil *n*
seated valve   Sitzventil *n*
seated-valve pump   druckgesteuerte (sitzventilgesteuerte) Pumpe *f*
seating member   *of a check valve* Schließelement *n*   *eines Sperrventils*
seating valve   Sitzventil *n*
secant bulk modulus   *FLUIDS* Sekan-

lltenkompressionsmodul *m*, mittlerer Kompressionsmodul *m*
second viscosity *FLUIDS* Dilatationsviskosität *f*, Sekundärviskosität *f*
sectional mounting Modulverkettung *f*, Batterieverkettung *f*
sectional valve Ventil *n* für Modulverkettung (Batterieverkettung), Modulventil *n*
seepage Leckage *f*, Leckverlust *m*
self-align fitting selbstausrichtende Verbindung *f*
self-bypassing filter Filter *m,n* mit automatischer Umgehung
self-cleaning filter element selbstreinigendes Filterelement *n*
self-energized seal selbstwirkende (druckgespannte) Dichtung *f*
self-flare fitting selbstbördelnde Verbindung *f*
self-guided weight-loaded accumulator gewichtsbelasteter Speicher *m* mit Innenführung
self-ignition temperature *FLUIDS* Selbstentzündungstemperatur *f*
self-lube (-lubricating) *FLUIDS* selbstschmierend
self-sealing coupling *FITT* selbstdichtende Kupplung *f*
self-warning filter Filter *m, n* mit Verstopfungsanzeige, Warnfilter *m,n*
semi-automatic seal vorgespannte Dichtung *f*
semi-rigid line halbstarre Leitung *f*
SEN *steam-emulsion number, measure of emulsibility FLUIDS* Wasserdampfemulsionszahl *m* *Maß der Emulgierbarkeit*
sensing element *s.* sensor
sensitive to contamination *FLUIDS* schmutzempfindlich, verschmutzungsempfindlich
sensitivity of adjustment *of an adjustable restrictor FCV* Auflösungsvermögen *n* *einer Verstelldrossel*
sensor Sensor *m*, Fühler *m*, Meßfühler *m*, Geber *m*, Aufnehmer *m*, Meßwandler *m*
sensor fitting Sensorverschraubung *f*
separable fitting lösbare Verbindung (Verschraubung) *f*
separate *v* *flow THEOR* [sich] ablösen, abreißen *Strömung*
separate circuits system Einkreissystem *n*, Einkreishydraulik *f*, Einzelhydraulik *f*
separate pump/motor transmission Getriebe *n* in aufgelöster Bauweise, Ferngetriebe *n*
separated accumulator Speicher *m* mit Trennwand
separation point *of flow THEOR* Ablösepunkt *m*, Abreißpunkt *m*
separation power *FLUIDS* Abscheidevermögen *n* *z. B. für Wasser*
separator *ACCUM* Trennglied *n*, Trennelement *n*, Trennwand *f*; *PREPAR* Wasserabscheider *m*
separator piston *ACCUM* Trennkolben *m*
separator tube Trennschlauch *m*
sequence diagram (plot) *CIRC* Funktionsablaufplan *m*, Ablaufdiagramm *n*
sequence valve Folge[schalt]ventil *n*, Zuschaltventil *n*
sequencer *s.* sequencing control
sequencing circuit Folgeschaltung *f*
sequencing control Folgesteuerung *f*
series: install (connect, pipe) *v* in series

hintereinanderschalten, in Serie (Reihe) *f* schalten
series circuit   Reihenschaltung *f*, Serienschaltung *f*
series flow-control valve   Zweiwege-Strombegrenzungsventil *n* (-Stromregelventil *n*)
series flow regulator   *s.* series flow-control valve
service indicator   *FILTERS* Verstopfungsanzeige *f*
servo   *VALV OP* Servosteuerung *f*; Servoregler *m* mit mechanischer Ausgangsgröße; *PU/MOT* *s.* servo stroker
servo brake   *APPL* Servobremse *f*
servo stroker   *PU/MOT* Servostelleinheit *f*
servoactuator   *VALV OP* Servostelleinheit *f*, Servostellglied *n*
servoamplifier   Servoverstärker *m*
servocontrol   *PU/MOT* Servostelleinheit *f*; *VALV OP* Servosteuerung *f*
servocontrolled   *VALV OP* servogesteuert
servocylinder   Servozylinder *m*
servodrive   Servoantrieb *m*
servomechanism   *VALV OP* Servoregler *m* mit mechanischer Ausgangsgröße
servomotor   *VALV OP* Servomotor *m*, Stellmotor *m*
servopump   Servopumpe *f*
servosolenoid   Steuermotor *m* mit linearer Bewegung, Proportionalmagnet *m*
servovalve   *DCV* Servoventil *n*
set pressure   Einstelldruck *m*
setting point   *FLUIDS* Stockpunkt *m*
shaft seal   Wellendichtung *f*

sharp-edged orifice   *THEOR* scharfkantige Blende *f*, blendenförmiger Drosselwiderstand *m*, Blende *f*
shear   *THEOR* Scherung *f*, Schub *m*, Schiebung *f*
shear breakdown   *of long polymer molecules* *FLUIDS* Scherungsversagen *n*   Bruch der langen Kettenmoleküle
shear rate   *THEOR* Schergeschwindigkeit *f*, Schergefälle *n*
shear stability (strength)   *FLUIDS* Scherfestigkeit *f*, Scherstabilität *f*
shear valve   Scherschlußventil *n*
shell   *of a hose fitting* Hülse *f* einer Schlauchverbindung
shell-and-tube heat exchanger   Röhrenwärmeübertrager *m* mit Mantel
shift *v*   *valve* schalten   *Ventil*; *valve spool* [sich] verschieben, schalten   *Ventilkolben*
shifting speed   *DCV* Schaltgeschwindigkeit *f*
shifting time   *DCV* Schaltzeit *f*
shifting velocity   *s.* shifting speed
shock absorber   Stoßdämpfer *m*, Stoßfänger *m*, Prallfänger *m*
shock pressure absorber   Druckstoßdämpfer *m*
shock suppressor   *s.* shock absorber
shock wave   Stoßwelle *f*
shoe   *PU/MOT* Gleitschuh *m*
shop air   *PREPAR* Betriebsdruckluft *f*
shop air mains   Werksdruckluftnetz *n*, Betriebsdruckluftnetz *n*
short-stroke cylinder   Kurzhubzylinder *m*
short-stroke valve   Kurzhubventil *n*
shutoff pressure   *PCV* Schließdruck *m*
shutoff pressure surge (shock)

Schließdruckstoß *m*, Schließdruckschlag *m*
shutoff stroke *PCV* Schließweg *m*
shutoff valve Absperrventil *n*
shuttle valve *DCV* Wechselventil *n*
side mount cylinder Zylinder *m* mit seitlicher Befestigung
sight gage (glass) *RESERVOIRS* Füllstandsglas *n*, Füllstandsauge *n*
signal port Signalanschluß *m*
silencer Schalldämpfer *m*, Geräuschdämpfer *m*
silica-gel desiccant Silicageltrocknungsmittel *n*, Kieselgeltrocknungsmittel *n*
silicate ester fluid Silicatesterflüssigkeit *f*
silicone fluid siliconbasische Flüssigkeit *f*
silt *contaminant particles less than about 5 μm in size FLUIDS* Schlamm *m* aus Schmutzpartikeln kleiner etwa 5 μm
silt *v* *small orifice* verstopfen, [sich] zusetzen, zuwachsen *enge Drosselquerschnitte*
silting *FLUIDS* Schlammbildung *f*
SIM *summing impact modulator LOG EL* Gegenstrahlelement *n* ohne Steuerdüsen, summierender Impaktmodulator *m*
simplified symbol vereinfachtes Symbol (Schaltzeichen) *n*
single-acting cylinder einfachwirkender Zylinder *m*
single-acting hand pump einfachwirkende Handpumpe *f*
single-acting intensifier einfachwirkender Druckübersetzer *m*
single-acting solenoid *VALV OP* Stellmagnet *m* mit einseitiger Wirkungsrichtung
single ball valve *CHECKS* Einkugelventil *n*
single banjo richtungseinstellbare Winkelverschraubung *f*
single circuit system Einkreissystem *n*, Einkreishydraulik *f*, Einzelhydraulik *f*
single-diaphragm element *LOG EL* Einmembranelement *n*
single-ear mount *CYL* Schwenkaugenbefestigung *f*
single-flow pump Pumpe *f* mit einem Förderstrom, Einstrompumpe *f*
single-fluid intensifier Druckübersetzer *m* für gleiche Medien
single-inlet screw pump einflutige Schraubenpumpe *f*
single-lip seal Einlippenring *m*
single-lip wiper Einlippenabstreifer *m*
single-path transmission Hydrogetriebe *n* mit einem *einzigen* Motor
single-plane swivel joint Einebenen-Rohrgelenk *n*
single poppet coupling *FITT* Einstrangsitzkupplung *f*
single pump Pumpe *f* mit einem Förderstrom, Einstrompumpe *f*
single-rack rotary actuator Schwenkmotor *m* mit einer Zahnstange
single-rod [end] cylinder Zylinder *m* mit einseitiger Kolbenstange, Einstangenzylinder *m*
single-screw pump Einschraubenpumpe *f*
single-shot booster (intensifier) Einzelhubdruckübersetzer *m*

single shut-off *FITT* Einstrangabsperrung *f*, Einwegabsperrung *f*
single shut-off coupling Einstrangabsperrkupplung *f*, Einwegabsperrkupplung *f*
single-stage compressor einstufiger Verdichter *m*
single-stage cylinder einstufiger Zylinder *m*
single-stage servovalve einstufiges Servoventil *n*
single-station subplate *VALV* Einzelunterplatte *f*, Einzelanschlußplatte *f*
single-suction screw pump einflutige Schraubenpumpe *f*
single-valved coupling Einstrangabsperrkupplung *f*, Einwegabsperrkupplung *f*
single-valve manifold Einzelunterplatte *f*, Einzelanschlußplatte *f*
single-vane rotary actuator Einflügelschwenkmotor *m*
single wire braid hose Schlauch *m* mit einfacher Drahtgeflechteinlage
sink *THEOR* Senke *f*
sintered bronze filter Sinterbronzefilter *m,n*
sintered fiber filter Sinterfaserfilter *m,n*
sintered metal-powder filter Sintermetallfilter *m,n*
sintered plastic filter Sinterplastikfilter *m,n*
sintered wire-cloth filter Sinterdrahtgewebefilter *m,n*
SIT *spontaneous ignition temperature* *FLUIDS* Selbstentzündungstemperatur *f*
six-port valve Ventil *n* mit sechs Anschlüssen

six-way valve Sechswegeventil *n*, 6-Wege-Ventil *n*
skid mount Kufenbefestigung *f*
skive *v* *hose cover* [ab]schälen Schlauchdecke, schärfen Schlauchende
skive-type fitting *hose fitting* Schälverbindung *f*, Schärfverbindung *f* Schlauchverbindung
slack diaphragm *VALV OP* Schlaffmembran *f*
slave cylinder Folgezylinder *m*, Stellzylinder *m*
slave spool *of a two-stage valve* *VALV OP* Hauptsteuerschieber *m*, Hauptkolben *m* *eines zweistufigen Ventils*
sleeve *VALV* Ventilbüchse *f*
sleeve-and-poppet coupling *FITT* Gleitmuffenkupplung *f*
slide block *of a radial piston pump* Gleitrahmen *m* *einer Radialkolbenpumpe*
slide-seal coupling *FITT* Gleitsitzkupplung *f*, Gleitdichtungskupplung *f*
slide shoe *PU/MOT* Gleitschuh *m*
sliding seal Gleitdichtung *f* Dichtung für hin- und hergehende Bewegung
sliding spool *DCV* Kolbenlängsschieber *m*
sliding spool valve *DCV* Kolbenlängsschieberventil *n*
sliding-vane compressor Zellenverdichter *m*, Lamellenverdichter *m*
sliding-vane motor Flügelzellenmotor *m*
sliding-vane pump Flügelzellenpumpe *f*
slip [flow] *PU/MOT* innerer Leckstrom *m*, Schlupfstrom *m*

slippage

slip ratio *of a pump* Förderstromschlupf *m*
slippage *s.* slip [flow]
slippage-compensation line *in a closed circuit* Nachfüllleitung *f*, Leckölergänzungsleitung *f in einem geschlossenen Kreislauf*
slippage pump *in a closed circuit* Spülpumpe *f*, Hilfspumpe *f in einem geschlossenen Kreislauf*
slipper pad *PU/MOT* Gleitschuh *m*
slipper piston *PU/MOT* Gleitschuhkolben *m*
slipper pump Gleitschuhpumpe *f*
slipper ring Gleitringdichtung *f* selbstschmierend
slope coefficient *viscosity-temperature characteristics FLUIDS* [Viskositäts-]-Richtungskonstante *f*, VRk *f*, m- Wert *m Viskositäts-Temperatur-Verhalten*
sloshing *of oil surface in the reservoir* Schwappen *n der Ölöberfläche im Behälter*
slot seal *CYL* Schlitzdichtung *f am geschlitzten Zylinder*
slotted cylinder geschlitzter Zylinder *m*, Schlitzzylinder *m*
slotted spear *CYL* Bremsansatz *m* mit Axialkerbe[n]
slow-closure valve schleichend schließendes Ventil *n*
slow-opening valve schleichend öffnendes Ventil *n*
sludge *contaminant particles less than about 5 µm in size FLUIDS* Schlamm *m aus Schmutzpartikeln kleiner etwa 5 µm*
sludge formation *FLUIDS* Schlammbildung *f*

small-bore cylinder Zylinder *m* mit kleinem Innendurchmesser
small size cylinder Miniaturzylinder *m*, Minizylinder *m*, Mikrozylinder *m*, Kleinzylinder *m*
smooth *PU/MOT* pulsationsfrei
smooth pipe *THEOR* glattes Rohr *n*
snap mount *CYL* Schnappbefestigung *f*
snubber *s.* cylinder cushion; *s.* gauge snubber; *s.* pressure snubber; *s.* shock absorber
socket *of a hose fitting* Hülse *f einer Schlauchverbindung*
socket end fitting Aufschraubverschraubung *f*
socket weld fitting with cone extension Schweißkegelverschraubung *f*, Schweißnippelverschraubung *f*
socket weld fitting with spherical sealing surface Schweißkugelverschraubung *f*
socket welding elbow Einsteckschweißwinkel *m*
socketless fitting *hose fitting* Aufsteckverbindung *f*, hülsenlose Verbindung *f Schlauchverbindung*
soft end *THEOR* freies Leitungsende *n*
soft seat *DCV* Weichsitz *m*
soft-seat valve *DCV* Weichsitzventil *n*
soft-shift directional valve weichschaltendes Wegeventil *n*
soldered fitting Lötverbindung *f*, Lötverschraubung *f*
solderless fitting lötlose Verbindung *f* (Verschraubung *f*)
solenoid *VALV OP* Stellmagnet *m*
solenoid-actuated valve elektro-

magnetisch betätigtes (magnetbetätigtes) Ventil *n*, Magnetventil *n*
solenoid actuator elektromagnetische Stelleinheit (Betätigungseinheit) *f*, Magnetstelleinheit *f*
solenoid armature Magnetanker *m*
solenoid coil Magnetspule *f*
solenoid control *PU/MOT* elektrohydraulische Servo-Stelleinheit *f*
solenoid drain valve Magnetablaßventil *n*
solenoid-operated drain valve Magnetablaßventil *n*
solenoid-operated valve *s.* solenoid-actuated valve
solenoid operator *s.* solenoid actuator
solenoid pin (pushpin) Magnetstößel *m*
solenoid stroker *s.* solenoid actuator
solenoid valve *s.* solenoid-actuated valve
solenoid vented relief valve elektrisch abschaltbares Druckbegrenzungsventil *n*
solid contamination *FLUIDS* Feststoffverschmutzung *f*, Festschmutz *m*
solid drawn tubing nahtlos gezogenes (nahtloses) Rohr *n*
solid-end cylinder Zylinder *m* mit nicht lösbar verbundenen Deckeln
solid flange mount *CYL* Befestigung *f* mit ungeteiltem Flansch
solidification point *FLUIDS* Stockpunkt *m*
solubility of air *FLUIDS* Luftlösungsvermögen *n*
solubility of gas *FLUIDS* Gaslösungsvermögen *n*
soluble in oil öllöslich
soluble oil *s.* oil-in-water emulsion
sonic flow *THEOR* Strömung *f* mit Schallgeschwindigkeit
sonic speed Schallgeschwindigkeit *f*
sound attenuator Schalldämpfer *m*, Geräuschdämpfer *m*
sound beam sensor *LOG EL* akustischer Oszillationssensor *m*
sound-proof enclosure *PREPAR* schalldichtes Gehäuse *n*, Schallschutzhaube *f*
sound velocity Schallgeschwindigkeit *f*
source *THEOR* Quelle *f*
source of contamination *FLUIDS* Verschmutzungsquelle *f*, Schmutzquelle *f*
spear *CYL* Bremsansatz *m*
special cut lip seal [form]gedrehter Dichtring *m*
special cylinder Zylinder *m* in Sonderausführung, Sonderzylinder *m*
special-function valve Sonderventil *n*, Spezialventil *n*
specialty cylinder *s.* special cylinder
speed-control muffler Entlüftungsdrossel *f*
speed fluctuation *PU/MOT* Drehzahlungleichförmigkeit *f*
speed governor *PU/MOT* Drehzahlregler *m*
speed loss *PU/MOT* Drehzahlschlupf *m*, Schlupfdrehzahl *f*
speed of sound Schallgeschwindigkeit *f*
speed pulsation *PU/MOT* Drehzahlungleichförmigkeit *f*
speed range *PU/MOT* Drehzahlbereich *m*
speed rating Nenndrehzahl *f*

spherical accumulator Kugelspeicher *m*
spherical mounting *CYL* Gelenkbefestigung *f*
spherical plug valve Kugelhahn *m*, Kugeldrehschieberventil *n*
spherical seal union fitting Schweißkugelverschraubung *f*
spherical swivel joint Universalrohrgelenk *n*, Kugelrohrgelenk *n*
spherical valve plate *PU/MOT* sphärischer Steuerspiegel *m*
spill-off flow-control valve Dreiwege-Stromregelventil *n*, Dreiwege-Strombegrenzungsventil *n*
spin *v* eg *air motor* durchdrehen, durchfallen z. B. Druckluftmotor
spin-on filter Wechselfilter *m,n*
spindle *VALV OP* Ventilspindel *f*
spiral Bourdon tube *MEASUR* Schneckenrohrfeder *f*
spiral wire wrap hose spiralarmierter Schlauch *m*
4-spiral wire wrap hose 4fach-spiralarmierter Schlauch *m*
split flange *FITT* geteilter Flansch *m*, Splitflansch *m*
split-flange fitting Verbindung *f* mit geteiltem Flansch, Splitflanschverbindung *f*
split-flow pump Zweistrompumpe *f*, Zweikreispumpe *f*, Zwillingspumpe *f*, Doppelpumpe *f*
split-ring seal geteilte Dichtung *f*
split-torque drive Getriebe *n* mit äußerer Leistungsverzweigung
split transmission Ferngetriebe *n*, Getriebe *n* in aufgelöster Bauweise
spontaneous ignition temperature *FLUIDS* Selbstentzündungstemperatur *f*

spool *DCV* Steuerschieber *m*, Ventilschieber *m*, im engeren Sinn Kolbenlängsschieber *m*
spool chamber *DCV* Schieberkammer *f*, Steuerkammer *f*
spool displacement *DCV* Steuerkolbenverschiebung *f*, Schieberweg *m*
spool element *LOG EL* Kolbenelement *n*
spool land *DCV* Steuerschieberkolben *m*, Steuerschieberbund *m*, Steuerschiebersteg *m*
spool position *DCV* Schaltstellung *f*, Schieberstellung *f*
spool stroke *DCV* Schieberhub *m*, Schieberweg *m*, Schaltweg *m* auch maximaler
spool travel (traverse) Steuerkolbenverschiebung *f*, Schieberweg *m*
spool valve *DCV* Schieberventil *n*, Kolbenventil *n*; in engerem Sinn Kolbenlängsschieberventil *n*
spool-valve pump schiebergesteuerte Pumpe *f*
spring-biased valve federbelastetes Ventil *n*
spring cavity *VALV OP* Federkammer *f*, Federraum *m*
spring-centered valve federzentriertes Ventil *n*
spring centering *VALV OP* Federzentrierung *f*
spring-energized seal federgespannte Dichtung *f*
spring-loaded accumulator federbelasteter Speicher *m*
spring-loaded valve federbelastetes Ventil *n*
spring-offset valve Ventil *n* mit Federabhub, federgeöffnetes Ventil *n*

spring-opposed valve federbelastetes Ventil *n*
spring pocket *VALV OP* Federkammer *f*, Federraum *m*
spring return *CYL* Federrückzug *m*
spring-return cylinder Zylinder *m* mit Federrückzug
spring-return piston *CYL* federrückgezogener Kolben *m*
spring-return valve Ventil *n* mit Federrückzug, federfixiertes Ventil *n*
spur gear motor Außenzahnradmotor *m*
spur gear pump Außenzahnradpumpe *f*
spur gear pump *or* **motor** Außenzahnradmaschine *f*, Außenzahnradeinheit *f*, Außenzahnradgerät *n*
square-bodied cylinder Zylinder *m* mit Quadratkörper (Quadratrohrmantel)
square-cut pipe end rechtwinklig getrenntes Rohrende *n*
square-edged orifice *THEOR* drosselartige Blende *f*
square-head cylinder Zylinder *m* mit Quadratflansch bzw. Quadratkopf
square-head piston *CYL* Rechteckkolben *m*
square land valve *DCV* Ventil *n* mit scharfkantigen Schieberbunden
square ring *SEALS* Quadrat[schnur]ring *m*
squeegee pump Walkverdrängerpumpe *f*
squeeze-type moulded seal Formweichdichtung *f*, Profilweichdichtung *f*
SSU *Saybolt Universal Second* Einheit der kinematischen Viskosität
stack *s.* valve stack

stack *vt* *valves* verketten *Ventile*
stack mounting Modulverkettung *f*, Batterieverkettung *f*
stack pump verkettbare Pumpe *f*
stack valve Ventil *n* für Modulverkettung (Batterieverkettung), Modulventil *n*
stackable *VALV OP* verkettungsfähig, verkettbar
stacked-diaphragm logic valve *LOG EL* Mehrmembranenlogikventil *n*, Membranpaketlogikventil *n*
stacking *VALV OP* verkettungsfähig, verkettbar
stacking subplate Verkettungsunterplatte *f*, verkettbare Unterplatte *f*
stage pressure ratio *COMPR* Stufendruckverhältnis *n*
staged pump Stufenpumpe *f*
stainless steel tubing Rohr *n* aus einem nichtrostenden Stahl
stall *vi* *hydromotor* blockieren, stehenbleiben infolge Überlastung *Hydromotor*
stall torque *PU/MOT* Blockiermoment *n*, Kippmoment *n*
stalled leakage flow *PU/MOT* Blockierleckstrom *m*, Leckstrom *m* im Kippunkt
standard air Luft *f* unter Norm[al]bedingungen (im Normzustand)
standard conditions *of temperature and pressure* Normzustand *m*, Normbedingungen *fpl*, Standardbedingungen *fpl* *von Temperatur und Druck*
standard cylinder Zylinder *m* in genormter Ausführung, Normzylinder *m*
standard pressure Normdruck *m*, Standarddruck *m*

standard reservoir (tank) Normbehälter *m*
standard temperature Normtemperatur *f*, Standardtemperatur *f*
standard volume Normvolumen *n*, Standardvolumen *n*
standardized fitting Normverschraubung *f*
stand-by pump Notpumpe *f*
standpipe senkrechtes Rohrstück *n*, Standrohr *n*
start-stop control *COMPR* Abschaltregelung *f*, Ausschaltregelung *f*, Stillsetzregelung *f*
starting characteristics *PU/MOT* Anfahrverhalten *n*, Anlaufverhalten *n*, Startverhalten *n*
starting force *CYL* Anfahrkraft *f*, Anlaufkraft *f*, Startkraft *f*
starting load *CYL, PU/MOT* Anfahrlast *f*, Anlauflast *f*, Startlast *f*
starting pressure *CYL, PU/MOT* Anfahrdruck *m*, Anlaufdruck *m*, Startdruck *m*
starting viscosity *PU/MOT* Startviskosität *f*
starvation *of a pump* unvollständige Füllung *f*, Abschnappen *n* *einer Pumpe*
starve *v pump* ungenügend Flüssigkeit *f* erhalten, ungenügend gefüllt werden, abschnappen *Pumpe*
static fluid sampling *FLUIDS* statische Flüssigkeitsprobenahme *f*
static pressure statischer Druck *m*
static seal ruhende (statische) Dichtung *f*
stationary-body (-rod) cylinder Zylinder *m* mit feststehendem Zylinderkörper

stationary-vanes motor Sperrschiebermotor *m*
stationary-vanes pump Sperrschieberpumpe *f*
stator *PU/MOT* Stator *m*
steady flow *THEOR* stationäre Strömung *f*
steady[-state] pressure *THEOR* stationärer Druck *m*, Druck *m* im Beharrungszustand
steam-emulsion number *measure of emulsibility FLUIDS* Wasserdampfemulsionszahl *m* *Maß der Emulgierbarkeit*
steel tubing Stahlrohr *n* Präzisionsrohr
steel-wool filter Stahlwollefilter *m,n*
steering booster (cylinder) *APPL* Lenkkraftverstärker *m*, Lenkhilfzylinder *m*
stem *VALV OP* Ventilspindel *f*
stem-operated valve Spindelventil *n*
stem-valve coupling *FITT* Stößelventilkupplung *f*
step vane *PU/MOT* Stufenflügel *m*
stepped spear *CYL* gestufter Bremsansatz *m*
stepper [motor] *VALV OP* Schrittmotor *m*
stepper-positioned (stepping) cylinder Schrittzylinder *m*
stepping sequence Schaltfolge *f*
stick *v eg valve spool* verklemmen, verkleben *z. B. Ventilkolben*
stiff end *of a conduit* festes Leitungsende *n*
stop tube *CYL* Hubbegrenzungsbuchse *f*, Anschlagbuchse *f*
stop valve *CYL* Stoppventil *n* *zum Anhalten des Kolbens in einer Zwischenstellung*

storage pressure *ACCUM* Speicherdruck *m*
store *v* *energy, pressure, fluid* speichern *Energie, Druck, Flüssigkeit*
straight connection (fitting) gerade Verschraubung *f* (Verbindung *f*)
straight spear *CYL* zylindrischer Bremsansatz *m*
straight thread *FITT* gerades Gewinde *n*
straight-through coupling *FITT* Durchgangskupplung *f*
straight-tube heat exchanger Geradrohrwärmeübertrager *m*
strain *v* filtern *mit Oberflächenfilter*
strainer Filter *m,n* *eher als Oberflächenfilter, für weniger kleine Schmutzteilchen; vgl* filter
strain gauge pressure transducer Dehnmeßstreifen-Druckwandler *m*, DMS-Druckwandler *m*
streak line *THEOR* Schlierenlinie *f*
stream Fluidstrom *m*, Flüssigkeitsstrom *m*, Strom *m*; *fluid motion* Flüssigkeitsströmung *f*, Strömung *f*
stream filament *THEOR* Stromfaden *m*
streamline *THEOR* Stromlinie *f*; Strombahn *f*
streamline flow *THEOR* laminare Strömung *f*, Laminarströmung *f*
streamtube *THEOR* Stromröhre *f*
street elbow Einschraubwinkelverschraubung *f*
street tee T-Verschraubung *f* mit Einschraubzapfen im durchgehenden Teil
stroke *CYL* Hub *m*; Hublänge *f*, Hub *m*; *of valve poppet* Hub *m* im Sitzventil

stroke: move *v* **off stroke** *variable pump* einschwenken *Verstellpumpe*
stroke: move *v* **on stroke .** *variable pump* ausschwenken *Verstellpumpe*
stroke *v* *piston* sich verschieben (bewegen) *Kolben*; *variable pump* ausschwenken, verstellen *Verstellpumpe*
stroke adjustment *CYL* Hubeinstellung *f*
stroke end *CYL* Hubende *n*
stroke length Hublänge *f*, Hub *m*
stroke limiter *of a pump control* Stellwegbegrenzer *m* einer Pumpenstelleinheit
stroke limiter cover *of a logic element DCV* Hubbegrenzerdeckel *m*, Hubeinstelldeckel *m*
stroke multiplier Hubvervielfältiger *m*
stroke positioner *CYL* Positioniereinrichtung *f*
stroke reversal *CYL* Hubumkehr *f*, Hubumsteuerung *f*
stroke speed *CYL* Hubgeschwindigkeit *f*
stroke volume *CYL* Hubraum *m*, Hubvolumen *n*
stroker *PU/MOT* Stelleinrichtung *f*, Stelleinheit *f*, Stellkopf *m*
stroker servo *PU/MOT* Servo-Stelleinheit *f*
stroking force *VALV OP* Stellkraft *f*, Betätigungskraft *f*, Verstellkraft *f*, Steuerkraft *f*, Schaltkraft *f*
stroking piston *displacement control PU/MOT* Stellkolben *m* Verdrängervolumenverstellung
stroking rate *CYL* Hubgeschwindigkeit *f*; *VALV OP* Schaltgeschwindigkeit *f*

stroking time   *VALV OP* Schaltzeit *f*
stud mount   *CYL* Schwenkzapfenbefestigung *f*, Zapfenbefestigung *f*
subbase   Unterplatte *f*, Anschlußplatte *f*; *single-valve manifold* Einzelunterplatte *f*, Einzelanschlußplatte *f*
subbase mount   Unterplattenbefestigung *f*, Unterplattenanbau *m*
subbase valve   Unterplattenventil *n*
submerged   getaucht, Tauch..., Unteröl...
submersed   *s.* submerged
subplate   *s.* subbase
subsonic flow   *THEOR* Unterschallströmung *f*
suction   *PU/MOT* Ansaugen *n*, Saugen *n*, Ansaugung *f*, Saugung *f*
suction capacity   *PU/MOT* Saugvermögen *n*, Ansaugvermögen *n*, Saugfähigkeit *f*
suction chamber   Saugraum *m*, Ansaugraum *m*, Einlaßkammer *f*, Eintrittskammer *f*, Zulaufkammer *f*, Zuflußkammer *f*
suction channel (duct)   Saugkanal *m*, Ansaugkanal *m*, Einlaßkanal *m*, Eintrittskanal *m*, Eingangskanal *m*, Zulaufkanal *m*, Zuflußkanal *m*
suction characteristics   *PU/MOT* Saugverhalten *n*, Ansaugverhalten *n*
suction cup   Saugnapf *m*
suction filter   Saugfilter *m,n*
suction flow [rate]   Saugstrom *m*, Ansaugstrom *m*, Einlaß[volumen]strom *m*, Eintrittsstrom *m*, Zulaufstrom *m*
suction head   Saug[druck]höhe *f*
suction horsepower   Saugleistung *f*

suction kidney   Saugniere *f*
suction line   Saugleitung *f*, Ansaugleitung *f*, Einlaßleitung *f*, Eintrittsleitung *f*, Eingangsleitung *f*, Zulaufleitung *f*, Zuführleitung *f*, Zuflußleitung *f*, Zuleitung *f*
suction port   Sauganschluß *m*, Ansauganschluß *m*, Einlaßanschluß *m*, Eintrittsanschluß *m*, Zulaufanschluß *m*, Zuführanschluß *m*, Zuflußanschluß *m*
suction power   Saugleistung *f*
suction pressure   Saugdruck *m*, Ansaugdruck *m*, Einlaßdruck *m*, Eintrittsdruck *m*, Eingangsdruck *m*, Zulaufdruck *m*
suction side   Saugseite *f*, Einlaßseite *f*, Eintrittsseite *f*, Eingangsseite *f*, Zulaufseite *f*, Zuflußseite *f*
suction stroke   Saughub *m*
suction valve   Saugventil *n*, Einlaßventil *n*
sudden contraction   *THEOR* plötzliche Verengung *f*
sudden enlargement   *THEOR* plötzliche Erweiterung *f*
sulfur content   *FLUIDS* Schwefelgehalt *m*
summation horsepower (torque) limiter control   *of a pump* Summenleistungsregelung *f einer Pumpe*
summer-grade oil   Sommeröl *n*
summing impact modulator   *LOG EL* Gegenstrahlelement *n* ohne Steuerdüsen, summierender Impaktmodulator *m*
sump filter   Behälterfilter *m,n*, Sumpffilter *m,n*
supercharge *v*   *pump* [vor]füllen *Pumpe*
supercharge pump   Speisepumpe *f*,

Füllpumpe *f*, Vorfüllpumpe *f*, Zuförderpumpe *f*, Ladepumpe *f*
superfiltration Feinstfilterung *f*, Mikrofilterung *f*
superpressure fluid power Höchstdruckhydraulik *f*
supersonic flow *THEOR* Überschallströmung *f*
supply Zufuhr *f*, Speisung *f*, Beaufschlagung *f*, Lieferung *f*, Versorgung *f*
supply *v* fördern, liefern, abgeben *Flüssigkeit in (an) das System*, beaufschlagen *Komponente mit Flüssigkeit*, speisen, beliefern, versorgen *System mit Flüssigkeit*; *with air* Druckluft *f* zuführen, belüften, mit Druckluft speisen (beaufschlagen)
supply air *LOG EL* Speiseluft *f*
supply jet *LOG EL* Versorgungsstrahl *m*, Speisestrahl *m*
supply line Speiseleitung *f*, Druckleitung *f*, Einlaßleitung *f*, Eintrittsleitung *f*, Eingangsleitung *f*, Zulaufleitung *f*, Zuführleitung *f*, Zuflußleitung *f*, Zuleitung *f*
supply nozzle *LOG EL* Versorgungsdüse *f*, Speisedüse *f*
supply port Einlaßanschluß *m*, Eintrittsanschluß *m*, Zulaufanschluß *m*, Zuführanschluß *m*, Zuflußanschluß *m*, *bei einer Pumpe auch* Sauganschluß *m*, Ansauganschluß *m*, *bei einem Verbraucher auch* Druckanschluß *m*
supply pressure *LOG EL* Versorgungsdruck *m*, Speisedruck *m*
supply tube *LOG EL* s. supply nozzle
support s. hose support
support ring *SEALS* s. female support ring; s. male support ring

supporting pressure *LOG EL* Stützdruck *m*
surface-active agent *FLUIDS* grenzflächenaktiver Stoff *m*, Tensid *n*
surface filter Oberflächenfilter *m,n*
surface filtration Oberflächenfilterung *f*
surface-mounted *VALV* flächenmontiert; *vgl* base-mounted
surface temperature sensor Oberflächentemperaturfühler *m*, Anlegetemperaturfühler *m*, Berührungstemperaturfühler *m*
surface tension *FLUIDS* Oberflächenspannung *f*
surfactant *FLUIDS* grenzflächenaktiver Stoff *m*, Tensid *n*
surge absorber Druckstoßdämpfer *m*
surge-damping valve Dämpfungsventil *n*
surge tank Nachsaugbehälter *m*
surplus flow Stromüberschuß *m*, Überstrom *m*
surplus-flow loss *PCV* Überströmverlust *m*
SUS *Saybolt Universal Second Einheit der kinematischen Viskosität*
swaged-on fitting Verpreßverbindung *f*, Quetschverbindung *f*, Preßverbindung *f*
swashplate *PU/MOT* Schrägscheibe *f*, Schiefscheibe *f*
swashplate angle *PU/MOT* Schwenkwinkel *m* *der Schrägscheibe*
swashplate axial piston motor Axialkolbenmotor *m* mit Schrägscheibe, Schrägscheibenmotor *m*
swashplate axial piston pump Axialkolbenpumpe *f* mit Schrägscheibe, Schrägscheibenpumpe *f*

swell characteristics *SEALS* Quellverhalten *n*
swept volume *CYL* Hubraum *m*, Hubvolumen *n*
switching-over filter Umschaltfilter *m,n*, Schaltfilter *m,n*, Doppelschaltfilter *m,n*
switching pressure *VALV OP* Schaltdruck *m*, Umschaltdruck *m*
switching pressure surge Schalt[druck]stoß *m*, Schaltschlag *m*
switching sequence Schaltfolge *f*
switching shock *s.* switching pressure surge
switching speed *DCV* Schaltgeschwindigkeit *f*
switching time *DCV* Schaltzeit *f*
swivel [connection] Rohrgelenk *n*
swivel housing *of a bent-axis pump* Schwenkgehäuse *n*, Schwenkkörper *m* einer Schrägachsenpumpe
swivel joint *s.* swivel connection
swivel mount *CYL* Schwenkbefestigung *f*
swivel nut elbow richtungseinstellbare Winkelverschraubung *f*
symbol Symbol *n*, Schaltzeichen *n*
symbolic diagram Schaltplan *m*
synchronism Gleichlauf *m*, Synchronlauf *m*
synchronization Synchronisierung *f*, Herstellung *f* von Gleichlauf
synchronize *v* synchronisieren, Gleichlauf *m* herstellen
synchronized motion Gleichlauf *m*, Synchronlauf *m*
synchronizing circuit Gleichlaufschaltung *f*, Synchronlaufschaltung *f*
synthetic [fluid] synthetische Flüssigkeit *f*
system System *n*, Anlage *f*, *s z. B.* fluid power system; *circuit* Kreislauf *m*, System *n*, Schaltung *f*
system air Systemluft *f* *die Luft im Druckluftsystem*
system pressure Druck *m* im System, Systemdruck *m*

T

T-ported filter Filter *m,n* mit T-Gehäuseausführung
T-ring (-seal) T-Ring *m*
T-type filter T-Gehäuse-Filter *m,n*
tailpiece *of a hose fitting* Tülle *f*, Nippel *m* einer Schlauchverbindung
take-off point Anzapfung *f*, Zapfstelle *f*, Entnahmestelle *f*, Abgriff *m*
tandem-center valve *DCV* Ventil *n* mit Umlaufstellung
tandem cylinder Tandemzylinder *m*
tandem position *DCV* Umlaufstellung *f*
tandem pump Zweistrompumpe *f*, Zweikreispumpe *f*, Zwillingspumpe *f*, Doppelpumpe *f*
tangent bulk modulus *FLUIDS* Tangentenkompressionsmodul *m*, wahrer Kompressionsmodul *m*
tank Flüssigkeitsbehälter *m*, Hydraulikbehälter *m*, Ölbehälter *m*, *umgangssprachlich noch üblich:* Tank *m*
tank: connected to tank *HY* mit dem Behälter *m* verbunden
tank bottom Behälterboden *m*
tank capacity Behältervolumen *n*, Behältergröße *f*
tank connection *DCV* Behälteranschluß *m*
tank cover Behälterdeckel *m*, Behälterdeckplatte *f*

tank drain äußere Leckölabführung *f*
tank filter Behälterfilter *m,n*, Sumpffilter *m,n*
tank mount filter Behälteraufbaufilter *m,n*, Behälteranbaufilter *m,n*
tank port *DCV* Behälteranschluß *m*
tank size Behältervolumen *n*, Behältergröße *f*
tank top plate Behälterdeckel *m*, Behälterdeckplatte *f*
tank volume Behältervolumen *n*, Behältergröße *f*
tank wall Behälterwand[ung] *f*
tap Anzapfung *f*, Zapfstelle *f*, Entnahmestelle *f*, Abgriff *m*
tap *v* anzapfen
tapered spear *CYL* kegliger Bremsansatz *m*
tapered thread *FITT* kegliges (konisches) Gewinde *n*
tapered-tube flowmeter (rotameter) Schwebekörper-Durchflußmesser *m*
tapped holes mounting *CYL* Gewindebohrungsbefestigung *f*
tapping plate *of a valve stack* Abzweigplatte *f*, Anzapfplatte *f einer Steuersäule*
tapping point Anzapfung *f*, Zapfstelle *f*, Entnahmestelle *f*, Abgriff *m*
tar number *FLUIDS* Teerzahl *f*
TDR *turndown ratio LOG EL* Volumenstromverhältnis *n*
tee T-Verschraubung *f*
tee *v to a line* anschließen *an eine Leitung*
tee off *v a flow* abzweigen *einen Strom*
tee branch Abzweig *m* der T-Verschraubung
tee fitting T-Verschraubung *f*

tee run gerader Teil *m* der T-Verschraubung *f*
telescopic cylinder Zylinder *m* mit Teleskopkolben, Teleskopzylinder *m*
telescopic damper Teleskopdämpfer *m*
telescopic line *LINES* Teleskoprohr *n*
telescopic plunger Teleskopkolben *m als Tauchkolben*
telescopic tube *CYL* Teleskoprohr *n*
temperature-compensated flow-control valve temperaturkompensiertes Stromventil *n*
temperature compensation *FCV* Temperaturkompensation *f*
temperature measurement Temperaturmessung *f*
temperature rating Nenntemperatur *f*
temperature recorder Temperaturschreiber *m*
temperature switch Temperaturschalter *m*
test bench Versuchsstand *m*, Prüfstand *m*
test contaminant *FILTERS* Prüfverschmutzung *f*
test dust *FILTERS* Prüfstaub *m*
test port Prüfanschluß *m*
test pressure gauge Prüfmanometer *m*
test stand Prüfstand *m*
thermal conductivity *FLUIDS* Wärmeleitfähigkeit *f*
thermal bypass valve thermisch ausgelöstes (temperaturbetätigtes) Umgehungsventil *n*
thermal flowmeter thermischer Durchflußmesser *m*
thermal stability *FLUIDS* thermi-

thermistor

sche Stabilität *f*, Wärmebeständigkeit *f*
thermal switch Temperaturschalter *m*
thermistor temperature sensor Thermistortemperaturfühler *m*, Halbleitertemperaturfühler *m*
thermocouple Thermoelement *n*, Thermopaar *n*
thermograph Temperaturschreiber *m*
thermometer Thermometer *n*
thick *FLUIDS* hochviskos, dickflüssig
thick-walled *LINES* dickwandig
thickener *FLUIDS* viskositätserhöhendes Additiv *n*, Verdicker *m*
thickplate orifice *THEOR* drosselartige Blende *f*
thin *FLUIDS* niedrigviskos, dünnflüssig
thin-walled *LINES* dünnwandig
threaded-end cylinder Zylinder *m* mit Schraubboden
threaded-end mount *CYL* Gewindeansatzbefestigung *f*
threaded fitting Gewindeverschraubung *f*, Gewindefitting *m*
threaded port Gewindeanschluß *m*
three-diaphragm element *LOG EL* Dreimembranenelement *n*
three-piece flare fitting *nut and sleeve* Bördelverbindung *f* mit Klemmring
three-port valve *DCV* Ventil *n* mit drei Anschlüssen; *sometimes used for* three-way valve
three-position valve *DCV* Dreistellungsventil *n*, 3-Stellungsventil *n*
three-screw pump Dreischraubenpumpe *f*
three-stage servovalve dreistufiges Servoventil *n*

three-vanes rotary actuator Dreiflügelschwenkmotor *m*
three-way ball valve *DCV* Dreiwegekugelhahn *m*, 3-Wege-Kugelhahn *m*
three-way valve *DCV* Dreiwegeventil *n*, 3-Wege-Ventil *n*
threshold pressure Schwellendruck *m*, Ansprechdruck *m*
throttle Drossel[stelle] *f*, Strömungswiderstand *m*; *FCV* Drosselventil *n*
throttle *v* *flow* drosseln *den Volumenstrom*
throttle characteristics Drosselcharakteristik *f*, Drosselverhalten *n*, Öffnungsverhalten *n*
throttle control *FCV* Drosselsteuerung *f*
throttle loss Drosselverlust *m*
throttle valve *FCV* Drosselventil *n*
throttling Drosselung *f*
throttling area Drosselquerschnitt *m*, Drosselfläche *f*
throttling element Drosselelement *n*
throttleing length Drossellänge *f*, Drosselstrecke *f*
through-shaft configuration *PU/MOT* Durchtriebausführung *f* mit durchgehender Welle
throw-away filter element Wegwerffilterelement *n*
thrust shoe *PU/MOT* Gleitschuh *m*
tie-rod cylinder Zugankerzylinder *m*
tie-rod mounting *CYL* Zugstangenbefestigung *f*
tightness *SEALS* Dichtheit *f*, Dichtigkeit *f*
tilt angle *of the swashplate* Schwenkwinkel *m* *der Schrägscheibe*
tilting block *of a radial piston unit* Schwenkrahmen *m*, Schwenkkörper *m* *einer Radialkolbenmaschine*

tilting cylinder *VALV OP* Schwenkzylinder *m*
tilting cylinder block *of a bent-axis axial piston unit* Schwenkkörper *m* der Schrägtrommelmaschine
TIM *transverse impact modulator* *LOG EL* Gegenstrahlelement *n* mit transversaler Steuerdüse, Querstrahlelement *n*, transversaler Impaktmodulator *m*
time-delay valve *FCV* Zeitventil *n*, Intervallventil *n*
timer-actuated drain zeitgesteuerter Ablaß *m*
timing valve *s.* time-delay valve
tiny cylinder *s.* miniature cylinder
tooth-mesh leakage *PU/MOT* Zahneingriffsverlust *m*
tooth number *PU/MOT* Zähnezahl *f*
tooth space *in a gear pump* Zahnkammer *f* in einer Zahnradpumpe
top blanking plate *of a valve stack* Sperrplatte *f*, Abschlußplatte *f*, Dichtungsplatte *f* einer Steuersäule
top crossover plate *of a valve stack* Umlenkplatte *f* einer Steuersäule
top mount filter Behälteraufbaufilter *m,n*, Behälteranbaufilter *m,n*
toroidal damper *pressure pulse* Ringwulstdämpfer *m* Druckpulsation
torque converter *APPL* Drehmomentwandler *m*
torque fluctuation *s.* torque pulsation
torque limiter control *of a pump* Leistungsregelung *f* einer Pumpe
torque motor *VALV OP* Torque-Motor *m*, Drehmomentmotor *m*
torque output Abtriebs[dreh]moment *n*, abgegebenes Moment *n*

torque pulsation *PU/MOT* Drehmomentpulsation *f*, Drehmomentungleichförmigkeit *f*
torque rating Nenn[dreh]moment *n*
torque ripple *s.* torque pulsation
total [pressure] head *THEOR* Gesamtdruckhöhe *f*, Gesamtdruck *m*
touch valve berührungssensitives Ventil *n*
toxicity *FLUIDS* Giftigkeitsprüfung *f*, Toxizitätsprüfung *f*
tracer valve *APPL* Kopierventil *n*
track *of the vane pump* Flügelleitkurve *f*
track ring *PU/MOT* Leitring *m*, Gehäusering *m*, Führungsring *m*
tractive magnet *VALV OP* Zugmagnet *m*
tramp oil *FLUIDS* unerwünschtes Öl *n*
transaxle *s.* hydrostatic transaxle
transfer barrier *ACCUM* Blasen- oder Membranschutzvorrichtung *f*
transfer-type accumulator Druckflüssigkeitsspeicher *m* mit einer oder mehreren nachgeschalteten Gasflaschen
transmission *s.* hydrostatic transmission
transonic flow *THEOR* schallnahe (transsonische) Strömung *f*
transport motion *CYL* Transportbewegung *f*
transverse impact modulator *LOG EL* Gegenstrahlelement *n* mit transversaler Steuerdüse, Querstrahlelement *n*, transversaler Impaktmodulator *m*
trap *LINES* Wasserfang *m*, Wasserablaß *m*; Öltasche *f*, Totölraum *m*
trapezoidal ring *SEALS* Trapezring *m*

trapped oil *PU/MOT* Quetschöl *n*
trapping relief groove *PU/MOT* Quetschölnut *f*
treadle-actuated (-operated) valve pedalbetätigtes (fußbetätigtes) Ventil *n*, Pedalventil *n*, Fußwippenventil *n*
treble pump *s.* triple pump
triple-plane swivel joint Dreiebenen-Rohrgelenk *n*
triple pump Dreistrompumpe *f*, Dreikreispumpe *f*
triple-vane rotary actuator Dreiflügelschwenkmotor *m*
triplex pump *s.* triple pump
true fluid *THEOR* reales (wirkliches) Fluid *n*
trunnion mount *CYL* Schwenkzapfenbefestigung *f*, Zapfenbefestigung *f*
tube *LINES* Rohr *n* Präzisionsrohr, Zusammensetzungen *s auch unter* line, pipe
tube bend Rohrkrümmer *m*, Rohrbogen *m*
tube bender Rohrbiegevorrichtung *f*
tube bundle Rohrbündel *n*, Rohrkabel *n*
tube-bundle heat exchanger Rohrbündelwärmeübertrager *m*
tube clamp Rohrschelle *f*, Rohrhalter *m*
tube cutter Rohrschneidegerät *n*, Rohrabschneider *m*
tube fitting Rohrverschraubung *f*, Rohrverbindung *f*
tube line Rohrleitung *f*, Leitung *f*
tube plug Rohrverschluß *m*
tube wall Rohrwand[ung] *f*
tubing Gesamtheit von Rohren, Rohr allgemein, Rohrnetz *n*, Rohrleitungssystem *n*, Leitungsnetz *n*, Verrohrung *f*, *s auch* tube; *Verlegen von Rohrleitungen*
tubing bundle Rohrbündel *n*, Rohrkabel *n*
turbine Turbine *f*
turbine flowmeter Flügelrad-Durchflußmesser *m*, Turbinen- Durchflußmesser *m*, Meßturbine *f*
turbine flow sensor Turbinen-Volumenstromsensor *m*
turbine motor Turbinenmotor *m*, Turbomotor *m*
turbo-compressor Strömungsverdichter *m*, Turboverdichter *m*
turbulator *s.* turbulence inducer
turbulence *THEOR* Turbulenz *f*
turbulence amplifier *LOG EL* Turbulenzverstärker *m*
turbulence inducer Turbulenzeinbauten *mpl*
turbulent *THEOR* turbulent
turbulent flow *THEOR* turbulente Strömung *f*, Turbulentströmung *f*
turbulent sampler *FLUIDS* Turbulenzprobenehmer *m*
turndown ratio *LOG EL* Volumenstromverhältnis *n*
Twedell accumulator Differentialkolbenspeicher *m*
twin filter Doppelfilter *m,n*, Zwillingsfilter *m,n*
two-ball valve *CHECKS* Zweikugelventil *n*, Doppelkugelventil *n*
two-bite flareless compression fitting Doppellippen-Schneidringverschraubung *f*
two-dimensional copying *APPL* zweidimensionales Kopieren *n*
two-jet flapper valve zweidüsiges Prallplattenventil *n*

two-piece flare fitting Bördelverbindung f ohne Klemmring
two-port valve Ventil n mit zwei Anschlüssen; *sometimes used for* two-way valve
two-position control *PU/MOT* Zweipunktverstellung f; Zweipunkt-Stelleinheit f
two-position valve Zweistellungsventil n, 2-Stellungsventil n
two-screw pump Zweispindelschraubenpumpe f
two-stage filter Zweistufenfilter m,n
two-stage pressure compensator vorgesteuerter (zweistufiger) Nullhubregler m
two-stage pump Zweistufenpumpe f
two-stage servovalve zweistufiges Servoventil n
two-stage valve zweistufiges Ventil n
two-step decompression cycle Entspannungszyklus m mit Vorentlastung
two-way ball valve *DCV* Zweiwegekugelhahn m, 2-Wege-Kugelhahn m
two-way seal coupling Zweistrangabsperrkupplung f, Zweiwegeabsperrkupplung f
two-way shut-off Zweistrangabsperrung f, Zweiwegeabsperrung f
two-way solenoid *VALV OP* Stellmagnet m mit zweiseitiger Wirkungsrichtung und Nullstellung, Umkehrmagnet m, Doppelhubmagnet m
two-way valve Zweiwegeventil n, 2-Wege-Ventil n
two wire braid hose Schlauch m mit zwei Drahtgeflechteinlagen

U

U-cup *SEALS* Nutring m, Doppellippenring m, Zweilippenring m
U-seal s. U-cup
U-tube heat exchanger U-Rohrwärmeübertrager m, Haarnadelwärmeübertrager m
ultrasonic flowmeter Ultraschall-Durchflußmesser m
unbalance fehlender Ausgleich m, Unausgeglichenheit f
unbalanced nicht [druck]entlastet, nicht [druck]ausgeglichen
uncentered spool exzentrisch liegender Kolben m
uncompensated s. unbalanced
underlap *DCV* negative Überdeckung f
undissolved air *FLUIDS* freie (ungelöste) Luft f
undoped oil *FLUIDS* unlegiertes Öl n, Öl n ohne Wirkstoffzusätze
uni-directional pump Pumpe f für eine Drehrichtung
uni-flow pump Pumpe f mit drehrichtungsunabhängiger Förderrichtung
uniform-pressure-drop valve Druckdifferenzventil n *für Differenz zwischen Ein- und Ausgangsdruck auch:* Druckgefälleventil n
uninflammability *FLUIDS* Nichtentflammbarkeit f, Unentflammbarkeit f, Unentzündbarkeit f, Unbrennbarkeit f, Flammbeständigkeit f
uninflammable *FLUIDS* nicht entflammbar, unentflammbar, unentzündbar, nicht brennbar, flammbeständig
uninhibited oil *FLUIDS* unlegiertes Öl n, Öl n ohne Wirkstoffzusätze

union

union *Rohrverschraubung nur mit Rohranschlüssen, ohne Rohrdrehung montierbar*
union cross Kreuzverschraubung *f* mit vierseitigem Rohranschluß
union elbow Winkelverschraubung *f* mit zweiseitigem Rohranschluß
union fitting *s.* union
union hose connector Schlauchverbindung *f* mit zweiseitigem Schlauchanschluß
union tee T-Verschraubung *f* mit dreiseitigem Rohranschluß
unload *v* mit dem Behälter verbinden, entlasten
unloader control *COMPR* Entlastungsregelung *f*
unloading valve Abschaltventil *n*, Entlastungsventil *n*; *ACCUM* Schaltventil *n*
unlock *v* *valve spool* losreißen, lösen *Steuerkolben*
unpressurized drucklos, nicht unter Druck *m*, nicht vorgespannt
unpressurized line drucklose Leitung *f*
unreinforced seal unverstärkte (unbewehrte, nicht armierte) Dichtung *f*
unseat *v* *seat valve* [sich] öffnen *Sitzventil*; *poppet* abheben *vi und vt* *Ventilelement von einem Sitz*
unsteady flow *THEOR* instationäre Strömung *f*
unthickened fluid *FLUIDS* nichtviskositätserhöhte Flüssigkeit *f*
unvalved coupling *FITT* ventillose Kupplung *f*
upstream line Leitung *f* vor *einem Element*, Leitung *f* oberhalb *eines Elements*, Leitung *f* gegen die Stromrichtung, *einem Element* vorgeschaltete Leitung *f*
upstream orifice *FCV* Vordrossel *f*
upstroke *CYL* Aufwärtshub *m*
used oil *FLUIDS* Gebrauchtöl *n*
user of compressed air equipment Pneumatikanwender *m*
user of hydraulic equipment Hydraulikanwender *m*

V

V-cup *SEALS* *s.* V-ring
V-notch *FCV* Dreieckkerbe *f*, V-Kerbe *f*
V-ring *SEALS* Dachring *m*, Dachmanschette *f*
V-ring packing *SEALS* Dachringsatz *m*, Dachmanschettensatz *m*
vacuum *reduced pressure* Unterdruck *m*, Vakuum *n*
vacuum lifter (lifting rig) *APPL* Vakuumhubeinrichtung *f*
vacuum pump Vakuumpumpe *f*
vacuum-relief valve Unterdruckbegrenzungsventil *n*
vacuum suction cup *APPL* Vakuum[saug]napf *m*
vacuum tight vakuumdicht
valve Ventil *n*
valve *v* mit Ventilen ausstatten
valve actuator Ventilstelleinheit *f*, Ventilbetätigungseinrichtung *f*
valve ball Ventilkugel *f*
valve bank Ventilbatterie *f*, wenn höhenverkettet auch Steuersäule *f*
valve block Blockventil *n*, Ventilblock *m*, Steuerblock *m*
valve body *DCV* Ventilgehäuse *n*, Ventilkörper *m*

valve card *combined microprocessor-air valves plug-in unit* VALV OP Ventilkarte *f* steckbare Mikroprozessor-Ventil-Kombination
valve cavity Ventilaufnahme[bohrung] *f*
valve characteristic Ventilcharakteristik *f*, Ventilverhalten *n*
valve chatter DCV Ventilschnarren *n*, Ventilrattern *n*, Ventilflattern *n*
valve control VALV OP Ventilstelleinheit *f*, Ventilbetätigungseinrichtung *f*; Ventilsteuerkreislauf *m*, Ventilsteuersystem *n*, Ventilsteuerung *f*; *s auch* valve element
valve control circuit Ventilsteuerkreislauf *m*, Ventilsteuersystem *n*, Ventilsteuerung *f*
valve control system *s.* valve control circuit
valve-controlled pump ventilgesteuerte Pumpe *f*
valve element DCV Ventilelement *m*, Steuerelement *n*
valve envelope *in symbolic diagrams* Ventilbaugruppenumrandung *f*; Ventilgehäuse *n*, Ventilkörper *m*
valve flutter DCV Ventilschnarren *n*, Ventilrattern *n*, Ventilflattern *n*
valve liner Ventilbüchse *f*
valve loss THEOR Ventil[druck]verlust *m*
valve mounting type Verkettungsart *f*, Anschlußart *f*
valve noise Ventilgeräusch *m*
valve operator Ventilstelleinheit *f*, Ventilbetätigungseinrichtung *f*
valve panel Ventilmontageplatte *f*, Ventilaufnahmeplatte *f*, Montagewand *f*

valve plate PU/MOT Steuerplatte *f*, Steuerscheibe *f*, Ventilplatte *f*
valve-plate controlled pump wegegesteuerte (ventillose, schlitzgesteuerte, flächengesteuerte) Pumpe *f*
valve plug DCV Kolbendrehschieber *m*
valve plunger Ventilstößel *m*
valve port Ventilanschluß *m*
valve pressure loss THEOR Ventil[druck]verlust *m*
valve response Ventilcharakteristik *f*, Ventilverhalten *n*
valve seat Ventilsitz *m*
valve sleeve Ventilbüchse *f*
valve spindle PU/MOT Mittelzapfen *m*, Steuerzapfen *m*; VALV OP Ventilspindel *f*
valve-spindle radial piston pump Radialkolbenpumpe *f* mit Steuerzapfen, wegegesteuerte (ventillose) Radialkolbenpumpe *f*
valve spool DCV Steuerschieber *m*, Ventilschieber *m*
valve spool with *n* metering edges DCV *n*kantenschieber *m*
valve stack Ventilbatterie *f*, *wenn höhenverkettet auch* Steuersäule *f*
valve stacking Modulverkettung *f*, Batterieverkettung *f*
valve stem VALV OP Ventilspindel *f*
valve unit Blockventil *n*, Ventilblock *m*, Steuerblock *m*
valving *Gesamtheit von Ventilen; Ventile allgemein; Ausstattung mit Ventilen*
valving cone PU/MOT Steuerwalze *f*, Steuerkonus *m*
valving element DCV Ventilelement *n*, Steuerelement *n*

valving surface *PU/MOT* Steuerspiegel *m*, Steuerfläche *f*
vane *PU/MOT* Flügel *m*, beim *Druckluftmotor* Lamelle *f*
vane block *PU/MOT* Flügelkörper *m*, Flügelträger *m*
vane compressor Zellenverdichter *m*, Lamellenverdichter *m*
vane motor Flügelzellenmotor *m*, als *Druckluftmotor* Lamellenmotor *m*
vane number *PU/MOT* Flügel[an]zahl *f*
vane pump Flügelzellenpumpe *f*
vane pump or **motor** Flügelzellenmaschine *f*, Flügelzelleneinheit *f*, Flügelzellengerät *n*
vane pump test *of anti-wear properties FLUIDS* Vickers-Pumpentest *m*, Flügelzellenpumpentest *m* der *Verschleißschutzeigenschaften*
vane rotary actuator Flügelschwenkmotor *m*, Drehflügelschwenkmotor *m*
vane slot Flügelschlitz *m*
vane throw Flügelhub *m*
vane tip Flügelkuppe *f*, Flügelkopf *m*
vane track Flügelleitkurve *f*
vane tracking force Flügelanpreßkraft *f*
vane transmission Flügelzellengetriebe *n*
vane-type damper Drehflügeldämpfer *m*
vane-within-a-vane-type pump Flügelzellenpumpe *f* mit Doppelflügeln
vanes-in-stator motor Sperrschiebermotor *m*
vanes-in-stator pump Sperrschieberpumpe *f*
vapour-phase inhibitor *to prevent corrosion of metals in contact with fluid vapour* Dampfphaseninhibitor, VPI-Stoff *m* *über die Dampfphase wirkender Korrosionsinhibitor*
vapour pressure *FLUIDS* Dampfdruck *m*
vapour-pressure (-tension) thermometer Dampfdruckthermometer *n*, Dampfspannungsthermometer *n*, Torsionsthermometer *n*
variable-angle swashplate *PU/MOT* Schwenkscheibe *f*
variable-delivery pump Verstellpumpe *f*, stellbare Pumpe *f*
variable-displacement motor Verstellmotor *m*, stellbarer Motor *m*
variable-displacement pump *s.* variable-delivery pump
variable orifice *FCV* Verstelldrossel *f*, einstellbare Drossel *n*; *in a compensated flow-control valve* Stelldrossel *f*, Passivdrossel *f* *im Stromregelventil*
variable-volume motor *s.* variable-displacement motor
variable-volume pump *s.* variable-delivery pump
vee notch *FCV* Dreieckkerbe *f*, V-Kerbe *f*
vee-ring *SEALS* Dachring *m*, Dachmanschette *f*
vee-ring packing (set) *SEALS* Dachringsatz *m*, Dachmanschettensatz *m*
vegetable-oil base fluid pflanzenölbasische Flüssigkeit *f*
velocity distribution *THEOR* Geschwindigkeitsverteilung *f*, Geschwindigkeitsprofil *n*
velocity head *THEOR* Geschwindigkeithöhe *f*
velocity profile *s.* velocity distribution

vena contracta *THEOR* Einschnürungsstelle *f*
vent *for entrained air* Entlüftungseinrichtung *f*, Entlüfter *m*, Entlüftung *f* *für ungelöste Luft*
vent *v* *entrained air* entlüften *ungelöste Luft entfernen*; *exhaust air* entlüften *Abluft ableiten*; *fluid im Nebenschluß m ableiten, umleiten*; *to tank* mit dem Behälter *m* verbinden, entlasten
vent port Entlüftungsöffnung *f*, Entlüftungsbohrung *f*
ventilate *v* *RESERVOIRS* belüften, entlüften
ventilating eyelet *RESERVOIRS* Belüftungsöffnung *f*, Entlüftungsöffnung *f*, Lüftungsöffnung *n*
ventilation *RESERVOIRS* Belüftung *f*, Entlüftung *f*, Lüftung *f*
venting *entrained air* Entlüftung *f*, *Entfernung ungelöster Luft*
venting line Entlüftungsleitung *f*
venting screw Entlüftungsschraube *f*
venting valve Entlüftungsventil *n*
Venturi [tube] *THEOR* Venturidüse *f*, Venturirohr *n*
Venturi-type lubricator Düsennebelöler *m*
vertical valve stacking Höhenverkettung *f*, Turmverkettung *f*
VI *s.* viscosity index
VI improver *s.* viscosity index improver
Vickers pump test *s.* vane pump test
viscometer Viskosimeter *n*, Zähigkeitsmesser *m*
viscosimeter *s.* viscometer
viscometry Viskosimetrie *f*, Zähigkeitsmessung *f*
viscosimetry *s.* viscometry

viscosity *FLUIDS* Viskosität *f*, Zähigkeit *f*
viscosity detector Viskositätsfühler *m*
viscosity force *FLUIDS* Viskositätskraft *f*, Zähigkeitskraft *f*
viscosity grade *SAE FLUIDS* Viskositätsklasse *f*, Viskositätsgrad *m* *SAE*
viscosity grade index *FLUIDS* Viskositäts-Gradexponent *m*, VGe, k-Wert *m*
viscosity improver *FLUIDS* viskositätserhöhendes Additiv *n*, Verdicker *m*
viscosity index *FLUIDS* Viskositätsindex *m*, VI *m*
viscosity index improver *FLUIDS* Viskositätsindexverbesserer *m*, VI-Verbesserer *m*
viscosity pole *FLUIDS* Viskositätspol *m*
viscosity pole height *FLUIDS* Viskositätspolhöhe *f*, Polhöhe *f*, VPh *f*
viscosity-pressure characteristics *FLUIDS* Viskositäts-Druck-Verhalten *n*, Vp-Verhalten *n*
viscosity rating *FLUIDS* Nennviskosität *f*
viscosity slope coefficient *viscosity-temperature characteristics FLUIDS* Viskositäts-Richtungskonstante *f*, Richtungskonstante *f*, VRk *f*, m-Wert *m*
viscosity-temperature characteristics *FLUIDS* Viskositäts-Temperatur-Verhalten *n*, VT-Verhalten *n*
viscosity-temperature chart (diagram) *FLUIDS* Viskositäts-Temperatur-Diagramm *n* (-Blatt *n*), VT-Diagramm *n*, Viskogramm *n*

viscous *FLUIDS* hochviskos, dickflüssig
viscous damper Flüssigkeitsdämpfer *m*, viskoser Dämpfer *m*
viscous damping (drag) force *THEOR* Widerstandskraft *f*
viscous drag windage *THEOR* dem Quadrat der Geschwindigkeit proportionale Reibung
viscous flow *THEOR* laminare Strömung *f*, Laminarströmung *f*
viscous friction *THEOR* viskose (innere, flüssige) Reibung *f*
viscous restriction *THEOR* Laminarwiderstand *m*, Drossel *f*
volume-controlled metering *PU/MOT* volumetrische Dosierung *f* (Zuteilung *f*)
volume rate of flow Volumenstrom *m*, Durchflußstrom *m*
volumetric efficiency *PU/MOT* volumetrischer Wirkungsgrad *m*
volumetric flow rate Volumenstrom *m*
volumetric losses *PU/MOT* volumetrische Verluste *mpl*
vortex (*pl:* vortices *or* vortexes) Wirbel *m*
vortex amplifier *LOG EL* Wirbelkammerverstärker *m*, Vortexverstärker *m*
vortex chamber *LOG EL* Wirbelkammer *f*
vortex diode *LOG EL* Wirbelkammerdiode *f*, Vortexdiode *f*
vortex element *LOG EL* Wirbelkammerelement *n*, Vortexelement *n*
vortex flow *THEOR* Wirbelströmung *f*
vortex noise *THEOR* Wirbelgeräusch *n*, Wirbellärm *m*

vortex sensor *LOG EL* Wirbelsensor *m*
VPI vapour-phase inhibitor, used to prevent corrosion of metals in contact with fluid vapour Dampfphaseninhibitor, VPI-Stoff *m* über die Dampfphase wirkender Korrosionsinhibitor
VT characteristics *s.* viscosity-temperature characteristics
VT chart (diagram) *s.* viscosity-temperature chart (diagram)

W

W-ring *SEALS* W-Ring *m*
wall-attached jet *LOG EL* Wandstrahl *m*
wall-attachment effect *LOG EL* Coanda-Effekt *m*, Wandhafteffekt *m*
wall-attachment element *LOG EL* Wandstrahlelement *n*, Haftstrahlelement *n*, Coanda-Element *n*
wall friction *THEOR* Wandreibung *f*
wall roughness *THEOR* Wand[ungs]rauheit *f*, Wand[ungs]rauhigkeit *f*
wall thickness Wanddicke *f*
warning device *FILTERS* Verschmutzungsanzeige *f*
washer pack *FILTERS* Scheibenpaket *n*, Lamellenpaket *n*, Plattenpaket *n*
washer seal Scheibendichtung *f*, Dicht[ungs]scheibe *f*
waste oil Gebrauchtöl *n*
water-base fluid wasserhaltige (wäßrige, wasserbasische) Flüssigkeit *f*
water-carrying capacity *FLUIDS*

Wasseraufnahmefähigkeit *f*, Wassertragvermögen *n*
water conditioning Wasseraufbereitung *f*, Wasserbehandlung *f*
water contamination *FLUIDS* Verschmutzung *f* durch Wasser, Wasserverschmutzung *f*
water content *FLUIDS* Wassergehalt *m*
water-cooled wassergekühlt
water-cooled heat exchanger Öl-Wasser-Wärmeübertrager *m*
water-cooled oil cooler wassergekühlter Ölkühler *m*, Wasser-Öl-Kühler *m*
water cooling Wasserkühlung *f*
water demulsibility *FLUIDS* Wasserabscheidevermögen *n*
water drain Wasserablaß *m*
water-glycol fluid *FLUIDS* wäßrige Polymerlösung *f*, Wasser-Glycol-Lösung *f*, glycolbasische Flüssigkeit *f*
water hammer *THEOR* hydraulischer Stoß *m*, Druckstoß *m*, Druckschlag *m*
water hammer absorber Druckstoßdämpfer *m*
water hydraulic press *APPL* Druckwasserpresse *f*
water hydraulic system Wasserhydraulikanlage *f*, Wasserhydrauliksystem *n*, Wasserhydraulik *f*, Druckwasserhydraulik *f*, Preßwasserhydraulik *f*
water hydraulics Druckwasserhydraulik *f*, Wasserhydraulik *f*
water-injected compressor wassereingespritzter Kompressor *m*
water-in-oil emulsion *FLUIDS* Wasser-in-Öl-Emulsion *f*

water in solution gelöstes Wasser *n*
water-oil heat exchanger *s.* oil-to-water heat exchanger
water preparation Wasseraufbereitung *f*, Wasserbehandlung *f*
water pump Druckwasserpumpe *f*, Preßwasserpumpe *f*, Wasserpumpe *f*
water-removing filter Wasserabscheidefilter *m,n*, Entwässerungsfilter *m,n*, Trocknungsfilter *m,n*
water-ring compressor Wasserringverdichter *m*
water separator Wasserabscheider *m*
water-soluble oil *s.* oil-in-water emulsion
water system *s.* water hydraulic system
water treatment Wasseraufbereitung *f*, Wasserbehandlung *f*
waterfree fluid nicht-wasserhaltige (wasserfreie) Flüssigkeit *f*
wave equation *THEOR* Wellengleichung *f*
wave transmission resistance *THEOR* Wellenwiderstand *m*
way *DCV gesteuerter Anschluß, innerer Strömungsweg des Wegeventils*
***n*-way valve** *DCV n*-Wegeventil *n*
2-way ball valve *DCV* Zweiwegekugelhahn *m*, 2-Wege-Kugelhahn *m*
3-way ball valve *DCV* Dreiwegekugelhahn *m*, 3-Wege-Kugelhahn *m*
wear *of the oil FLUIDS* Ölverschleiß *m*
wear plate *of a pump or motor* Seitenplatte *f* *einer Verdrängermaschine*
wear ring *CYL* Führungsring *m*, Gleitring *m*
weight flow rate *THEOR* Gewichts-

weighted

strom *m*   in Gewichtskrafteinheiten je Zeiteinheit
weight-loaded accumulator   gewichtsbelasteter Speicher *m*
weight return   *CYL* Schwerkraftrückzug *m*
weight-returned cylinder   Zylinder *m* mit Schwerkraftrückzug
weight-returned ram   schwerkraftrückgezogener Tauchkolben *m*
weighted accumulator   gewichtsbelasteter Speicher *m*
weld fitting   Schweißverbindung *f*, Schweißverschraubung *f*
weld flange   *FITT* Vorschweißflansch *m*
welded cylinder   geschweißter Zylinder *m*, Schweißzylinder *m*
welded reservoir   geschweißter Behälter *m*, Schweißbehälter *m*
welded steel pipe   geschweißtes Stahlrohr *n*
welded tank   *s.* welded reservoir
welding neck flange   Vorschweißflansch *m*
weldless fitting   schweißlose Verbindung (Verschraubung)
wet   *s.* wet mount
wet air   Feuchtluft *f*
wet filter   Naßfilter *m,n*
wet mount   getaucht, Tauch ... , Unteröl ...
wet pin solenoid   *VALV OP* Ölbadmagnet *m* nicht druckdicht
wetted perimeter   *FCV* benetzter Umfang *m*
wheel motor   *APPL* Rad[naben]motor *m*
wick test   persistence of burning *FLUIDS* Dochtprüfung *f* Brennbeständigkeit

windage   *THEOR* dem Quadrat der Geschwindigkeit proportionale Reibung
winter-grade oil   Winteröl *n*
wipeage   *CYL* Abstreifverlust *m*
wiper   *CYL* Schmutzabstreifring *m*, Abstreifring *m*, Abstreifer *m*
wiper-scraper [seal]   Abstreifer *m* mit weicher und harter Lippe
wiper seal   *s.* wiper
wire braid hose   Schlauch *m* mit Drahtgeflechteinlage[n], drahtgeflechtarmierter Schlauch *m*
n-**wire braid hose**   Schlauch *m* mit *n* Drahtgeflechteinlagen
wire-cloth (-gauze, -mesh, -screen) filter   Drahtgewebefilter *m,n*, Drahtsiebfilter *m,n*
wire-screen disk filter   Siebscheibenfilter *m,n*
wirewound filter   Drahtbandfilter *m,n*
withdraw *vi and vt*   cylinder, piston zurückziehen, einfahren   *Zylinder, Kolben*
withdrawal   *CYL* Rückzug *m*, Rücklauf *m*, Einfahren *n*
withdrawal stroke   *CYL* Rückzug *m*, Rücklauf *m*, Rückhub *m*, Einfahrhub *m*
wobble plate   *PU/MOT* Taumelscheibe *f*
wobble-plate axial piston motor   Taumelscheibenmotor *m*
wobble-plate axial piston pump   Taumelscheibenpumpe *f*
wobble-plate flowmeter   Treibscheibenzähler *m*, Scheibenzähler *m*
work[ing] cycle   Arbeitszyklus *m*, Arbeitsspiel *n*, Betriebszyklus *m*, Betriebsspiel *n*
working fluid   Arbeitsflüssigkeit *f*, Arbeitsmedium *n*, Arbeitsmittel *n*,

Betriebsflüssigkeit *f*, Druck[übertragungs]mittel *n*, Energieübertragungsmittel *n*, Druckflüssigkeit *f*,
vgl hydraulic fluid
working line   Arbeitsleitung *f*, Hauptleitung *f*
working port   *DCV* Verbraucheranschluß *m*
working pressure   Betriebsdruck *m*, Arbeitsdruck *m*
working speed   Betriebsdrehzahl *f*, Arbeitsdrehzahl *f*
working stroke   *CYL* Arbeitshub *m*, Nutzhub *m*
working temperature   Betriebstemperatur *f*, Arbeitstemperatur *f*
working viscosity   Betriebsviskosität *f*
wound-ribbon paper filter   Papierbandfilter *m,n*
wound-wire filter   Drahtbandfilter *m,n*
woven-cloth filter   Gewebefilter *m,n*, Textilfilter *m,n*
woven-screen filter   Siebfilter *m,n*, Gewebefilter *m,n*

X

X-ring   *SEALS* X-Ring *m*

Y

Y-fitting   Y-Verschraubung *f*
yoke rotary actuator   Kulissenschwenkmotor *m*

Z

ZDDP   *s.* zinc dialkyldithiophosphate
ZDP   *s.* zinc dithiophosphate
zero delivery   *PU/MOT* Nullförderstrom *m*
zero displacement   *PU/MOT* Nullverdrängungsvolumen *n*
zero lap   *DCV* Nullüberdeckung *f*
zero-leakage   Leck[age]freiheit *f*
zero-load flow   *PU/MOT* Nullastvolumenstrom *m*
zero pressure   Druck *m* mit Wert Null
zero-speed torque   *PU/MOT* Drehmoment *n* bei Drehzahl Null
zero-stroke position   *CYL* Einfahrstellung *f*
zinc dialkyldithiophosphate   *FLUIDS* Zinkdialkyldithiophosphat *n*
zinc dithiophosphate   *FLUIDS* Zinkdithiophosphat *n*

German/English
Deutsch/Englisch

A

abdichten seal *v*
abfallen *z. B. Druck* decay *v*, drop *v* *eg pressure*
Abfallen *n* droop [of an characteristic]
Abfluid *n* return (exhaust) fluid
Abfluß *m* outlet, exhaust, outflow
   Abfluß *m*: **im Abfluß steuern** *STROMV* meter-out *v*
Abflußkanal *m* outlet (exhaust) channel, *also, in a pump* discharge (delivery, output, pressure) channel
Abflußleitung *f* outlet (exhaust) line, *also, at a pump* discharge (delivery, output, pressure) line
abflußseitige Stromsteuerung *f* *STROMV* meter-out flow control, metering-out
Abflußsteuerung *f* *s.* abflußseitige Stromsteuerung *f*
Abflußsteuerventil *n* meter-out valve
Abführanschluß *m* outlet (exit) port, *also, of a pump* discharge (delivery, output, pressure) port
Abführleitung *f* *s.* Abflußleitung *f*
Abführstrom *m* *s.* Abführvolumenstrom *m*
Abführung *f* outlet, exhaust, outflow
Abführvolumenstrom *m* outlet (exit) flow [rate], outlet (exit) rate
Abgabeleistung *f* output horsepower (power), horsepower (power) output
abgeben *z. B. Flüssigkeit an eine Komponente* deliver *v*, supply *v*, provide *v* *eg fluid to a componente*; *Flüssigkeit, Energie* release *v* *fluid, energy*
abgegebene Leistung *f* *s.* Abgabeleistung *f*

abgegebenes Moment *n* output torque, torque output
abgerundete Kante *f* *s.* Blende *f* mit abgerundeter Kante
abgewickelte Rohrlänge *f* developed pipeline length, length of straight pipe
Abgriff *m* tap, tapping (take-off) point
abheben *vt und vi* *Ventilelement von einem Sitz* unseat *v*, lift *v*, move *v* *poppet off (from) a seat*
Ablaß *m* *BEHÄLT* drainage; *s.* Kondensatablaß *m*; *s.* Wasserablaß *m*
ablassen *vt* *Behälter* drain *vt* *tank*
Ablaßleitung *f* drain leg
Ablaßöffnung *f* *BEHÄLT* drain [opening]
Ablaßschraube *f* *BEHÄLT* drain plug
Ablaßventil *n* dump (bottom, foot) valve; *Entleerungsventil FILTER, BEHÄLT* drain valve
Ablauf *m* outlet, exhaust, outflow
   Ablauf *m*: **im Ablauf steuern** meter-out *v*
Ablaufdiagramm *n* [cycle] sequence diagram (plot), functional circuit
Ablaufdosierpumpe *f* meter-out pump
Ablaufdrosselung *f* meter-out flow control, metering-out
Ablaufdruck *m* outlet (exhaust) pressure, *also, in a pump* delivery (discharge, output) pressure
ablaufdruckentlastet *WEGEV* externally drained
ablaufdruckentlastetes Druckbegrenzungsventil *n* externally drained relief valve, balanced relief valve

Ablaufdruckentlastung *f*   *WEGEV*
external drainage
ablaufgesteuert   metered-out, *auch als Verb:* meter-out *v*
Ablaufleitung *f*   outlet (exhaust) line, also, at a pump discharge (delivery, output, pressure) line
Ablaufplatte *f*   *einer Ventilbatterie* outlet (return) section, return plate of ganged valves
Ablaufseite *f*   outlet (exhaust) side, also, of a pump discharge (output, pressure) side
ablaufseitige Stromsteuerung *f*   meter-out flow control, metering-out
Ablaufsteuerventil *n*   meter-out valve
Ablauf[volumen]strom *m*   outlet (exit) flow [rate], outlet (exit) rate
ableiten *Abluft*   bleed *v*, vent *v*; *im Nebenschluß ableiten* bypass *v*, divert *v*, bleed [off] *v*, vent *v*
Ableiter *m*   *Kondensatableiter FILTER* condensate drain (discharge)
Ableitung *f*   outlet (exhaust) line, also, at a pump discharge (delivery, output, pressure) line
Ablenkverlust *m*   *THEOR* deflection loss
ablösen/sich   *Strömung* separate *v* flow
Ablösepunkt *m*   *Strömung* separation point *flow*
Abluft *f*   exhaust air
  Abluft *f* **ableiten**   bleed *v*, vent *v* exhaust air
Abluftaufbereitungsgerät *n*   kombinierter Schalldämpfer-Filter muffler-reclassifyer
Abluftdrossel *f*   bleed (exhaust) throttle

Abluftdrosselung *f*   *s.* Abluftsteuerung *f*
Abluftleitung *f*   exhaust [air] line
Abluftsteuerung *f*   metering-out, exhaust air metering
Abmeßkreislauf *m*   meter-out circuit
Abmeßventil *n*   meter-out valve
Abnutzung *f*   *des Öls* oil wear
Aböl *n*   *Rücköl* return (exhaust) oil
abreißen   *Strömung* separate *v* flow
Abreißpunkt *m*   *einer Strömung* separation point *of a flow*
Abreißsicherung *f*   *s.* Schlauchkupplung *f* mit Abreißsicherung
abschälen   *Schlauchdecke* skive *v* hose cover
Abschaltregelung *f*   *KOMPR* start-stop control
Abschaltventil *n*   *DRUCKV* unloading valve
Abscheider *m*   *s.* Wasserabscheider *m*
Abscheidevermögen *n*   *FLÜSS* separation power
Abschlußplatte *f*   *einer Steuersäule* port (top) blanking plate, blanking (end) plate, end cover   *of a valve stack*
abschnappen   *Pumpe* starve *v* pump
Abschnappen *n*   *einer Pumpe* pump (intake) starvation
Absenkventil *n*   lowering valve
absetzverhinderndes Mittel *n*   *FLÜSS* detergent
Absolutdruck *m*   absolute pressure
Absolutdruckmesser *m*   absolute pressure gauge
Absolutdruckwandler *m*   absolute pressure transducer
absolute Filterfeinheit *f*   absolute filtration (filter) rating (fineness)

Absorptionsschalldämpfer *m* absorptive muffler
Absorptionstrockner *m* deliquescent (absorption) dryer
absperren *z. B. Strom, Leitung* block [off] *v*, isolate *v* *eg flow, line*
Absperrgröße *f* *s.* Filterfeinheit *f*
Absperrventil *n* *WEGEV* shutoff (isolating) valve
Abstreifer *m* *DICHT* wiper [seal]; *gering nachgiebig:* scraper [seal]; Spalträumer *FILTER* scraper, knife
Abstreifring *m* *s.* Abstreifer *m*
Abstreifverlust *m* *ZYL* wipeage
Abströmkante *f* *WEGEV* efflux edge
Abtriebsdrehmoment *n* output torque, torque output
Abtriebsdrehzahl *f* output speed (rpm)
Abtriebsleistung *f* output horsepower (power), horsepower (power) output
Abtriebsmoment *n* *s.* Abtriebsdrehmoment *n*
Abtriebswelle *f* output shaft
Abwärtshub *m* *ZYL* downstroke
Abwasser *n Rückwasser* return (exhaust) water
abzweigen *einen Strom* tee [off] *v*, branch *v* *a flow*
Abzweigplatte *f* *einer Steuersäule* tapping plate *of a valve stack*
Abzweigstromsteuerung *f* bleed-off (bypass) flow control
AD *Außendurchmesser* outside (outer) diameter, OD, O. D.
Additiv *n* *FLÜSS* additive, agent
additivieren *Hydraulikflüssigkeit* dope *v*, inhibit *v* *hydraulic fluid*
additiviertes Öl *n* doped (inhibited) oil
Admittanz *f* admittance *Volumen-strom/Druckabfall* *THEOR* flow/pressure drop
ADR-Verschraubung *f* *Aufdornringverschraubung* expansion fitting
Adsorptionsfilter *m,n* adsorption filter
Adsorptionstrockner *m* desiccant (adsorptive) dryer
Aerodynamik *f* aerodynamics, gas dynamics
aerodynamisch aerodynamic, gas-dynamic
Aeroemulsion *f* *FLÜSS* air-in-oil emulsion, air emulsion
Aeromechanik *f* aeromechanics, gas mechanics
Aerosol *n* *FLÜSS* aerosol
Aerostatik *f* aerostatics
Akku *m* *s.* Akkumulator *m*
Akkumulator *m* *Druckflüssigkeitsspeicher* [hydraulic] accumulator, *compounds s with* Speicher *m*
Aktionsturbinenmotor *m* impulse-type turbine motor
aktive Vorsteuerung *f* *STELL* external piloting
Aktivdrossel *f* *Differenzdruckregler im Stromregelventil* controlled orifice, pressure compensator, hydrostat *in a compensated flow-control valve*
Aktivkohlefilter *m,n* activated carbon filter
Aktivtonerde-Trockenmittel *n* activated alumina desiccant
akustischer Oszillationssensor *m* *LOGIKEL* interruptible sound beam sensor
akustischer Sensor *m* *LOGIKEL* acoustic wave sensor, fluidic ear
akustisches Filter *m,n* pulse (acoustic) filter

allmähliche Erweiterung *f*   *THEOR*
gradual enlargement
allmähliche Verengung *f*   *THEOR*
gradual contraction
altern   *FLÜSS* age *v*
Alterung *f*   *FLÜSS* ag[e]ing
Alterungsbeständigkeit *f*   *FLÜSS*
ag[e]ing resistance (stability)
Alterungsgeschwindigkeit *f*   *FLÜSS*
ag[e]ing rate
Alterungsstabilität *f*   *s.* Alterungsbeständigkeit *f*
Aluminiumrohr *n*   aluminium tubing   *GB*, aluminum tubing *USA*
Alu-Rohr *n*   *s.* Aluminiumrohr *n*
analoges Fluidikelement *n*   fluid analog device, analog fluidic device
Anbaufilter *m,n*   Behälteraufbaufilter tank mount (top) filter, reservoir mount filter
Anbauzahnradpumpe *f*   flange-mount (flanged) gear pump
Anfahrdrehmoment *n*   *PU/MOT*
starting torque
Anfahrkraft *f*   *ZYL* starting force
Anfahrlast *f*   *PU/MOT, ZYL* starting load
Anfahrmoment *n*   *s.* Anfahrdrehmoment *n*
Anfahrverhalten *n*   starting characteristics
Anfangsverschmutzung *f*   *FLÜSS* initial contamination
Anflanschpumpe *f*   flange-mount (flanged) pump
Anflanschzahnradpumpe *f*   flange-mount (flanged) gear pump
Anilinpunkt *m*   *FLÜSS* aniline point
Anker *m*   *s.* Magnetanker *m*

Anlage *f*   system, *s eg* Hydrauliksystem *n*
Anlaufdrehmoment *n*   *PU/MOT*
starting torque
Anlaufdruck *m*   *PU/MOT, ZYL*
starting pressure
Anlaufkraft *f*   *PU/MOT, ZYL*
starting force
Anlauflast *f*   *PU/MOT, ZYL* starting load
Anlaufmoment *n*   *s.* Anlaufdrehmoment *n*
Anlaufstrecke *f*   *bis zur vollständigen Ausbildung der Strömungsform THEOR* entrance length   *required to attain fully developed flow mode*
Anlaufverhalten *n*   starting characteristics
Anlegetemperaturfühler *m*   surface temperature sensor
anliegen   *Druck, z. B. an einem Ventil* act *v* on (upon)   *pressure, on a valve*
Anordnung *f* **im Nebenstrom**   bypass installation
Anpaßstück *n*   *VERBIND* adaptor
Anpreßkraft *f*   *der Flügel* vane tracking force
Ansatz *m*   *Bremsansatz ZYL* cushion spear (plunger)
Ansauganschluß *m*   suction (input, inlet, intake, supply) port
Ansaugdrosselung *f*   *KOMPR* inlet (intake) throttling
Ansaugdruck *m*   suction (input, inlet) pressure
ansaugen   *aus dem Behälter* draw *v from reservoir*; *erstmalig ansaugen, anzusaugen beginnen* prime *v*
Ansaugen *n*   *PU/MOT* suction, sucking, drawing

Ansaugkanal *m* suction (input, inlet, intake) channel
Ansaugleitung *f* suction (input, inlet, intake) line
Ansaugluft *f* *KOMPR* intake air
Ansaugraum *m* suction (input, inlet, intake) chamber
Ansaugstrom *m* s. Ansaugvolumenstrom *m*
Ansaugung *f* suction, sucking, drawing
Ansaugverhalten *n* *PU/MOT* suction (intake) characteristics
Ansaugvermögen *n* *PU/MOT* suction (intake) capacity
Ansaugvolumenstrom *m* suction (input, inlet, intake) flow [rate]
Ansaugzustand *m* s. Luft *f* im Ansaugzustand
Anschlagbuchse *f* *zur Hubbegrenzung ZYL* stop tube
anschließen connect *v*, port *v*; *an eine Leitung* tee *v* *to a line*
Anschluß *m* port; *s auch ED* way
Anschluß *m*: **mit Anschluß versehen** port *v*
Anschlußart *f* valve mounting type
Anschluß[loch]bild *n* port (hole) configuration
Anschlußplatte *f* *mit den Anschlüssen für Zu- und Ableitung PU/MOT* port plate *connecting with intake and outlet pipes*; *VENTILE* subplate, subbase, manifold, mounting plate, *s auch ED* manifold *v*; *als Verbindungsplatte* connector plate
Anschlußquerschnitt *m* port area (cross-section, size)
Anschlußverschraubung *f* port fitting (connection)

Anschlußweite *f* port area (cross-section, size)
ANSI = **American National Standards Institute**
Ansprechdruck *m* response pressure; *DRUCKV* cracking (response) pressure; *einer Nullhubsteuerung PU/MOT* cutoff pressure *of a pressure-compensator control*; Schwellendruck *m* *THEOR* threshold pressure
Ansprechdurchflußstrom *m* s. Ansprechvolumenstrom *m*
ansprechen *DRUCKV* crack *v*, respond *v*, open *v*, blow [off] *v*
Ansprechstrom *m* s. Ansprechvolumenstrom *m*
Ansprechverhalten *n* *DRUCKV* cracking (response) characteristics
Ansprechvolumenstrom *m* *DRUCKV* cracking (minimum opening) flow [rate]
Anstieg *m* *einer im Idealfall horizontalen Kennlinie* negative droop
antiparallel geschaltete Rückschlagventile *npl* *KREISL* back-to-back check valves
Anti-Schaum-Additiv *n* *FLÜSS* antifoaming additive, defoamer, foam (froth) inhibitor (depressant)
Antriebsdrehmoment *n* input (driving) torque
Antriebsdrehzahl *f* input (driving) speed (rpm)
Antriebsläufer *m* *KOMPR* male rotor
Antriebsleistung *f* input horsepower (power), driving horsepower
Antriebsmaschine *f* *PU/MOT* prime mover
Antriebsmoment *n* s. Antriebsdrehmoment *n*

Antriebsmotor *m*   *PU/MOT* prime mover
Antriebswelle *f*   drive (driving, input) shaft
anzapfen   tap *v*
Anzapfplatte *f*   *einer Steuersäule* tapping plate *of a valve stack*
Anzapfung *f*   tap, tapping (take-off) point
anzugsbegrenzte Dichtung *f*   confinement-controlled seal
AP   *Anilinpunkt FLÜSS* aniline point
APC, APZ   *automatischer Partikelzähler* automatic particle counter, APC
äquivalente Rohrlänge *f*   *THEOR* equivalent pipeline length
äquivalenter Drosselquerschnitt *m*   *THEOR* equivalent orifice (restrictive) area
Arbeitsdrehzahl *f*   operating (working) speed (rpm)
Arbeitsdruck *m*   operating (working) pressure
Arbeitsflüssigkeit *f*   working (operating, pressure) fluid, fluid power medium, *cf* Hydraulikflüssigkeit *f*
Arbeitshub *m*   *ZYL* power (working, operating) stroke
Arbeitsleitung *f*   working (operating, main) line
Arbeitsmedium *n*   *s.* Arbeitsflüssigkeit *f*
Arbeitsmittel *n*   *s.* Arbeitsflüssigkeit *f*
Arbeitspumpe *f*   *eines hydrostatischen Getriebes* main (primary) pump *of a hydrostatic transmission*
Arbeitsspiel *n*   *s.* Arbeitszyklus *m*
Arbeitstemperatur *f*   operating (working) temperature

Arbeitszyklus *m*   work[ing] (operating, duty) cycle
Arbeitszyklusdiagramm *n*   cycle plot (diagram)
Arbeitszylinder *m*   cylinder, *Weiterbildungen s unter* Zylinder *m*; *als Kraftzylinder, Leistungszylinder ANWEND* power cylinder
Armatur *f*   *s.* Rohrverschraubung *f*; *s.* Schlauchverschraubung *f*
armierte Dichtung *f*   reinforced seal
Armierung *f*   *s.* Schlaucharmierung *f*
Armierungslage *f*   reinforcement layer
Aschegehalt *m*   *FLÜSS* ash content
ATF-Öl *n*   *für Drehmomentwandler* automatic transmission fluid, ATF
Atmosphäre *f*: **mit der Atmosphäre verbunden**   *PN* open to atmosphere
atmosphärische Luft *f*   *Luft im Ansaugzustand* atmospheric (free) air
atmosphärischer Druck *m*   atmospheric pressure
Aufbereitung *f*   *s.* Luftaufbereitung *f*
Aufdornringverschraubung *f*   expansion fitting
Aufenthaltszeit *f*   *des Öls im Behälter* dwell (rest) time *of the oil in the reservoir*
Aufladedruck *m*   *SPEICH* precharge (charging, preload, inflation) pressure
aufladen   *mit Flüssigkeit SPEICH* charge *v*, load *v* *with fluid*; *mit Gas SPEICH* charge *v*, precharge *v*, load *v*, pressurize *v* *with gas*
Aufladung *f*   *SPEICH* filling, loading, charging
Auflösungsvermögen *n*   *einer Verstelldrossel* adjustment sensitivity *of an adjustable restriction*
Aufnehmer *m*   *s.* Sensor *m*
Aufschraubverbindung *f*   Schlauch-

verbindung screw- together fitting
hose fitting
Aufschraubverschraubung *f* female connector (end fitting), socket end fitting
Aufschraubwinkelverschraubung *f* female elbow
Aufsteckverbindung *f* *hülsenlos, für Schlauch* socketless fitting
Aufwärtshub *m* *ZYL* upstroke
Auge *n* *Füllstandsglas* level sight glass, sight gage, gage glass
Augenbefestigung *f* *ZYL* eye (ear, single-ear) mount
ausbilden/sich *Strömung* develop *v flow*
Ausdehnungsthermometer *n* expansion thermometer
Ausdehnungsviskosität *f* *FLÜSS* expansional viscosity
ausfahren *vt und vi* *Zylinder* extend *v cylinder*
Ausfahren *n* *der Kolbenstange* *übermäßiges Ausfahren ZYL* rod over-extension
Ausfahrgeschwindigkeit *f* *ZYL* extension speed
Ausfahrhub *m* *ZYL* extend (out, extension, outward) stroke
Ausfahrlänge *f* *ZYL* extended length
Ausfahrseite *f* *ZYL* rod (front, head) end
ausfahrseitige Befestigung *f* *ZYL* head (front) mount
ausfahrseitiger Druck *m* *ZYL* rod-end (head-end, front-end) pressure
ausfahrseitiges Zylinderende *n* rod (front, head) end
Ausfahrstellung *f* *ZYL* extended position

Ausfall *m* *der Hydraulik* *ANWEND* hydraulic failure
Ausflockungsmittel *n* *FLÜSS* flocculating agent, flocculant
Ausfluß *m* outlet, exhaust, outflow
Ausflußviskosimeter *n* efflux (capillary) viscosimeter
Ausführung *f* *mit durchgehender Welle* *PU/MOT* through-shaft configuration
Ausgang *m* outlet, exhaust, outflow
Ausgangsdruck *m* outlet (exhaust) pressure, *also, in a pump* delivery (discharge, output) pressure
Ausgangsleistung *f* *abgegebene Leistung* output horsepower (power), horsepower (power) output
Ausgangsleitung *f* outlet (exhaust) line, *also, at a pump* discharge (delivery, output, pressure) line
Ausgangsraum *m* outlet (exhaust) chamber, *also, in a pump* discharge (delivery, output, pressure) chamber
Ausgangsseite *f* outlet (exhaust) side, *also, of a pump* discharge (output, pressure) side
Ausgangs[volumen]strom *m* outlet (exit) flow [rate], outlet rate
ausgasen *Luft ausscheiden* release *v* air, outgas *v*
ausgeglichene Pumpe *f* *druckausgeglichen* pressure-compensated (-balanced) pump, compensated (balanced) pump
ausgleichen *Druck* pressure-balance (-compensate) *v*; *Leckverluste* compensate *v*, make up *v* *for leakage*
Ausgleichleitung *f* *VENTILE* balance (compensation) line
Auslaß *m* outlet, exhaust, outflow
Auslaßanschluß *m* outlet (exhaust)

Auslaßdruck

port, *also, of a pump* discharge (delivery, output, pressure) port
Auslaßdruck *m*   outlet (exhaust) pressure, *also, in a pump* delivery (discharge, output) pressure
Auslaßkammer *f*   outlet (exhaust) chamber, *also, in a pump* discharge (delivery, output, pressure) chamber
Auslaßkanal *m*   outlet (exhaust) channel, *also, in a pump* discharge (delivery, output, pressure) channel
Auslaßleitung *f*   outlet (exhaust) line, *also, at a pump* discharge (delivery, output, pressure) line
Auslaßseite *f*   outlet (exhaust) side, *also, of a pump* discharge (output, pressure) side
Auslaßventil *n*   *einer Pumpe* outlet valve *of a pump*
Auslaufverlust *m*   *THEOR* exit loss
Auslaufviskosimeter *n*   efflux (capillary) viscosimeter
Auslegungsdruck *m*   design pressure
auspumpen   evacuate *v*
Ausschaltregelung *f*   *KOMPR* start-stop control
ausschieben   displace *v*
ausschwenken   *Nullhubpumpe* move *vi* on stroke   *pressure-compensated pump*;   *Verstellpumpe* stroke *v* variable pump
außenbeaufschlagte Radialkolbenpumpe *f*   peripherally ported radial-piston pump, radial-piston pump with exterior admission
Außenbeaufschlagung *f*   *PU/MOT* exterior admission
Außendruck *m*   external pressure
Außendurchmesser *m*   outside (outer) diameter, OD, O. D.

Außenfilterung *f*   *Spülung AUFBER* external (bulk) filtration
Außenläufer *m*   *eines Magnetkolbenzylinders* external yoke *of a magnetic rodless cylinder*
Außenlippenring *m*   *Topfmanschette* cup seal (ring)
außenschrägverzahnte Zahnradpumpe *f*   helical gear pump
Außenzahnradmaschine *f*   external gear (spur gear, gear-on-gear) pump *or* motor
Außenzahnradmotor *m*   external gear (spur gear, gear-on-gear) motor
Außenzahnradpumpe *f*   external gear (spur gear, gear-on-gear) pump
äußere Flüssigkeitsrückführung *f*   *Ablaufdruckentlastung WEGEV* external drainage
äußere Leckölabführung *f*   external (reservoir, tank) drain [line]
äußere Leistungsverzweigung *f*   *s. Getriebe* n *mit äußerer Leistungsverzweigung*
äußerer Leckverlust *m*   external leakage
ausstoßen   displace *v*
Austauschfiltereinsatz *m*   replacement filter cartridge
Austauschfilterpatrone *f*   *s. Austauschfiltereinsatz m*
Austritt *m*   outlet, exhaust, outflow
Austrittsanschluß *m*   outlet (exit) port, *also, of a pump* discharge (delivery, output, pressure) port
Austrittsdeckel *m*   *einer Ventilbatterie* outlet (return) section, return plate *of ganged valves*
Austrittsdruck *m*   outlet (exhaust) pressure, *also, in a pump* delivery (discharge, output) pressure

Austrittskammer *f* outlet (exhaust) chamber, *also, in a pump* discharge (delivery, output, pressure) chamber
Austrittskanal *m* outlet (exhaust) channel, *also, in a pump* discharge (delivery, output, pressure) channel
Austrittsleitung *f* outlet (exhaust) line, *also, at a pump* discharge (delivery, output, pressure) line
Austrittsseite *f* outlet (exhaust) side, *also, of a pump* discharge (output, pressure) side
Austrittsstrom *m* *s.* Austrittsvolumenstrom *m*
Austrittsverlust *m* exit loss
Austrittsvolumenstrom *m* outlet (exit) flow [rate], outlet (exit) rate
automatischer Ablaß *m* *FILTER* automatic drain
automatischer Partikelzähler *m* *FLÜSS* automatic particle counter, APC
axiale Dichtung *f* axial (face) seal
Axialkerbe *f* *STROMV* axial notch
Axialkolbengetriebe *n* axial piston transmission
Axialkolbenmaschine *f* axial piston pump *or* motor
Axialkolbenmotor *m* axial piston motor
Axialkolbenmotor *m* **mit abgewinkelter Hauptachse** bent-axis (angled) axial piston motor
Axialkolbenmotor *m* **mit Schrägscheibe** swashplate axial piston motor
Axialkolbenmotor *m* **mit Taumelscheibe** wobble-plate axial piston motor
Axialkolbenpumpe *f* axial piston pump

Axialkolbenpumpe *f* **mit abgewinkelter Hauptachse** bent-axis (angled) piston pump
Axialkolbenpumpe *f* **mit antriebsachsparallelem Kolbenträger** in-line axial piston pump
Axialkolbenpumpe *f* **mit Schrägscheibe** swashplate axial piston pump
Axialkolbenpumpe *f* **mit Taumelscheibe** wobble-plate axial piston pump
Axialverdichter *m* axial[-flow] compressor
Axko ... *s.* Axialkolben ...

B

Back-up-Ring *m* *DICHT* anti-extrusion (back-up) ring
Bajonettverschlußkupplung *f* bayonet-type coupling
Balgdruckschalter *m* bellows pressure switch
Balgdruckwandler *m* bellows pressure transducer
Balgfedermanometer *n* bellows pressure gauge
Balgspeicher *m* bellows accumulator
Balgzylinder *m* flexible-wall (bellows) cylinder
Bandzylinder *m* band cylinder
bar *unit of pressure: 1 bar = 14.5 lbf/in^2*
Batterie *f* *s.* Filterbatterie *f*; *s.* Ventilbatterie *f*
Batterieplatte *f* custom (circuit, cross drilled) manifold, manifold block, drilled plate
Batterieverkettung *f* sectional (stack,

modular, gang, sandwich) mounting, valve stacking
Baumwollfilter *m,n*   cotton filter
Bauteil *n*   *s.* Hydraulikbauteil *n*; *s.* Pneumatikbauteil *n*
Bauteillebensdauer *f*   component life
beaufschlagen mit Druck   pressurize *v*, apply *v* pressure   *to a component*, pressure-load *v*, charge *v* with pressure, expose *v* to a pressure;   *Druck, eine Fläche* act *v* on (upon)   *pressure, on an area*;   *mit Flüssigkeit*   deliver *v*, supply *v*, provide *v* *system; fluid to component*
Beaufschlagung *f*   supply, admission, delivery
Befestigung *f*   *s.* Zylinderbefestigung *f*
Befestigung *f* **an ausfahrseitigem Gewindeansatz**   *ZYL* neck (nose) mount
Befestigung *f* **an der Ausfahrseite**   *ZYL* head (front) mount
Befestigung *f* **an der Bodenseite**   *ZYL* cap mount
Befestigung *f* **an einer Stirnfläche**   *boden- oder ausfahrseitig ZYL* face mounting   *cap or front end*
Befestigung *f* **an Laschen in Höhe der Zylinderachse**   centerline lugs mount
Befestigung *f* **in einer Ebene mit der Zylinderachse**   centerline mount
Befestigungsart *f*   mounting style (pattern, configuration)
Befestigungselement *n*   mount[ing] [element]
Befüllgerät *n*   oil conditioner (filtration unit, reconditioner), filter cart, portable filter unit, fluid transfer cart, kidney machine

Behälter *m*   fluid reservoir (tank), hydraulic reservoir (tank)
Behälter *m*:   **mit dem Behälter verbinden**   vent *v* [to tank], exhaust *v*, dump *v*, release *v*, relieve *v*, connect *v* to reservoir (tank), *HY mit dem Behälter verbunden auch:* open to atmosphere
Behälter *m* **ohne Trennmittel**   air-over-oil tank
Behälteranschluß *m*   *WEGEV* reservoir (tank) port (connection)
Behälteraufbaufilter *m,n*   tank mount (top) filter, reservoir mount filter
Behälterboden *m*   reservoir (tank) bottom
Behälterdeckel *m*   reservoir (tank) cover (top plate)
Behälterfilter *m,n*   in-tank (in-reservoir, tank, reservoir, sump, submerged, immersion) filter
Behältergröße *f*   reservoir (tank) volume (capacity, size)
Behältervolumen *n*   reservoir (tank) volume (capacity, size)
Behälterwand[ung] *f*   reservoir (tank) wall
beid[er]seitige Kolbenstange *f*   *ZYL* double-end rod
... belasteter Speicher *m*   ...-loaded accumulator
Belastungszahl *f*   *LOGIKEL* fan-out
beliefern   deliver *v*, supply *v*, provide *v*   *system; fluid to component*
belüften   *Druckluft zuführen* supply *v* with air, provide *v* air;   *entlüften BEHÄLT* breathe *v*, ventilate *v*, aerate *v*
Belüftung *f*   *BEHÄLT* breathing, ventilation, aeration
Belüftungs-Einfüll-Kombination *f*

BEHÄLT filler-breather [assembly], breather/filler
Belüftungsfilter *m,n* BEHÄLT air breather filter
Belüftungsöffnung *f* BEHÄLT breather hole, ventilating eyelet, aeration opening
Belüftungssteuerung *f* LOGIKEL bleed-on system
Bemessungsdruck *m* design pressure
benetzter Umfang *m* STROMV wetted perimeter
beölen *Druckluft* lubricate *v* compressed air
Bernoulli-Gleichung *f* THEOR Bernoulli's equation
Bernoullisches Gesetz *n* *Energieerhaltung bei strömenden Medien* THEOR Bernouilli's law *of energy conservation in a flowing fluid*
Berstdruck *m* burst pressure
Berstmembran *f* DRUCKV rupture (blow-out) disk
Berstscheibe *f* *s.* Berstmembran *f*
Berührungsdichtung *f* contact seal
berührungssensitives Ventil *n* touch valve
Berührungstemperaturfühler *m* surface temperature sensor
Beschleunigungsdruck *m* THEOR acceleration head (pressure)
Beschleunigungsverträglichkeit *f* acceleration tolerance
Beständigkeit *f* **gegen Emulgieren** FLÜSS emulsification resistance, resistance to emulsification
betätigen STELL operate *v*, actuate *v*, control *v*
Betätigung *f* *eines Ventils* valve operation (actuation, control); *s also* Betätigungseinrichtung *f*

Betätigungsdruck *m* STELL operating (actuating, control) pressure
Betätigungseinrichtung *f* *für Ventile* valve operator (actuator, control mechanism, control)
Betätigungskolben *m* STELL operating (actuating, control) piston
Betätigungskraft *f* STELL operating (actuating, control, stroke) force
Betätigungsmoment *n* STELL operating (actuating, control) torque
Betätigungsstellung *f* DRUCKV actuated (offset) position
Beta-Verhältnis *n* **(-Wert** *m***)** *ein Maß für den Filterwirkungsgrad* beta ratio, β ratio, *(β for X µm particle size = particles > X µm upstream / particles > X µm downstream)*
Betriebsdrehzahl *f* operating (working) speed (rpm)
Betriebsdruck *m* operating (working) pressure
Betriebsdruckluft *f* shop (plant) air
Betriebsdruckluftnetz *n* shop (plant) air mains
Betriebsflüssigkeit *f* working (operating, pressure) fluid, fluid power medium, *cf* Hydraulikflüssigkeit *f*
Betriebsmanometer *n* industrial pressure gauge; *geringerer Genauigkeit* commercial pressure gauge
Betriebsspiel *n* *s.* Betriebszyklus *m*
Betriebstemperatur *f* operating (working) temperature
Betriebsverschmutzung *f* *erzeugte Verschmutzung* FLÜSS generated contamination
Betriebsviskosität *f* operating (working) viscosity
betriebswarm warmed-up

Betriebswirkungsgrad m running efficiency

Betriebszyklus m work[ing] (operating, duty) cycle

bewegen/sich *Kolben* stroke *vi piston*

bewegte Teile npl *s.* Steuertechnik *f* mit bewegten Teilen

Bewegungsdichtung f dynamic seal

bewehrte Dichtung f reinforced seal

Bewehrungslage f *LEIT* reinforcement layer

Bezugsbedingungen fpl reference conditions

Bezugsdruck m reference pressure

Bezugsfläche f *Fluiddynamik* control surface *fluid dynamics*

Bezugsflüssigkeit f reference fluid

Bezugsgebiet n *Fluiddynamik* control area *fluid dynamics*

Bezugstemperatur f reference temperature

Bezugsvolumen n *Fluiddynamik* control volume *fluid dynamics*

Bezugszustand m reference conditions *pl*

Biegeradius m *LEIT* bend radius

Biegevorrichtung f *LEIT* pipe (tube) bender

biegsame Leitung f flexible line

Bimetallthermometer n bimetall[ic] thermometer

biochemisch (biologisch) abbaubar *FLÜSS* biodegradable

Blase f *SPEICH* accumulator bag (bladder)

Blasendruckprüfung f *FILTER* bubble[-point] test

Blasenleckage f *bei Tauchprüfung DICHT* bubble leakage *when immersion-testing*

Blasenschutzvorrichtung f *SPEICH* transfer barrier

Blasenschutzventil n *SPEICH* bladder (bag) protection valve

Blasenpunkttest m *FILTER s.* Blasendruckprüfung *f*

Blasenschwenkmotor m bladder rotary actuator

Blasenspeicher m bladder (bag) accumulator

Blasenstrom m *bei Tauchprüfung DICHT* bubble leakage *when immersion-testing*

Blasiussches Gesetz n *Strömungswiderstand* Blasius' law *of fluid resistance*

Blende f *THEOR* nonviscous restriction, orifice

Blende f mit abgerundeter Kante *THEOR* round-edged orifice

Blendendurchflußmesser m orifice meter

blendenförmiger Drosselwiderstand m *THEOR* sharp-edged orifice

Blindabschnitt m *einer Rohrleitung* blind run

Blockanbau m *VENTILE* manifold mounting

blockieren *Hydromotor* stall *v hydromotor*; *z. B. Strom, Leitung* block [off] *v*, isolate *v eg flow, line*

Blockierleckstrom m *PU/MOT* stalled leakage flow

Blockiermoment n *PU/MOT* stall torque

Blockschaltbild n block diagram

Blockventil n valve unit (block)

Boden m *s.* Zylinderboden *m*

Bodenseite *ZYL* cap (rear, blind, blank) end

bodenseitige Befestigung *f*   *ZYL* cap mount
bodenseitiger Druck *m*   cap (rear, blind, blank) end pressure
bodenseitiges Zylinderende *n*   cap (rear, blind, blank) end
Bodenventil *n*   dump (bottom, foot) valve
Bogenverlust *m*   bend loss
Bohrbild *n*   *VENTILE* port (hole) configuration
Bohrung *f*   *s.* Zylinderbohrung *f*
Bohrungseinbau *m*   *s.* Ventil *n* für Bohrungseinbau
bördellose Verbindung *f*   *VERBIND* flareless (non-flared) fitting
Bördelmaschine *f*   *VERBIND* pre-flaring device
bördeln   *VERBIND* flare *v*
Bördelverbindung *f*   flare (flared) fitting
   Bördelverbindung *f* **mit Klemmring**   3-piece flare fitting
   Bördelverbindung *f* **ohne Klemmring**   2-piece flare fitting
Bördelverschraubung *f*   *s.* Bördelverbindung *f*
Bördelwinkel *m*   *VERBIND* flare angle
Bourdonrohr *n*   *s.* Rohrfeder *f*
Bremsansatz *m*   *ZYL* cushion spear (plunger)
   Bremsansatz *m* **mit Axialkerbe[n]**   *ZYL* slotted spear
Bremse *f*   Endlagenbremse *ZYL* cylinder cushion
Bremsbetrieb *m*   *PU/MOT* braking duty (operation)
Bremsdichtung *f*   *ZYL* cushion seal
Bremsdruck *m*   *ZYL* cushioning pressure
bremsen   *am Hubende ZYL* cushion *v* at the end of stroke
Bremsfläche *f*   *ZYL* cushion area
Bremskraft *f*   *ZYL* cushioning force
Bremsraum *m*   *ZYL* cushion cavity (chamber, dashpot)
Bremsventil *n*   *STROMV* braking (brake) valve
Bremsweg *m*   *ZYL* cushion stroke, length of cushion
Brennpunkt *m*   *FLÜSS* fire point
Brillenpumpe *f*   gear pump with double bearing *with pressure-dependent axial and radial clearance*
Bruch-/Berstfestigkeitsprüfung *f*   *FILTER* collapse-burst test
Brückenschaltung *f*   *KREISL* bridge circuit
BS-Rohrgewinde *n*   *VERBIND* BSP (British Standard pipe) thread
Büchse *f*   *s.* Ventilbüchse *f*
Bündelstrahlverstärker *m*   *LOGIKEL* focused-jet amplifier
Bunsenscher Lösungskoeffizient *m*   *FLÜSS* Bunsen absorption coefficient
Bypass *m*   bypass
Bypassfilter *m,n*   *Filter mit Umgehung* integral bypass filter
Bypassschaltung *f*   bleed-off (bypass) circuit
Bypass-Stromsteuerung *f*   bleed-off (bypass) flow control
Bypassventil *n*   *FILTER* bypass valve

C

C-förmige Rohrfeder *f*   C-type Bourdon tube
C-Ring *m*   *DICHT* C-ring
Cartridge *n*   cartridge (hydraulic) logic

valve, logic element (cartridge valve), but s ED cartridge valve
CETOP = **Comité Européen des Transmissions Oléohydrauliques et Pneumatiques**
chlorierte Kohlenwasserstoffe *mpl* s. Flüssigkeit *f* auf Basis chlorierter Kohlenwasserstoffe
Coanda-Effekt *m* *LOGIKEL* Coanda (wall-attachment) effect
Coanda-Element *n* *LOGIKEL* Coanda effect element, wall-attachment element

D

D-Ring *m* *DICHT* D-ring
Dachmanschette *f* *DICHT* V-ring, V-cup, chevron (vee) ring
Dachmanschettensatz *m* *DICHT* V-ring (chevron) packing, vee set
Dachring *m* s. Dachmanschette *f*
Dampfdruck *m* vapour pressure
Dampfdruckthermometer *n* vapour-pressure (-tension) thermometer
Dampfemulsionszahl *f* *Maß der Emulgierbarkeit FLÜSS* steam-emulsion number, SEN *measure of emulsibility*
dämpfen *Schwingungen* attenuate *v*, damp[en] *v*; *bremsen, am Hubende ZYL* cushion *v* *at the end of stroke*
Dampfphaseninhibitor *m* *über die Dampfphase wirkender Korrosionsinhibitor FLÜSS* vapour-phase inhibitor, VPI *to prevent corrosion of metals in contact with fluid vapour*
Dampfspannungsthermometer *n* vapour-pressure (-tension) thermometer

Dämpfungseinsatz *m* damping plug
Dämpfungsschlauch *m* damping hose
Dämpfungsventil *n* [surge-]damping (cushion) valve
Dauerarbeitsdruck *m* continuous (permanent) operating (working) pressure
Dauerbetriebsdruck *m* s. Dauerarbeitsdruck *m*
Dauerdrucksignal *n* continuous pressure signal
Dauerentlüftung *f* permanent exhaust
Dauerermüdungsprüfung *f* *FILTER* flow fatigue test
Dauerstrom *m* *STELL* holding current
Dauerstromdruckverstärker *m* continuous (reciprocating) intensifier
Deckel *m* *BEHÄLT* reservoir (tank) cover (top plate); *ZYL* cylinder [end] cover (cap, closure)
Dehnmeßstreifen-Druckwandler *m* *strain gauge pressure transducer*
Dehnschlauch *m* *für Druckspitzenabbau* expanding hose *used to smoothen pressure peaks*
Dekompression *f* *allmählicher Druckabbau* decompression
Deltaring *m* *Dreieckring DICHT* delta ring
Demulgator *m* *FLÜSS* demulsifier, demulsifying agent
demulgierbar *FLÜSS* demulsible
Demulgierbarkeit *FLÜSS* demulsibility
demulgieren *FLÜSS* demulsify *v*
Demulgierung *f* *FLÜSS* demulsification
Depressionswelle *f* depression wave
Detergens *n* *absetzverhinderndes Mittel FLÜSS* detergent

Diagonalverdichter *m* mixed-flow compressor
Dicarbonsäureester *m* *s.* Flüssigkeit *f* auf Basis von Dicarbonsäureestern
dicht *flüssigkeitsdicht* leakproof, fluid-tight, leaktight
Dichte-Druck-Verhalten *n* *FLÜSS* density-pressure characteristics
Dichte-Temperatur-Verhalten *n* *FLÜSS* density- temperature characteristics
Dichteinrichtung *f* seal, sealing device, packing
Dichtelement *n* sealing member (element), seal
dichten seal *v*
Dichtfläche *f* sealing surface
Dichtheit *f* tightness
Dichtigkeit *f* tightness
Dichtkante *f* sealing edge
Dichtkeil *m* *PU/MOT* crescent-shape separator, crescent
Dichtkraft *f* *DICHT* sealing force
Dichtlinie *f* sealing band, seal line
Dichtlippe *f* seal[ing] lip
Dichtmittel *n* *s.* Dichtstoff *m*
Dichtring *m* seal ring, ring seal
Dichtsatz *m* *Packung* packing [seal], seal assembly
Dichtscheibe *f* seal washer, washer seal
Dichtschnur *f* sealing profile
Dichtspalt *m* seal clearance (gap)
Dichtstoff *m* sealant
Dichtung *f* seal, sealing device, packing; *Dichtelement* sealing member (element), seal; *Flachdichtung* gasket
Dichtung *f* gegen Ausströmen inclusion (internal) seal

Dichtung *f* gegen Einströmen exclusion (external, protective) seal
Dichtung *f* in Sonderausführung custom seal
Dichtung *f* mit Verschleißring capped seal
Dichtung *f* zwischen Kolben und Kolbenstange *ZYL* piston-to-rod seal
Dichtungsgehäuse *n* *s.* Zylinderkopfdeckel *m*
Dichtungskitt *m* sealant
Dichtungsnut *f* seal groove
Dichtungsplatte *f* *einer Steuersäule* port (top) blanking plate, end plate (cover) *of a valve stack*
Dichtungsraum *m* seal cavity (pocket)
Dichtungsreibung *f* seal friction
Dichtungsring *m* seal ring, ring seal
Dichtungssatz *m* packing [seal], seal assembly
Dichtungsscheibe *f* seal washer, washer seal
Dichtungsschnur *f* sealing profile
Dichtungsstandzeit *f* seal life
Dichtungsstoff *m* seal material
Dichtungsverträglichkeit *f* *FLÜSS* seal compatibility
Dichtungsverträglichkeitsindex *m* *FLÜSS* seal compatibility index
Dichtungswerkstoff *m* seal material
dickes Wasser *n* *s.* hochwasserhaltige Flüssigkeit *f*
dickflüssig *FLÜSS* [high] viscous, thick, high-viscosity ...
dickwandig thick- (heavy-)walled
Differentialkolben *m* *ZYL* differential-area piston
Differentialkolben-Druckübersetzer *m* differential-piston intensifer

Differentialschaltung 154

Differentialkolben-Speicher *m* dual diameter (differential, Twedell) accumulator
Differentialschaltung *f* regenerative (differential) circuit
Differentialzylinder *m* differential cylinder
Differenzdruck *m* *Meßgröße* pressure difference, differential pressure
Differenzdruckmanometer *n* differential pressure gauge
Differenzdruckregler *m* *Druckwaage im Stromregelventil* controlled orifice, pressure compenator, hydrostat *in a compensated flow-control valve*
Differenzdruckschalter *m* differential pressure (pressure differential) switch
Differenzdruckstabilität *f* *der Filterfeinheit* differential pressure stability *of filter stability*
Differenzdruckwandler *m* differential pressure transducer
Diffusor *m* *Rückstromteiler in einem Behälter* diffuser *in a reservoir*
Digitalfluidik *f* digital fluidics
Digitalmanometer *n* digital pressure gauge
Digitalventil *n* *WEGEV* digital (digitally controlled) valve, microprocessor-controlled valve
Digitalzylinder *m* positional (digital) cylinder
Dilatationsviskosität *f* *FLÜSS* dilatational (second) viscosity
DIN-Flansch *m* DIN flange *4-bolt-type flange*
direkte Betätigung (Steuerung) *f* *STELL* direct operation (actuation, control)
direkter Impaktmodulator *m* *LOGIKEL* direct impact modulator, DIM
direktgesteuerte Pumpe *f* direct operated pump
direktgesteuertes Druckbegrenzungsventil *n* direct-acting relief valve
direktgesteuertes Ventil *n* direct-acting (one-stage, direct) valve
Dismulgator *m* *FLÜSS* demulsifier, demulsifying agent
dismulgierbar *FLÜSS* demulsible
Dismulgierbarkeit *FLÜSS* demulsibility
dismulgieren *FLÜSS* demulsify *v*
Dismulgierung *f* *FLÜSS* demulsification
Dispergens *n* *FLÜSS* dispersing agent, dispersant
Dispergierungsmittel *m* *s.* *Dispergens* *n*
dissipativer Schalldämpfer *m* dissipating muffler
Ditherantrieb *m* *zur Umgehung der Haftreibung*) *STELL* dither drive *to eliminate breakaway friction*
Ditherfrequenz *f* *VENTILE* dither frequency
DMS-Druckwandler *m* *mit Dehnmeßstreifen* strain gauge pressure transducer
Dochtprüfung *f* *auf Brennbeständigkeit FLÜSS* wick test *persistence of burning*
DOM-Rohr *n* *stangengezogen* DOM tubing *drawn over mandrel*
Doppelfilter *m,n* duplex (twin) filter
Doppelflügelschwenkmotor *m* double-vane rotary actuator
doppelflutige Schraubenpumpe *f* *axial ausgeglichen* double-suction (-inlet) screw pump

Doppelhub *m* *Weglänge ZYL* extension and return stroke length
Doppelhubmagnet *m* *Umkehrmagnet STELL* double-acting (dual-operation, push-pull type, two-way) solenoid
Doppelkegel *m* *SPERRV* double end poppet
Doppelkugelventil *n* *SPERRV* two ball valve
Doppellippenabstreifer *m* *DICHT* double-lip wiper
Doppellippenring *m* U-seal, U-cup, double lip seal
Doppellippen-Schneidringverschraubung *f* two-bite flareless compression fitting
Doppelmagnetventil *n* double-solenoid valve
Doppelmantelzylinder *m* double-barreled cylinder
Doppelmembranrelais *n* *LOGIKEL* double-diaphragm element
Doppelpumpe *f* tandem (dual, double, split-flow) pump
doppelreihige Radialkolbenpumpe *f* double-piston row radial pump
Doppelrückschlagventil *n* *wechselseitig entsperrbar* double (dual) check valve
Doppelschaltfilter *m,n* switch-over filter
Doppelsitzventil *n* double-seat valve
Doppelstangenzylinder *m* double-rod [end] cylinder
doppeltwirkende Flügelzellenpumpe *f* *mit Ovalring* elliptical cam ring-type vane pump
 doppeltwirkende Handpumpe *f* double-acting hand pump
 doppeltwirkender Zylinder *m* double-acting cylinder
 doppeltwirkender Druckübersetzer *m* double-acting intensifier (booster)
Doppelzahnstangenschwenkmotor *m* double-rack rotary actuator
Dose *f* *s.* Kupplungsdose *f*
Dosierpumpe *f* metering pump
Dosierverschraubung *f* flow control fitting
Drahtbandfilter *m,n* wound-wire (wirewound, metall ribbon) filter
drahtgeflechtarmierter Schlauch *m* wire braid hose
Drahtgewebefilter *m,n* wire-cloth (-gauze, -screen, -mesh) filter, metal screen filter
Drahtsiebfilter *m,n* *s.* Drahtgewebefilter *m,n*
Drehantrieb *m* *PN* rotary actuator; *s also* Pneumatikmotor *m*
Drehdurchführung *f* *s.* Drehverbindung *f*
Drehflügeldämpfer *m* vane-type damper
Drehflügelschwenkmotor *m* vane rotary actuator
Drehgelenk *n* swivel joint (connection), swivel
Drehkolbenmotor *m* lobed-element (lobe) motor
Drehkolbenpumpe *f* lobed-element (lobe) pump
Drehlängsschieber *m* combination sliding and rotary spool valve
Drehmoment *n* **bei Drehzahl Null** *PU/MOT* zero-speed torque
Drehmoment *n* **bei Nenndrehzahl** running torque
Drehmomentmotor *m* *STELL* torque motor

Drehmomentpulsation *f*   *PU/MOT*
torque pulsation (ripple, fluctuation)
Drehmomentungleichförmigkeit *f*
  s. Drehmomentpulsation *f*
Drehmomentwandler *m*   *ANWEND*
torque converter
Drehrichtung *f*   direction of rotation
Drehrichtungsumkehr *f*   *PU/MOT*
change of direction of rotation,
reversing the rotation
Drehrichtungswechsel *m*   s. Drehrichtungsumkehr *f*
Drehschiebersteuerung *f*   *PU/MOT*
rotary valve control
Drehschieberventil *n*   *WEGEV* plug
(rotary) valve
Drehübertrager *m*   rotary joint (connection, union), rotating distributor
drehungsfreie Strömung *f*   *THEOR*
irrotational (nonrotational) flow
Drehventil *n*   *WEGEV* rotary valve
Drehverbindung *f*   rotary joint (connection, union), rotating distributor
Drehverschraubung *f*   s. Drehverbindung *f*
Drehwinkelmotor *m*   rotary actuator, *compounds s with* Schwenkmotor *m*
Drehzahlbereich *m*   *PU/MOT* speed range
Drehzahlregler *m*   *PU/MOT* [speed] governor
Drehzahlschlupf *m*   *PU/MOT* speed loss
Drehzahlungleichförmigkeit *f*
PU/MOT speed pulsation (fluctuation)
Dreiebenen-Rohrgelenk *n*   triple-plane swivel joint
Dreieckkerbe *f*   *STROMV* vee notch, V-notch

Dreieckring *m*   *DICHT* delta ring
Dreiflügelschwenkmotor *m*   three-vanes (triple-vane) rotary actuator
Dreikreispumpe *f*   triple (treble, triplex) pump
Dreimembranenelement   *LOGIKEL*
three-diaphragm element
Dreiplattenpumpe *f*   sandwiched gear pump
Dreispindelschraubenpumpe *f*   three-screw pump
Dreistellungsventil *n*   *WEGEV* three-position (3-position) valve
Dreistrompumpe *f*   triple (treble, triplex) pump
dreistufiges Servoventil *n*   *STELL*
three-stage servovalve
Dreiwege-Druckminderventil *n*   *mit Sekundärdruckbegrenzung* relieving pressure-reducing valve
Dreiwegekugelhahn *m*   *WEGEV* three-way (3-way) ball valve
Dreiwege-Strombegrenzungsventil (-Stromregelventil) *n*   bypass (spill-off) flow-control valve, bypass (spill-off) flow regulator
Dreiwegeventil *n*   *WEGEV* three-way (3-way) valve
Drossel *f*   *THEOR* orifice, throttle, choke, restrictor, restriction; *Laminarwiderstand* laminar-type (viscous) restriction, choke; *STROMV* restrictor (restriction, restrictive, throttle, throttling) valve
Drosselblende *f*   *STROMV* metering orifice
Drosselcharakteristik *f*   *STROMV*
restriction (throttle, metering, flow, area) characteristics
Drosselelement *n*   restrictive (throttling) element

Drosselfläche *f* throttling (orifice, restriction) area
Drosselgleichung *f* orifice equation
Drosselkerbe *f* STROMV metering groove (notch)
Drosselkolben *m* STROMV metering piston
Drossellänge *f* THEOR throttling (restrictive, choke) length
drosseln *Volumenstrom* throttle *v*, restrict *v*, choke *v*, orifice *v*, meter *v* flow
Drosselplatte *f* *einer Steuersäule* restrictor module, choke block *of a valve stack*
Drosselquerschnitt *m* throttling (orifice, restriction) area
Drosselreihenstoßdämpfer *m* multiple-orifice shock absorber
Drossel-Rohrverschraubung *f* orifice union, *cf ED* union
Drosselrückschlagventil *n* return-orifice (metering, orifice) check valve
Drosselstelle *f* THEOR orifice, throttle, choke, restrictor, restriction
Drosselsteuerung *f* STROMV restrictive (throttle) control
Drosselstrecke *f* THEOR throttling (restrictive, choke) length
Drosselung *f* restriction, throttling
Drosselventil *n* STROMV restrictor (restriction, restrictive, throttle, throttling) valve
Drosselverhalten *n* restriction (throttle, metering, flow, area) characteristics
Drosselverlust *m* restriction (throttle) loss
Druck *m* *Druckhöhe, nutzbarer Druck* THEOR [pressure] head; *Druckwert* pressure [level]; *Überdruck* [gauge] pressure
Druck *m*: **Druck ausgleichen** *s.* Druck kompensieren
Druck *m*: **Druck kompensieren** pressure-balance *v*, pressure-compensate *v*
Druck *m*: **Druck zuführen** pressurize *v*, apply *v* pressure to *a component*, pressure-load *v*, charge *v* with pressure, expose *v* to a pressure
Druck *m*: **mit Druck beaufschlagen** *s.* Druck zuführen
Druck *m*: **nicht unter Druck** non-pressurized, unpressurized, atmospheric, pressureless
Druck *m*: **unter Druck** pressurized, under pressure
Druck *m*: **unter Druck einsetzbar** pressure-resistant
Druck *m*: **unter Druck setzen** pressurize *v*
Druck *m*: **unter Druck stehender Raum** *m* pressurized volume
Druck *m*: **vom Druck entlasten** pressure-balance *v*, pressure-compensate *v*
Druck *m* **bei maximalem Strom** DRUCKV full flow pressure
Druck *m* **bei Nullförderung** PU/MOT deadhead pressure, *s also* ED deadhead *v*
Druck *m* **im Beharrungszustand** steady[-state] pressure
Druck *m* **im System** system pressure
Druck *m* **in der Entlüftungsleitung** PN exhaust pressure
Druck *m* **mit Wert Null** zero pressure
Druckabbau *m* *allmählich* decompression

Druckabfall

Druckabfall *m* pressure drop; *Drucktal* pressure decay
Druckabfall *m* **im Neuzustand** *FILTER* clean pressure drop
druckabhängig pressure-dependent (-sensitive)
Druckabnahme *f* pressure decrease
Druckabschaltung *f* pressure relief (unloading)
Druckabschneidung *f* *Pumpenregelung* pressure limitation *pump control*
Druckänderung *f* pressure change
Druckänderungsgeschwindigkeit *f* rate of pressure change
Druckanschluß *m* discharge (delivery, output, pressure, outlet, exit) port; *WEGEV* pressure port
Druckanstieg *m* pressure rise
Druckanzeiger *m* pressure indicator
Druckaufbau *m* pressure build-up
Druckaufnahmeleitung *f* pressure-sensing line
Druckausfall *m* pressure failure
druckausgeglichen pressure-compensated (-balanced), *also as a verb:* pressure-compensate (-balance) *v*
druckausgeglichene Pumpe *f* pressure-compensated (-balanced) pump, compensated (balanced) pump
Druckausgleich *m* pressure compensation (balance)
Druckausgleichnut *f* *WEGEV* balancing (centering) groove
druckbeaufschlagt *s.* Druck *m*: mit Druck beaufschlagen
Druckbeaufschlagung *f* pressurization, pressure supply
druckbedingter Schlupfstrom *m* *PU/MOT* pressure slip [flow], pressure slippage

Druckbegrenzungsventil *n* relief [valve]
Druckbegrenzungsventil *n* **mit äußerer Leckflüssigkeitsrückführung** externally drained relief valve, balanced relief valve
druckbelastetes Ventil *n* pressure-loaded (-biased) valve
Druckbereich *m* pressure range
druckbetätigt *STELL* pressure-operated (-actuated, -controlled), *also as a verb:* pressure-actuate (-operate, -control) *v*
Druckbetätigung *f* *STELL* pressure control
Druckdauerfestigkeitsprüfung *f* pressure fatigue test
Druckdiagramm *n* *s.* Druck-Weg-Diagramm *n*
druckdicht pressure- (bubble-)tight
druckdichter Behälter *m* sealed pressurized reservoir (tank)
druckdichter Magnet *m* *STELL* [oil-]immersed solenoid
druckdichter Torque-Motor *m* *STELL* [oil-]immersed torque motor
Druckdifferenz *f* *Wert* pressure differential
Druckdifferenzmesser *m* differential pressure gauge
Druckdifferenzventil *n* differential pressure regulator
Druckdifferenzwandler *m* differential pressure transducer
Druckeinstellung *f* *Wert DRUCKV* pressure setting, preset pressure level
Druckenergie *f* pressure energy
druckentlastet pressure-compensated (-balanced), *also as a verb:* pressure-compensate (-balance) *v*
druckentlastete Pumpe *f* pressure-

compensated (-balanced) pump, compensated (balanced) pump
Druckentlastung *f* pressure compensation (balance); *Druckabschaltung* pressure relief (unloading)
druckentlastungsgesteuert *STELL* bleed-operated
Druckerhöhung *f* pressure rise
Druckfeld *n* pressure field
druckfest pressure-resistant
druckfester Sensor *m* *ZYL* pressure-resistant sensor
Druckfilter *m,n* *Hochdruck* pressure[-line] filter, high-pressure filter
Druckflüssigkeit *f* pressurized (pressure) fluid, fluid under pressure; *Druckmittel* working (operating, pressure) fluid, fluid power medium, power transmitting agent
Druckflüssigkeit *f*: **mit Druckflüssigkeit beaufschlagen** *z. B. einen Verbraucher, eine Fläche* apply *v* pressure fluid *eg to an actuator, at an area*
Druckflüssigkeit *f*: **Druckflüssigkeit zuführen** *z. B. einem Verbraucher, einer Fläche* apply *v* pressure fluid *eg to an actuator, at an area*
Druckflüssigkeitsspeicher *m* [hydraulic] accumulator, *compounds s with* Speicher *m*
Druckflüssigkeitsspeicher *m* **mit nachgeschalteter Gasflasche (nachgeschalteten Gasflaschen)** transfer-typ accumulator
druckführende Leitung *f* line under pressure, pressurized (pressure) line
Druckgas *n* pressurized (pressure) gas, gas under pressure
Druckgefälle *n* *Neigung bzw.* *Steigung einer Druckverlaufskurve* pressure gradient; *Druckabfall* pressure drop
Druckgefälleventil *n* differential pressure regulator
druckgespannte Dichtung *f* pressure-energized (-actuated, -activated) seal, self-energized (automatic) seal
druckgesteuert *STELL* pressure-operated (-actuated, -controlled), *also as a verb:* pressure-actuate (-operate, -control) *v*
druckgesteuerte Pumpe *f* check-valve (seated-valve) pump
druckgesteuerter Ablaß *m* *FILTER* pressure-actuated drain
Druckgradient *m* pressure gradient
Druckhaltekreislauf *m* pressure-holding circuit
Druckhaltestellung *f* *PU/MOT* dead-head position
Druckhöhe *f* nutzbarer Druck [pressure] head
Druckhub *m* *ZYL* push stroke
Druckimpuls *m* pressure pulse
Druckkammer *f* *in einem Verbraucher* pressure (input, inlet, intake) chamber *in an actuator*
Druckkanal *m* *in einer Pumpe* discharge (delivery, output, pressure, outlet, exhaust) channel; *in einem Verbraucher* pressure (input, inlet, intake) channel
druckknopfbetätigt *STELL* [push]button-actuated
druckkompensiert pressure-compensated (-balanced), *also as a verb:* pressure-compensate (-balance) *v*
druckkompensierte Pumpe *f* pressure-compensated (-balanced) pump, compensated (balanced) pump

Druckleitung f *an einer Pumpe* discharge (delivery, output, pressure, outlet, exhaust) line; *an einem Verbraucher* pressure (input, inlet, intake, supply, feed) line; *unter Druck stehende Leitung* line under pressure, pressurized (pressure) line
drucklos nonpressurized, unpressurized, atmospheric, pressureless
drucklos ableiten (machen), mit dem Behälter verbinden vent *v* [to tank], exhaust *v*, dump *v*, release *v*, relieve *v*, unload *v*; depressurize *v*
drucklose Leitung f nonpressurized (unpressurized) line
drucklose Rücklaufleitung f atmospheric return line
Druckluft f pressurized (pressure) air, air under pressure; *in der Werkstattumgangssprache auch* Preßluft *f* compressed air, *compounds s also with* pneumatisch, Pneumatik..., Pneumo..., Luft...
Druckluft f: mit Druckluft beaufschlagen (speisen) supply *v* with air, provide *v* air to
Druckluft f: Druckluft zuführen supply *v* with air, provide *v* air to
Druckluftanlage f pneumatic (compressed air) system
Druckluftanschluß m pneumatic (compressed air, air) connection, air port
Druckluftantrieb m ANWEND pneumatic (compressed air, air) drive
Druckluftaufbereitung f air conditioning, compressed air preparation
Druckluftbedarf m [compressed] air demand
Druckluftbehälter m receiver
Druckluftbetätigungseinrichtung f STELL pneumatic operator (actuator, control), air operator
druckluftbetrieben pneumatically powered, air-powered (-operated)
Druckluftfilter m,n air [line] filter, pneumatic filter
druckluftgetrieben pneumatically powered, air-powered (-operated)
Druckluxthydraulikpumpe f hydropneumatic (air/hydraulic, air-powered, air-operated) pump
Druckluftkupplung f ANWEND pneumatic (air) clutch
Druckluftleitung f airline, [compressed-]air line, pneumatic line
Druckluftmotor m [compressed-]air motor
Druckluftnetz n [compressed-]air mains
Drucklufttöler m AUFBER air line lubricator
Druckluftregler m pressure regulator
Druckluftschaltung f pneumatic (compressed-air, air) circuit (system)
Druckluftschlauch m [compressed-]air hose, pneumatic hose
Druckluftspeicher m receiver
Druckluftstelleinheit f STELL pneumatic operator (actuator, control), air operator
Druckluftsteuerung f ANWEND pneumatic (compressed-air, air) control
Druckluftsystem n pneumatic (compressed-air, air) system
Drucklufttrockner m AUFBER air dryer
Druckluftventil n pneumatic (compressed-air, air) valve
Druckluftverbrauch m [compressed-]air consumption

Druckluftverteilungssystem *n* [compressed-]air distribution system

druckluftvorgesteuert *STELL* pneumatically (air-)piloted

Druckluftwartungseinheit *f* air conditioner [unit], filter- regulator-lubricator, FRL (*pl* FRLs)

Druckluftwerkzeug *n* *ANWEND* air (pneumatic) tool

Druckluftzufuhr *f* [compressed-]air supply

Druckluftzylinder *m* pneumatic linear actuator, pneumatic (compressed-air, air) cylinder,

Druckmeßabzweig *m* pressure tap, gauge tapping point

Druckmeßdose *f* *Membrandruckwandler* diaphragm pressure transducer

Druckmeßstelle *f* *s.* Druckmeßabzweig *m*

Druckmessung *f* pressure measurement

Druckminderventil *n* [pressure-]reducing valve, reducer; *Reduzierventil PN* pressure regulator

Druckminderventil *n* **mit Sekundärdruckbegrenzung (Überlastsicherung)** relieving pressure-reducing valve

Druckmittel *n* working (operating, pressure) fluid, fluid power medium, *cf* Hydraulikflüssigkeit *f*

Druckmittelwandler *m* dual-fluid intensifier (booster)

Druckniere *f* *PU/MOT* pressure (outlet) kidney

Drucköl *n* pressurized (pressure) oil, oil under pressure

Druckprofil *n* *THEOR* pressure distribution (profile)

druckproportional proportional to pressure

Druck-Proportionalventil *n* proportional pressure control valve

Druckpulsation *f* pressure pulsation (ripple, fluctuation)

Druckpulsationsdämpfer *m* pressure snubber, pulsation (ripple) damper

Druckquelle *f* *KREISL* constant-pressure source (supply)

Druckraum *m* *in einer Pumpe* discharge (delivery, output, pressure, outlet, exhaust) chamber; *in einem Verbraucher* pressure (input, inlet, intake) chamber; *unter Druck stehender Raum* pressurized volume

Druckreduzierventil *n* [pressure-]-reducing valve, reducer

Druckregelventil *n* *für den Ausgangsdruck s.* Druckminder-, Druckdifferenz- *oder* Druckverhältnisventil *n*

Druckring *m* *eines Dichtungssatzes* female (back) support ring, female adaptor *of a V-ring assembly*

Druckringverschraubung *f* compression (grip) fitting

Druckrückführung *f* *STELL* pressure feedback

Druckrückgewinn *m* pressure recovery

Druckschalter *m* pressure switch

Druck-Schieberweg-Verhalten *n* *VENTILE* pressure-displacement characteristics

Druckschlag *m* hydraulic (fluid) shock, water (fluid) hammer, pressure surge

Druckschlauch *m* pressure hose

Druckschreiber *m* pressure recorder

Druckschwankung *f* pressure variation

Druckschwingung *f* pressure vibration (oscillation)
Druckschwingungsdämpfer *m* pressure snubber, pulsation (ripple) damper
Druckseite *f* *einer Pumpe* discharge (output, pressure, outlet, exhaust) side; *eines Verbrauchers* pressure (input, inlet, intake) side
Drucksenkung *f* pressure decrease; *auf Unterdruck* depression
Drucksensor *n* pressure sensor
Drucksignal *n* pressure signal
Druckspitze *f* pressure peak (spike)
Druckstau *m* *s.* Gegendruck *m*
Drucksteigerung *f* pressure rise
Drucksteuereinrichtung *f* pressure control
Drucksteuerkreislauf *m* pressure control circuit (system), pressure control
Drucksteuerung *f* pressure control; *s.* Drucksteuereinrichtung *f*; *s.* Drucksteuerkreislauf *m*
Druckstoß *m* hydraulic (fluid) shock, water (fluid) hammer, pressure surge
Druckstoßdämpfer *m* desurger, water hammer (shock pressure) absorber, surge suppressor, pressure snubber
Druckstrom *m* flow of fluid under pressure, pressurized flow
Druckstromerzeuger *m* THEOR pressure [flow] generator; *s also* Pumpe *f*
Druck-Strom-Steuerplatte *f* VENTILE PQ manifold block, pressure/flow control section
Druckstromverbraucher *m* THEOR actuator, *but s* ED actuator
Druckstufe *f* *s.* Nenndruck *m*
Drucktal *n* pressure decay

drucktastenbetätigt STELL fingertip actuated
Drucktaupunkt *m* FLÜSS pressure dewpoint
Druckteilventil *n* pressure-dividing valve
Druckträger *m* *eines Schlauchs* pressure carrier *of a hose*
Drucktransiente *f* pressure transient
Druckübergang[sabschnitt] *m* pressure transient
Drucküberschwingweite *f* pressure overshoot
Druckübersetzer *m* [pressure] intensifier, booster
Druckübersetzer *m* **für gleiche Medien** single-fluid intensifier (booster)
Druckübersetzer *m* **für verschiedene Medien** dual-fluid intensifier (booster)
druckübersetzergespeist intensifier-(booster-)operated
Druckübersetzungsverhältnis *n* boost pressure ratio
Druckübertragungsmittel *n* working (operating, pressure) fluid, fluid power medium, power transmitting agent, *cf* Hydraulikflüssigkeit *f*
druckunabhängig pressure-independent (-insensitive)
Druckungleichförmigkeit *f* pressure pulsation (ripple, fluctuation)
Druckungleichgewicht *n* pressure imbalance
Druckventil *n* DRUCKV pressure-control valve; *Auslaßventil einer Pumpe* outlet valve *of a pump*
Druckverhältnis *n* pressure ratio
Druckverhältnisventil *n* proportional

pressure-reducing valve, proportioning pressure regulator
Druckverlust *m*   head (pressure) loss
Druckverminderung *f*   pressure decrease; *auf Unterdruck* depression
Druckverringerung *f*   *s.* Druckverminderung *f*
Druckversorgung *f*   pressurization, pressure supply
Druckverstärker *m*   *STELL* pressure amplifier; *s also* Druckübersetzer *m*
Druckverstärkung *f*   pressure intensification (boost, amplification, gain)
Druckverteilung *n*   *THEOR* pressure distribution (profile)
Druckviskosität *f*   *FLÜSS* compressional viscosity
Druck-Viskositäts-Verhalten *n*   *FLÜSS* viscosity-pressure characteristics
Druckwaage *f*   *Differenzdruckregler im Stromregelventil* controlled orifice, pressure compenator, hydrostat *in a compensated flow-control valve*
Druckwächter *m*   pressure indicator
Druckwasser *n, in der Werkstattumgangssprache auch* Preßwasser *n* compressed water
Druckwasseranlage *f*   water [hydraulic] system, water hydraulic
Druckwasserhydraulik *f*   water hydraulics
Druckwasserpresse *f*   *ANWEND* water hydraulic press
Druckwasserpumpe *f*   [compressed] water pump
Druckwassersystem *n*   *s.* Druckwasseranlage *f*
Druckwechselregenerierung *f*   *von Trockenmittel* pressure-swing regeneration *of a desiccant*

Druck-Weg-Diagramm *n*   pressure-displacement (-pattern) plot, pressure plot
Druckwelle *f*   pressure wave
Druckwert *m*   pressure [level], pressure value
Druckwirkungsgrad *m*   pressure efficiency
druckzentriertes Ventil *n*   pressure-centered valve
Druckzentrierung *f*   *STELL* hydraulic (pressure) centering
Druckzuführung *f*   pressurization, pressure supply
Druckzusammenbruch *m*   pressure collapse
Druckzyklusprofil *n*   *s.* Druck-Weg-Diagramm *n*
Druckzylinder *m*   *ZYL* push-action cylinder
Drückzylinder *m*   *überwindet Pressenkolbenreibung ANWEND* kicker cylinder *to overcome ram friction in presses*
dünnflüssig   *FLÜSS* low viscous, thin, low-viscosity ...
dünnwandig   thin-walled
durchdrehen, durchfallen   *z. B. Druckluftmotor* spin *v*, overspeed *v* *eg air motor*
durchfließen   *z. B. Öl durchfließt eine Öffnung, fließt durch eine Öffnung* pass *v* *eg oil passes a restriction, through a restriction*, flow *v* *eg through a restriction*
durchfließen lassen   *z. B. Ventil läßt Öl zum Verbraucher fließen* pass *v* *eg valve passes fluid to the actuator*
Durchfluß *m*   *s.* Durchflußstrom *m*
Durchflußbeiwert *m*   *THEOR* flow

Durchflußkapazität

(discharge, loss) coefficient, flow resistance value
Durchflußkapazität *f*   [flow] capacity
Durchflußleistung *f*   [flow] capacity
Durchflußmesser *m*   flowmeter
Durchflußmessung *f*   flow [rate] measurement
Durchflußquerschnitt *m*   passage area
Durchflußrichtung *f*   direction of passing fluid
Durchflußschreiber *m*   flow recorder
Durchflußsensor *m*   flow sensor
Durchflußstrom *m*   oft nur *Durchfluß* flow [rate], *in a more exact context* volumetric flow rate, volume rate of flow
Durchflußstrombereich *m*   range of flow [rate], flow range
Durchführung *f*   *VERBIND* rotary joint (connection, union), rotating distributor
Durchgangsdrehzahl *f*   *PU/MOT* runaway (free) speed
Durchgangskupplung *f*   *VERBIND* free-flow (free-passage, straight-through, high-flow) coupling
durchgehende Last *f*   overrunning load
durchgehende Welle *f*   *PU/MOT* s. *Durchtriebausführung f*
durchlässig: in Ruhestellung durchlässig   *LOGIKEL* normally passing
Durchlässigkeit *f*   *DICHT* permeability
Durchlaßquerschnitt *m*   *THEOR* passage area
Durchlaufregelung *f*   *KOMPR* on-line/off-line control
durchströmen   z. B. *Öl durchströmt eine Öffnung, strömt durch eine Öffnung* pass *v*   eg *oil passes a restriction, through a restriction*, flow *v* eg *through a restriction*
durchströmen lassen   z. B. *Ventil läßt Öl zum Verbraucher strömen* pass *v* eg *valve passes fluid to the actuator*
Durchtriebausführung *f*   *PU/MOT* through-shaft configuration
Düse *f*   *THEOR* nozzle; *des Düse-Prallplatten-Ventils* *STELL* nozzle *of the flapper-and-nozzle valve*
Düsendurchflußmesser *m*   flow nozzle
Düseneinschnürung *f*   *THEOR* nozzle throat
Düsengeräusch *n*   nozzle noise
Düsengleichung *f*   *THEOR* nozzle equation
Düsenhals *m*   *THEOR* nozzle throat
Düsenlärm *m*   nozzle noise
Düsennebelöler *m*   venturi-type lubricator
Düsenverengung *f*   *THEOR* nozzle throat
Düse-Prallplatten-Ventil *n*   *STELL* flapper[-and-nozzle] valve
DVI *m*   *Dichtungsverträglichkeitsindex* *FLÜSS* seal compatibility index
Dynamik *f* **flüssiger und gasförmiger Körper**   fluid dynamics
dynamische Dichtung *f*   dynamic seal
dynamische Flüssigkeitsprobenahme *f*   dynamic fluid sampling
dynamische Viskosität *f*   *FLÜSS* dynamic (absolute) viscosity
dynamischer Druck *m*   *Staudruck THEOR* dynamic pressure
dynamisches Element *n*   *LOGIKEL* fluidic

E

E-Ring *m*   DICHT E-ring
Eckleistung *f*   PU/MOT corner horsepower, CHP
Eck-Rückschlagventil *n*   angle check valve
Eckventil *n*   WEGEV angle valve
effektiver Förderstrom *m*   PU/MOT actual pump output (delivery, discharge rate, flow rate)
effektiver Schluckstrom *m*   PU/MOT actual motor input (inlet, intake) flow rate
Effektivdruck *m*   mean effective pressure, MEP
Eichflüssigkeit *f*   calibration fluid
eigengesteuertes Ventil *n*   STELL internally piloted valve
Eigensteuerdruck *m*   STELL internal pilot pressure
Eigensteuerung *f*   STELL internal piloting
Eilgangdruckübersetzer *m*   double-pressure intensifier
Eilgangkolben *m*   ZYL rapid (quick, fast) traverse piston
Eilgangpumpe *f*   rapid approach (advance, traverse) pump
Eilgangschaltung *f*   mit Differentialzylinder regenerative (differential) circuit
Eilgangstellung   WEGEV regenerative position
Eilhub *m*   rapid (quick, fast) stroke
Eilrückgang *m*   rapid (quick, fast) return
Eilrückhub *m*   ZYL rapid (quick, fast) return stroke
Eilrücklauf *m*, **Eilrückzug** *m*   rapid (quick, fast) return
Eilvorlaufschaltung *f*   mit Differentialzylinder regenerative (differential) circuit
Eilvorschub *m*   rapid (quick, fast) feed
Einbau *m* im Nebenstrom   bypass installation
Einbaudichtsatz *m*, **Einbaudichtung** *f*   seal cartridge
Einbauflanschventil *n*   flanged cartridge valve
Einbaulänge *f*   ZYL retracted (collapsed) length
Einbaupumpe *f*   cartridge (plug-in) pump
Einbauraum *m*   seal cavity (pocket)
Einbaurückschlagventil *n*   cartridge check [valve]
Einbaustromventil *n*   cartridge flow valve, flow (restrictor) cartridge
Einbauventil *n*   für Bohrungseinbau cartridge [insert] valve; s. Wegesitzventil *n*
eindringen lassen   z. B. Schmutzteilchen ingress *v*   eg particles
Einebenen-Rohrgelenk *n*   single-plane swivel joint
einfachwirkende Flügelzellenpumpe *f*   mit Exzenterring excentric cam ring-type vane pump
einfachwirkende Handpumpe *f*   single-acting hand pump
einfachwirkender Druckübersetzer *m*   single-acting intensifier (booster)
einfachwirkender Zylinder *m*   single-acting cylinder
einfahren *vi und vt*   z. B. Kolben retract *v*, withdraw *v*, draw-back *v*   eg the piston
Einfahrgeschwindigkeit *f*   ZYL retrac-

Einfahrhub

tion (withdrawal, draw-back, return) speed
Einfahrhub *m*   *ZYL* return (in, inward, retract, retraction, withdrawal, draw-back) stroke
Einfahrlänge *f*   *ZYL* retracted (collapsed) length
Einfahrstellung *f*   *ZYL* retracted position
Einflügelschwenkmotor *m*   single-vane rotary actuator
einflutige Schraubenpumpe *f*   single-suction (-inlet) screw pump
Einfüll-Belüftungs-Kombination *f*   *BEHÄLT* filler-breather [assembly], breather/filler
Einfüllfilter *m,n*   *BEHÄLT* filler-strainer
Einfüllöffnung   *BEHÄLT* filler [opening], refill opening
Eingang *m*   input, inlet, intake, inflow
Eingangsdruck *m*   input (inlet) pressure, *also, at a pump* suction pressure, *also, at a motor or actuator* supply (feed) pressure
Eingangskanal *m*   input (inlet) channel, *also, in a pump* suction channel, *also, in a motor* pressure channel
Eingangsleistung *f*   input horsepower (power), driving horsepower
Eingangsleitung *f*   input (inlet, intake) line, *also, to a pump* suction line, *also, to a motor or actuator* supply (feed, pressure) line
Eingangsseite *f*   input (inlet, intake) side, *also, of a pump* suction side, *also, of a motor* pressure side
Eingangsstrahl *m*   *LOGIKEL* input jet
Eingangs[volumen]strom *m*   input (inlet, intake) flow [rate],   *also, into a pump* suction flow [rate]
Eingangszahl *f*   *LOGIKEL* fan-in
eingebaute Verschmutzung *f*   *FLÜSS* built-in contamination
eingebautes Rückschlagventil *n*   integral (built-in) check valve
eingedrungene Verschmutzung *f*   *FLÜSS* ingressed contamination
eingefaßte Dichtung *f*   cased (captive) seal
eingeprägter Druck *m*   *KREISL* impressed pressure
eingeprägter Volumenstrom *m*   *KREISL* impressed flow [rate]
Eingriff *m   in einen automatischen Ablauf* override, emergency control
Eingriffsleckverlust *m*   *PU/MOT* [tooth-]mesh leakage
Einheit *f s eg* Radialkolbeneinheit *f*
Einkantenfühler *m*   *STELL* spool-type servovalve with one metering orifice
Einkantenschieber   *WEGEV* valve spool with one metering edge
Einkreishydraulik *f*   *s.* Einkreissystem *n*
Einkreissystem *n*   separate circuits (single circuit) system
Einkugelventil *n*   single-ball valve
Einlaß *m*   input, inlet, intake, inflow
Einlaßanschluß *m*   input (inlet, intake, supply) port, *also, of a pump* suction port, *also, of a motor* pressure port
Einlaßdruck *m*   input (inlet) pressure, *also, at a pump* suction pressure, *also, at a motor or actuator* supply (feed) pressure
Einlaßkammer *f*   input (inlet, intake) chamber, *also, in a pump* suction

chamber, *also, in a motor* pressure chamber
Einlaßkanal *m*   input (inlet) channel, *also, in a pump* suction channel, *also, in a motor* pressure channel
Einlaßleitung *f*   input (inlet, intake) line, *also, to a pump* suction line, *also, to a motor or actuator* supply (feed, pressure) line
Einlaßseite *f*   input (inlet, intake) side, *also, of a pump* suction side, *also, of a motor* pressure side
Einlaßstrom *m*   s. Einlaßvolumenstrom *m*
Einlaßventil *n*   PU/MOT suction (inlet) valve
Einlaßvolumenstrom *m*   input (inlet, intake) flow [rate], *also, into a pump* suction flow [rate]
Einlaufverlust *m*   THEOR entrance loss
Einlippenabstreifer *m*   ZYL single-lip wiper
Einlippenring *m*   single-lip seal
Einmembranelement   LOGIKEL single-diaphragm element
einregeln   *z. B. Strom, Druck* set *v*, meter *v*   *eg flow, pressure*
Einsatz *m*   FILTER filter cartridge
Einschaltstrom *m*   STELL inrush current
einsaugen   *durch Leckstellen* ingest *v*, draw in *v*
Einschnürungsstelle *f*   THEOR vena contracta   (*pl* venae contractae)
Einschraubenpumpe *f*   single-screw pump
Einschraubventil *n*   screw-in cartridge valve
Einschraubverschraubung *f*   male connector, male (plug) end fitting

Einschraubwinkelverschraubung *f*   male (street) elbow
Einschwenkcharakteristik *f*   *einer Nullhubpumpe* destroking characteristics   *of a pressure-compensated pump*
einschwenken   *Verstellpumpe* destroke *v*, move *v*, off stroke   *variable pump*
einseitige Kolbenstange *f*   s. Einstangenzylinder *m*
Einspeisedruckbegrenzungsventil *n*   *in einem geschlossenen Kreislauf* charge pressure relief valve   *in a closed circuit*
einspritzgekühlter Verdichter *m*   injected compressor
Einspritzöler *m*   AUFBER oil injector
Einspritzpumpe *f*   injection (injector) pump, injector
Einspritzschmierung *f*   injection lubrication
Einstangenzylinder *m*   single-rod [end] cylinder
Einsteckschweißwinkel *m*   socket welding elbow
Einsteckverbindung *f*   plug-in (push-in) fitting (connection)
einstellbare Drossel *f*   STROMV variable (adjustable) restrictor (restriction)
einstellbare Endlagenbremse (Hubendebremse) *f*   ZYL adjustable cushion
einstellbares Druckbegrenzungsventil *n*   pressure-adjustment (adjustable) relief valve
Einstelldruck *m*   preset (set) pressure
einstellen   *z. B. Strom, Druck* set *v*, meter *v*   *eg flow, pressure*
Einstrangabsperrkupplung *f*

Einstrangabsperrung *VERBIND* one-way seal (single shut-off, single-valved) coupling
Einstrangabsperrung *f* *VERBIND* one-way (single) shut-off
Einstrangsitzkupplung *f* *VERBIND* single poppet coupling
einströmige Pumpe *f* single[-flow] pump
Einstrompumpe *f* *s.* einströmige Pumpe *f*
einstufiger Verdichter *m* single-stage compressor
einstufiger Zylinder *m* single-stage cylinder
einstufiges Servoventil *n* single-stage servovalve
einstufiges Ventil *n* direct[-acting] (one-stage) valve
Eintauchthermometer *n* immersion thermometer
Eintritt *m* input, inlet, intake, inflow
Eintrittsanschluß *m* input (inlet, intake, supply) port, *also, of a pump* suction port, *also, of a motor* pressure port
Eintrittsdeckel *m* *einer Ventilbatterie* inlet (intake) section *of ganged valves*
Eintrittsdruck *m* input (inlet) pressure, *also, at a pump* suction pressure, *also, at a motor or actuator* supply (feed) pressure
Eintrittskammer *f* input (inlet, intake) chamber, *also, in a pump* suction chamber, *also, in a motor* pressure chamber
Eintrittskanal *m* input (inlet) channel, *also, in a pump* suction channel, *also, in a motor* pressure channel
Eintrittsleitung *f* input (inlet, intake) line, *also, to a pump* suction line, *also, to a motor or actuator* supply (feed, pressure) line
Eintrittsseite *f* input (inlet, intake) side, *also, of a pump* suction side, *also, of a motor* pressure side
Eintrittsstrom *m* *s.* Eintrittsvolumenstrom *m*
Eintrittsverlust *m* *THEOR* entrance loss
Eintrittsvolumenstrom *m* input (inlet, intake) flow [rate], *also, into a pump* suction flow [rate]
Einwegabsperrkupplung *f* *VERBIND* one-way seal (single shut-off, single-valved) coupling
Einwegabsperrung *f* *VERBIND* one-way (single) shut-off
Einwegdrossel *f* return-orifice (metering, orifice) check valve
Einzelanschlußplatte *f* [single-valve] manifold, [single-station] subplate, subbase
Einzelhubdruckübersetzer *m* single-shot intensifier (booster)
Einzelhydraulik *f* separate circuits (single circuit) system
Einzelunterplatte *f* [single-valve] manifold, [single-station] subplate, subbase
Eisabstreifer *m* *ZYL* ice scraper
elastische Dichtung *f* flexible (elastic, resilient) seal
Elastizitätsmodul *m* *FLÜSS* bulk modulus [of elasticity]; *s also* Sekantenkompressionsmodul *m*
Elastomerdichtung *f* elastomeric seal
elektrisch abschaltbares Druckbegrenzungsventil *n* solenoid vented relief valve

elektrisch-fluidischer Wandler *m*
MESS electrical-to-fluid transducer
elektrisch-hydraulischer Wandler *m*
MESS electrohydraulic transducer
elektrisch-pneumatischer Wandler *m*
MESS electropneumatic transducer
**elektrische Betätigungseinrichtung
(Stelleinheit)** *f STELL* electric operator (actuator, control), *PU/MOT*
also electric controller (stroker)
elektrischer Druckwandler *m MESS*
electrical pressure transducer
Elektrohydraulik *f* electrohydraulics
elektrohydraulisch electrohydraulic
elektrohydraulische Achse *f eines
Roboters ANWEND* electrohydraulic
axis *of a robot*
elektrohydraulische Servo-Stelleinheit *f PU/MOT* solenoid (electrohydraulic) control (controller, stroker)
elektrohydraulischer Schrittmotor *m*
electrohydraulic pulse motor, EHPM
elektrohydraulischer Wandler *m*
MESS electrohydraulic transducer
elektrohydraulisches Servoventil *n*
electrohydraulic servovalve
elektromagnetisch betätigtes Ventil *n*
solenoid-operated (-actuated,
-controlled) valve, solenoid valve
elektromagnetische Betätigungseinrichtung (Stelleinheit) *f STELL*
solenoid operator (actuator)
elektromagnetischer Druckwandler *m*
induction (inductive, electromagnetic)
pressure transducer
elektromechanische Betätigungseinrichtung (Stelleinheit) *f STELL* electromechanical operator (actuator, control)
elektromotorbetätigtes Ventil *n*
electromotor-operated (motorised)
valve

elektronische Stelleinheit *f PU/MOT*
electronic control (controller, stroker)
Elektropneumatik *f* electropneumatics
elektropneumatisch electropneumatic
elektropneumatischer Wandler *m*
electropneumatic transducer
elektrorheologische Flüssigkeit *f mit
elektrisch beeinflußbarer Viskosität*
electro-rheological fluid, ER fluid
*with electrostatically controlled
viscosity*
elektrostatische Filterung *f* electrostatic filtration
elektrostatischer Flüssigkeitsreiniger
m electrostatic liquid cleaner, ELC
Element *n FILTER* filter element;
LOGIKEL fluidic; *s.* Hydraulikelement *n*; *s.* Pneumatikelement *n*
Element *n* **mit bewegten Teilen**
LOGIKEL moving-part element
Element *n* **ohne bewegte Teile, reinfluidisches Element** *n LOGIKEL*
pure fluid (non-moving part)
element
Emulgator *m FLÜSS* emulsifier,
emulsifying agent
emulgierbar *FLÜSS* emulsible, emulsifiable
Emulgierbarkeit *f FLÜSS* emulsibility, emulsifiability
emulgieren *FLÜSS* emulsify *v*
Emulgierung *f FLÜSS* emulsification
Emulgierwiderstand *m FLÜSS* emulsification resistance, resistance to
emulsification
Emulsion *f FLÜSS* emulsion
Emulsionsbeständigkeit *f FLÜSS*
emulsion stability
Emulsionsbildner *m FLÜSS* emulsifier, emulsifying agent

Emulsionsbildung *f* *FLÜSS* emulsification
Ende *n* *s.* Zylinderende *n*
Endlage *f* *ZYL* end-of-stroke position
Endlage *f* **vollständig erreichen** *ZYL* bottom *v* out
Endlagenbremse *f* *ZYL* cylinder cushion
Endlagenbremse *f* **mit konstanter Verzögerung** *ZYL* constant-deceleration cushion
Endlagenbremsung (Endlagendämpfung) *f* *ZYL* cushioning
Endlagenschaltventil *n* *WEGEV* limit valve
Endlagenverriegelung *f* *ZYL* end-of-stroke locking
Endstück *n* end fitting (connection), connector
Energieübertragungsmittel *n* power transmitting agent, *cf* Hydraulikflüssigkeit *f*
Engler-Grad *n* *veraltete Einheit der kinematischen Viskosität* degree Engler *conventional unit of kinematic viscosity*
Engler-Viskosimeter *n* *ein Auslaufviskosimeter* Engler's viscometer *efflux-type*
Enteisungsschmierstoff *m* *AUFBER* anti-freeze lubricant
entflammbar *FLÜSS* [in]flammable
Entflammbarkeit *f* *FLÜSS* [in]flammability
Entflammbarkeitsprüfung *f* *FLÜSS* [in]flammability (ignition) test
Entflammbarkeitsprüfung *f* **an heißer Fläche** *FLÜSS* manifold ignition test
entgasen *Luft ausscheiden FLÜSS* release *v* air, outgas *v*

entkuppeln *Schlauchkupplung* disconnect *v* hose coupling
Entladung *f* *SPEICH* discharge
entlasten *vom Druck* pressure-compensate (-balance) *v*; *mit dem Behälter verbinden* vent *v* [to tank], exhaust *v*, dump *v*, release *v*, relieve *v*, unload *v*
entlastete Pumpe *f* pressure-compensated (-balanced) pump, compensated (balanced) pump
Entlastungsanschluß *m* *WEGEV* relief port (connection); *ZYL* bypass port
Entlastungsleitung *f* *VENTILE* balance (compensation) line
Entlastungsregelung *f* *KOMPR* unloader control
Entlastungsventil *n* *DRUCKV* unloading valve
entleeren *BEHÄLT* drain *v*
Entleerung *f* *BEHÄLT* drainage; *SPEICH* discharge
Entleerungsöffnung *f* *BEHÄLT* drain [opening]
Entleerungsschraube *BEHÄLT* drain plug
Entleerungsventil *n* drain valve
entlüften *Abluft ableiten* bleed *v*, vent *v* exhaust air; *ungelöste Luft entfernen* deaerate *v*, bleed *v*, vent *v* entrained air; *BEHÄLT* breathe *v*, ventilate *v*, aerate *v*
entlüftend: nach außen entlüftend *PN* open to atmosphere
Entlüfter *m* *für ungelöste Luft* deaerator, [air] bleeder, vent *for entrained air*
Entlüftung *f* *Entfernung ungelöster Luft* deaeration, bleeding, venting *of entrained air*; *BEHÄLT*

breathing, ventilation, aeration; *s also* Entlüfter *m*

Entlüftungsblech *n* BEHÄLT air-detrainer plate

Entlüftungsbohrung *f* *für ungelöste Luft* bleeder hole (port), vent port

Entlüftungsdrossel *f* speed-control muffler, exhaust-speed controller, bleed (exhaust) throttle

Entlüftungsfilter *m,n* BEHÄLT air breather filter

Entlüftungsleitung *f* bleed (venting, vent) line

Entlüftungsöffnung *f* BEHÄLT breather hole, ventilating eyelet, aeration opening; *für ungelöste Luft* bleeder hole (port), vent port

Entlüftungsorgan *n* BEHÄLT air breather

Entlüftungsschraube *f* *für ungelöste Luft* bleeder (venting) screw *for entrained air*

Entlüftungssteuerung *f* LOGIKEL bleed-off system

Entlüftungsventil *n* deaerator (bleeding, venting) valve

Entnahmestelle *f* tap, tapping (take-off) point

Entnahmeventil *n* *s.* Probenahmeventil *n*

entölen *Druckluft* remove *v* the oil from the air

Entöler *m* AUFBER oil remover (scrubber)

entregt *Elektromagnet* STELL de-energized *solenoid*

Entschäumadditiv *n* anti-foaming additive, defoamer, foam inhibitor (depressant)

entschäumen FLÜSS defoam *v*

Entspannung *f* allmählicher Druck-abbau decompression; *Druckentlastung* pressure relief (unloading)

Entspannungsarbeit *f* THEOR expansion work

Entspannungsdruckstoß *m* decompression pressure surge (shock)

Entspannungsschlag *m* *s.* Entspannungsdruckstoß *m*

Entspannungsventil *n* DRUCKV decompression valve

Entspannungszyklus *m* **mit Vorentlastung** two-step decompression cycle

entsperrbares Rückschlagventil *n* pilot[-operated] check valve, pilot check

entsperren SPERRV open *v*

Entwässerung *f* drain (line, pipe) trap

Entwässerungsfilter *m,n* water-removing filter

Entwässerungsleitung *f* drain leg

entzündbar FLÜSS [in]flammable

Entzündbarkeit *f* FLÜSS [in]flammability

Entzündungsprüfung *f* FLÜSS [in]flammability (ignition) test

EP-Additiv *n* FLÜSS extreme-pressure (EP) additive

EP-Wandler *m* *elektrisch-pneumatisch* electropneumatic transducer

ER-Flüssigkeit *f* *elektrorheologische Flüssigkeit: mit elektrisch beeinflußbarer Viskosität* electro-rheological fluid, ER fluid *with electrostatically controlled viscosity*

erforderliche Saughöhe *f* PU/MOT net positive suction head, NPSH

Ergänzungsflüssigkeit *f* make-up (replenishing) fluid

Ergänzungsstrom *m* replenishment flow [rate]

erregen *einen Elektromagneten* STELL energize v a solenoid
erzeugte Verschmutzung *f* FLÜSS generated contamination
Erweiterungsverlust *m* THEOR enlargement loss
esterbasische Flüssigkeit *f* ester-base fluid
Eulersche Gleichung *f* *Bewegung von Flüssigkeiten* Euler's (Eulerian) equation
evakuieren evacuate v
Expansionsarbeit *f* expansion work
Expansionsthermometer *n* expansion thermometer
Expansionsviskosität *f* FLÜSS expansional viscosity
Expansionswelle *f* expansion wave
extern vorgesteuertes Ventil *n* STELL externally piloted valve
externe Vorsteuerung *f* STELL external piloting
Extrusion *f* DICHT extrusion
extrusionsfest DICHT non-extruding, extrusion-resistant
Extrusionsfestigkeit *f* DICHT extrusion resistance
Exzentermaschine *f* centrally-guided radial piston pump
Exzenterring-Flügelzellenpumpe *f* excentric cam ring-type vane pump
exzentrisch liegender Kolben *m* WEGEV uncentered (off-centered) spool
EY-Wandler *m* *elektrisch-hydraulisch* MESS electrohydraulic transducer

F

4fach-spiralarmierter Schlauch *m* 4-spiral wire wrap hose
Fahrhydraulik *f* ANWEND propel (drive) hydraulic circuit
Fahrzeughydraulik *f* ANWEND automotive hydraulics; mobile hydraulics
Fallbehälter *m* overhead reservoir (tank)
Fallkörper *m* falling (dropping) body
Fallkugel *f* falling (dropping) ball
Fallviskosimeter *n* falling-body viscometer
Faltenbalg *m* DICHT bellows, [accordion] boot, gaiter
Faltenbalgspeicher *m* bellows accumulator
Fangdruck *m* LOGIKEL collection pressure
Fangdüse *f* LOGIKEL collector [tube], output tube
Fangraum *m* LOGIKEL collection chamber
Fangrohr *n* s. Fangdüse *f*
Fan-in *m* LOGIKEL fan-in
Fan-out *m* LOGIKEL fan-out
Faser[stoff]filter *m,n* fiber filter
federbelasteter Speicher *m* spring-loaded accumulator
federbelastetes Ventil *n* spring-loaded (-biased) valve
federfixiertes Ventil *n* spring-return valve
federgeöffnetes Ventil *n* spring-offset valve
federgespannte Dichtung *f* spring-energized seal
Federkammer *f* STELL spring pocket (cavity)

federloses Rückschlagventil *n* no-spring check [valve]
Federraum *m* STELL s. Federkammer *f*
federrückgezogener Kolben *m* ZYL spring-return piston
Federrückzug *m* ZYL spring return
federzentriertes Ventil *n* spring-centered valve
Federzentrierung *f* STELL spring centering
fehlender Ausgleich *m* unbalance, imbalance
Feinfilter *m,n* fine filter, *s also* Feinstfilter *m,n*
Feinfilterung *f* fine filtration, *s also* Feinstfilterung *f*
feinmaschig FILTER fine-meshed
Feinmeßmanometer *n* high-grade (precision) pressure gauge
Feinstelldrossel *f* STROMV precision restriction (throttle, choke)
Feinsteuernut *f* PU/MOT *groove decompressing captive oil in crossover position*; zur Verminderung von Schaltdruckspitzen WEGEV *metering notch to reduce crossover pressure peaks*
Feinstfilter *m,n* micronic filter
Feinstfilterung *f* micronic filtration, superfiltration
fernbetätigte Kupplung *f* remote-type coupling
ferngesteuertes Wegeventil *n* remote-operated (-actuated, -controlled) directional valve
Ferngetriebe *n* *Getriebe in aufgelöster Bauweise* split (non-integral, separate pump/motor) transmission
Fernmanometer *n* remote (distance) pressure gauge

fernvorgesteuert STELL remote-piloted, remote pilot-operated (-actuated)
Fernvorsteuerung *f* STELL remote pilot actuation (piloting)
Fernvorsteuerventil *n* remote pilot-control valve
Festdrossel *f* STROMV fixed (non-adjustable) restrictor (restriction)
festes Leitungsende *n* THEOR closed termination, closed (stiff, dead) end
Festflügelmotor *m* s. Sperrschiebermotor *m* oder Rollflügelmotor *m*
Festhälfte *f* *einer Schlauchkupplung:* coupling half attached to rigid piping or component
Festschmutz *m* FLÜSS solid contamination
Feststelleinheit *f* *Feststellung in beliebiger Hubposition* ZYL locking device *at any midstroke position*
Feststoffverschmutzung *f* FLÜSS solid contamination
fetten *Druckwasser* lubricate *v* compressed water
Feuchtluft *f* AUFBER wet air
Filter *m,n* allgemein filter; *eher als Oberflächenfilter, für weniger kleine Schmutzteilchen* strainer; *eher als Tiefenfilter, für kleine Schmutzteilchen* filter
Filter *m,n* **für Behälteranbau (Behälteraufbau)** reservoir (tank) [top] mount filter
Filter *m,n* **für Rohrleitungseinbau** in-line (line) filter
Filter *m,n* **im Abzweig zum Behälter** bleed-off filter
Filter *m,n* **in L-Gehäuseausführung** L-type (-ported) filter

Filteranordnung 174

Filter *m,n* **in T-Gehäuseausführung** T-type (-ported) filter
Filter *m,n* **mit automatischer Umgehung** self-bypassing filter
Filter *m,n* **mit gefilterter Umgehung** filtered-bypass filter
Filter *m,n* **mit Umgehung** *m,n* [integral] bypass filter
Filter *m,n* **mit Verstopfungsalarm** *m* self-warning filter
Filter *m,n* **ohne Umgehungsmöglichkeit** non-bypass filter
Filteranordnung *f*   *im Kreislauf* filter location *in the system*
filterbar filt[e]rable
Filterbatterie *f*   filter assembly
Filterbypassfilter *m,n*   *s.* Filter *m,n* mit gefilterter Umgehung
Filterdruckabfall *m*   filter pressure drop
Filter-Druckminderventil *n*   filter-regulator
Filterdruckverlust *m*   filter [pressure] loss
Filtereinsatz *m*   filter cartridge
Filterelement *n*   filter element
filterfähig filt[e]rable
Filterfeinheit *f*   filtration (filter) rating (fineness)
Filterfläche *f*   filtration (filtering) area, filter surface
Filtergehäuse *n*   filter housing (body)
Filtergrad *m*   filtration ratio, *cf ED* beta ratio
Filterkorb *m*   bag-type strainer
Filterkörper *m*   filter housing (body)
Filterlamelle *f*   filter disk (plate, washer)
Filtermittel *n*   filter[ing] material (media, medium)
filtern   *auf X µm* filter *v* *to X µm;*

mit Oberflächenfilter screen *v,* strain *v*
Filterpatrone *f*   filter cartridge
Filterplatte *f*   *s.* Filterscheibe *f*
Filter-Reduzierventil *n*   filter-regulator
Filter-Schalldämpfer *m*   muffler-reclassifyer
Filterscheibe *f*   filter disk (plate, washer)
Filterspalt *m*   filter[ing] gap
Filterstandzeit *f*   filter life
Filterung *f*   filtering, filtration
Filterungsgrad *m*   filtration ratio, *cf ED* beta ratio
Filterverlust *m*   filter [pressure] loss
Filterwartung *f*   filter maintenance
Filterwehr *n*   *BEHÄLT* dirt dam
Filterwirkungsgrad *m*   filter efficiency; *s also ED* beta ratio
Filtration *f*   filtering, filtration
filtrationsfähig fit[e]rable
Filtrationsquotient *m*   *s.* Filterungsgrad *m*
filtrieren   *auf X µm* filter *v* *to X µm; mit Oberflächenfilter* screen *v,* strain *v*
Filtrierung *f*   *s.* Filtration *f*
Filzfilter *m,n*   felt filter
Fitting *m*   *s.* Rohrverschraubung *f*
Flachdichtung *f*   gasket
Flachdrehschieber *m*   *WEGEV* rotary plate (disk)
Flachdrehschieberventil *n*   *WEGEV* rotary-plate (-disk) valve
flächengesteuerte Pumpe *f*   port (valve) plate controlled pump
flächenmontiert   *VENTILE* surface- (gasket-) mounted, *cf* unterflächenmontiert
Flächenverhältnis *n*   *ZYL* area ratio

Flachschieberventil *n* WEGEV flat-slide (plate) valve
Flachzylinder *m* oval cylinder
flammbeständig FLÜSS non-[in]flammable, non-flam, uninflammable, flame-proof
Flammbeständigkeit *f* FLÜSS non-[in]flammability, uninflammability, flameproofness
Flammpunkt *m* FLÜSS flash point
Flanschanschluß *m* flange[d] port
Flanschbefestigung *f* flange[d] mount
Flansch-Einbauventil *n* flanged cartridge valve
Flanschhälfte *f* flange half
Flanschpumpe *f* flange-mount (flanged) pump
Flanschverbindung *f* flange connection
flattern *z. B. Ventile* chatter *v*, flutter *v* *eg valves*
Fleckvergleichsprüfung *f* *des Schmutzgehalts des Öls* patch test *of oil contamination*
flexible Leitung *f* flexible line
fliegender Kolben *m* SPEICH floating (free) piston
fließen *z. B. durch eine Öffnung* pass *v* *eg a restriction, through a restriction,* flow *v* *eg through a restriction;* *im Nebenschluß fließen* bypass *v*
Fließpunkt *m* *als Pourpoint angegeben* FLÜSS pour (flow) temperature
Fließpunkterniedriger *m* FLÜSS pour-point depressant
Fließvermögen *n* *Kehrwert der Viskosität* FLÜSS fluidity *reciprocal of viscosity*
Flockungsmittel *n* FLÜSS flocculating agent, flocculant

Flügel *m* PU/MOT vane
Flügelanpreßkraft *f* PU/MOT vane tracking force
Flügelanzahl *f* PU/MOT number of vanes, vane number
Flügelhub *m* PU/MOT vane throw
Flügelkopf *m* s. Flügelkuppe *f*
Flügelkörper *m* PU/MOT vane block
Flügelkuppe *f* PU/MOT vane tip
Flügelleitkurve *f* PU/MOT vane track
Flügelrad-Durchflußmesser *m* turbine (propeller) flowmeter
Flügelschlitz *m* PU/MOT vane slot
Flügelschwenkmotor *m* vane rotary actuator
Flügelträger *m* PU/MOT vane block
Flügelzahl *f* PU/MOT number of vanes, vane number
Flügelzellengetriebe *n* vane transmission
Flügelzellenmaschine *f* vane pump *or* motor
Flügelzellenmotor *m* [sliding-]vane motor
Flügelzellenpumpe *f* [sliding-]vane pump
Flügelzellenpumpe *f* mit Doppelflügeln vane-within-a-vane-type pump
Flügelzellenpumpe *f* mit druckentlastetem Rotor balanced-rotor (pressure-compensated) vane pump
Flügelzellenpumpentest *m* *der Verschleißschutzeigenschaften* FLÜSS Vickers (vane) pump test *of antiwear properties*
Flugkolben *m* SPEICH floating (free) piston
Flugzeugflüssigkeit *f* aircraft fluid
Flugzeughydraulik *f* aircraft hydraulics

Fluid n (*pl:* -e *oder* -s) fluid, *compounds s with* hydraulisch, Hydraulik ... , Hydro ... , fluidtechnisch, Fluidtechnik ... , fluidisch, Flüssigkeits ... , Öl ... ; *s.* Arbeitsflüssigkeit *f*
Fluid n: aus dem Fluid stammend fluidborne
Fluid n: im Fluid enthalten fluidborne
Fluidadmittanz *Volumenstrom/Druckabfall THEOR* fluid admittance *flow rate/pressure drop*
Fluiddynamik *f* *THEOR* fluid dynamics
Fluidic n *LOGIKEL* fluidic
Fluidik *f* fluid logic, fluidics; *meist im engeren Sinn: Einsatz von Logikelementen ohne bewegte Teile* fluidics *as the application of no-moving part elements*
Fluidikelement n *s.* Fluidic *n*
Fluidimpedanz *f* *Druckabfall/Volumenstrom THEOR* fluid impedance *pressure drop/flow*
Fluidinduktivität *f* *THEOR* fluid inductivity
Fluidingenieur m fluid power engineer (designer)
fluidisch *LOGIKEL* fluidic; *sometimes used in the sense of* fluidtechnisch
fluidisch betätigt *STELL* fluidic-operated (-actuated, -controlled)
fluidisch-elektrischer Wandler fluidic-to-electrical transducer
fluidische Admittanz *s.* Fluidadmittanz *f*
fluidische Digitaltechnik *f* *LOGIKEL* fluid logic, fluidics
fluidische Impedanz *f* *s.* Fluidimpedanz *f*
fluidische Induktivität *f* *s.* Fluidinduktivität *f*
fluidische Kapazität *f* *s.* Fluidkapazität *f*
fluidische Konduktanz *f* *s.* Fluidkonduktanz *f*
fluidischer Leitwert m *s.* Fluidleitwert *m*
fluidischer Oszillator m *LOGIKEL* fluidic oscillator
fluidischer Widerstand m *s.* Fluidwiderstand *m*
fluidischer Zeitgeber m *LOGIKEL* fluidic timer
Fluidität *f* *Kehrwert der Viskosität FLÜSS* fluidity *reciprocal of viscosity*
Fluidkapazität *f* *THEOR* fluid capacitance
Fluidkonduktanz *f,* Fluidleitwert m *reeller Leitwert THEOR* fluid conductance
Fluidlogik *f* *LOGIKEL* fluid logic, fluidics
Fluidmechanik *f* fluid mechanics
Fluidstatik *f* fluid statics
Fluidtechnik *f* fluid power technology (engineering), fluid power, *s also* Hydraulik *f*, Pneumatik *f*, *compounds s also with* hydraulisch, Hydraulik ... , Hydro ... , fluidtechnisch, Flüssigkeits ... , Öl ...
Fluidtechniker m fluid power engineer (designer)
Fluidtechnikhersteller m manufacturer of fluid power equipment
Fluidtechnik-Instandhaltungsingenieur m fluid power maintainer
fluidtechnisch fluid power ... , *s also*

hydraulisch, pneumatisch, *compounds s also with* hydraulisch, Hydraulik ... , Hydro ... , fluidtechnisch, Flüssigkeits ... , Öl ...
fluidtechnisch betätigt *STELL* fluid power-operated (-actuated, -controlled)
fluidtechnische Anlage *f* fluid power system, hydraulic *or* pneumatic system
fluidtechnischer Antrieb *m* fluid drive, hydraulic *or* pneumatic drive
fluidtechnischer Kreislauf *m* fluid power circuit (system), hydraulic *or* pneumatic circuit (system)
fluidtechnisches Bauelement *n* (**Gerät** *n*) fluid power component (element), hydraulic *or* pneumatic component (element)
Fluidwiderstand *m* ohmscher Widerstand, Eigenschaft *THEOR* fluid resistance
Fluktuation *f* *des Förderstroms* *PU/MOT* output pulsation (ripple, fluctuation), pump pulsation (ripple)
Flüssigdichtstoff *m* liquid sealant
flüssige Reibung *f* *THEOR* viscous (fluid) friction
Flüssigkeit *f* liquid, *most frequently in fluid power technology:* fluid; *Kehrwert der Viskosität FLÜSS* fluidity *reciprocal of viscosity*
Flüssigkeit *f*: aus der Flüssigkeit stammend fluidborn
Flüssigkeit *f*: in der Flüssigkeit enthalten fluidborn
Flüssigkeit *f* auf Basis chlorierter Kohlenwasserstoffe chlorinated hydrocarbon fluid
Flüssigkeit *f* auf Basis halogenierter Kohlenwasserstoffe halogenated hydrocarbon fluid
Flüssigkeit *f* auf Basis von Dicarbonsäureestern di-basic acid ester fluid
Flüssigkeit *f* auf Basis von Pflanzenölen vegetable-oil base fluid
Flüssigkeit *f* auf Basis von Polyglycolestern polyglycol ester fluid
Flüssigkeit *f* auf Basis von Siliconölen silicone fluid
Flüssigkeit *f* auf Esterbasis ester-base fluid
Flüssigkeit *f* auf Kohlenwasserstoffbasis hydrocarbon-base fluid
Flüssigkeit *f* auf Mineralölbasis petroleum[-based] fluid, mineral-oil base fluid
Flüssigkeit *f* auf Ölbasis oil-base fluid
Flüssigkeit *f* auf Phosphatesterbasis phosphate ester base fluid
Flüssigkeit *f* auf Rizinusbasis castor-base fluid
Flüssigkeit *f* auf Wasserbasis aqueous (water-base) fluid
Flüssigkeit *f* unter Druck pressurized (pressure) fluid, fluid under pressure
Flüssigkeitsabnutzung *f* fluid wear
Flüssigkeitsanschluß *m* *SPEICH* hydraulic fluid port (connection)
Flüssigkeitsaufbereitung *f* fluid maintenance (conditioning)
Flüssigkeitsbehälter *m* fluid (hydraulic) reservoir (tank)
Flüssigkeitsbremse *f* *ANWEND* hydraulic (oil) brake
Flüssigkeitsdämpfer *m* fluid absorber, viscous damper
flüssigkeitsdicht fluid-tight, leaktight
Flüssigkeitsfeder *f* liquid spring

Flüssigkeitsfilter *m,n* hydraulic filter
Flüssigkeitsgebrauchsdauer *f,*
Flüssigkeitsgrenznutzungsdauer *f* fluid life
Flüssigkeitskörper *m* bulk liquid
Flüssigkeitskupplung *f* ANWEND hydrodynamic (hydraulic, fluid) coupling (clutch)
Flüssigkeitsmechanik *f* hydromechanics
flüssigkeitsmechanisch, hydromechanisch hydromechanic
Flüssigkeitspflege *f* fluid maintenance (conditioning)
Flüssigkeitsprobe *f* fluid sample
Flüssigkeitsraum *m* SPEICH hydraulic fluid chamber
Flüssigkeitsreibung *f* fluid friction
Flüssigkeitsringverdichter *m* liquid-ring compressor
Flüssigkeitsseite *f* SPEICH hydraulic fluid side
Flüssigkeitsspeicher *m* [hydraulic] accumulator
Flüssigkeitsstand *m* BEHÄLT fluid level
Flüssigkeitsstandanzeiger *m* BEHÄLT level indicator (gage)
Flüssigkeitsstandschalter *m* BEHÄLT level[-control] switch
Flüssigkeitsstrom *m* flow, flow stream, stream, current; liquid flow (stream, current), *s also* Strom *m*
Flüssigkeitsstromanzeiger *m* flow indicator
Flüssigkeitsstrommesser *m* flowmeter
Flüssigkeitsstrommessung *f* flow [rate] measurement
Flüssigkeitsstromschreiber *m* flow recorder
Flüssigkeitsstromsensor *m* flow sensor
Flüssigkeitsströmung *f* flow, fluid flow (motion), flow stream, stream
Flüssigkeitsstromwandler *m* flow transducer
Flüssigkeitstasche *f* fluid trap
Flüssigkeitsthermometer *n* liquid-expansion thermometer
Flüssigkeitsventil *n* SPEICH hydraulic fluid [discharge] valve
Flüssigkeitsverbrauch *m* fluid consumption
Flüssigkeitsverschleiß *m* fluid wear
Flüssigkeitsverträglichkeit *f* DICHT fluid compatibility
Flüssigkeitswartung *f* fluid maintenance (conditioning)
Flüssigkeitswiderstand *m* THEOR resistance to flow, flow (fluid) resistance
Flüssigmetall *n* FLÜSS liquid metal
Flüssigverschmutzung *f* liquid contamination
Folgekolbenrückführung *f* STELL hydraulic follower feedback
Folgekolbenverstärker *m* STELL hydraulic-follower amplifier
Folgeschaltung *f* sequencing circuit
Folgeschaltventil *n* DRUCKV sequence valve
Folgesteuerung *f* sequencing control, sequencer; *Nachfolgesteuerung* follower control
Folgeventil *n* *s.* Folgeschaltventil *n*
Folgezylinder *m* slave cylinder
Folienelement *n* LOGIKEL flexing-diaphragm element
Förderanschluß *m* discharge (delivery, output, pressure, outlet, exit) port
Förderdruck *m* delivery (discharge, output, outlet, exhaust) pressure

Förderhub *m* PU/MOT delivering (discharge, pumping) stroke
Förderkanal *m* discharge (delivery, output, pressure, outlet, exhaust) channel
Förderkennlinie *f* delivery (discharge, output) characteristics
Förderleistung *f* PU/MOT hydraulic (pump horsepower) output
Förderleitung *f* discharge (delivery, output, pressure, outlet, exhaust) line
Förderluft *f* KOMPR discharge (delivered) air
fördern *z. B. Öl zum Verbraucher* deliver *v*, supply *v*, provide *v* eg *fluid to actuator*
Förderpumpe *f* im Gegensatz zur Hochdruckpumpe ANWEND high-flow pump
Förderrad *n* einer Zahnradpumpe pump gear
Förderraum *m* discharge (delivery, output, pressure, outlet, exhaust) chamber
Förderrichtung *f* PU/MOT direction of delivery (discharge, output flow), output flow direction
Förderrichtungsumkehr *f,* **Förderrichtungswechsel** *m* change of delivery (discharge, output) direction
Förderschwankung *f* s. Förderstrompulsation *f*
Förderseite *f* discharge (delivery, pressure, outlet, exhaust) side
Förderstrom *m* einer Pumpe delivery (discharge) rate, discharge flow, output flow [rate], delivery, discharge, output; *frequently used for* Volumenstrom *m*
Förderstrombedarf *m* flow (fluid) requirements

Förderstrombereich *m* PU/MOT delivery (discharge, output flow) range
Förderstromfluktuation *f* s. Förderstrompulsation *f*
förderstromgeregelte Pumpe *f* flow-compensated pump
Förderstrompulsation *f* PU/MOT output pulsation (ripple, fluctuation), pump pulsation (ripple)
Förderstromregler *m* *Pumpenregeleinheit* flow compensator *pump control*
Förderstromschlupf *m* slip ratio *of a pump*
Förderstromverstellung *f* delivery (discharge rate, output flow) control
Förderung *f* delivery, discharge, output
Förderungleichförmigkeit *f* s. Förderstrompulsation *f*
Förderverhalten *n* delivery (discharge, output) characteristics
Fördervolumen *n* [pump] displacement, output [volume]
Fördervolumen *n* je Umdrehung delivery (discharge, output volume) per cycle
Formdichtung *f* moulded (preformed) seal
formgedrehter Dichtring *m* special cut lip (SCL) seal, lathe-cut ring
Formlippe *f* DICHT moulded (preformed) lip
Formverlust *m* THEOR pressure loss due to local change of shape
Formweichdichtung *f* squeeze-type moulded seal
freie Luft *f* FLÜSS entrained (free, undissolved) air
freier Rücklauf *m* free[-flow] return

freies Leitungsende *n* open termination, open (soft) end
freies Wasser *n* *im Öl* free water in the oil
Freikolben *m* floating (free) piston
Freistrahl *m* *LOGIKEL* free jet
fremdgesteuertes Ventil *n* *STELL* externally piloted valve
Fremdsteuerung *f* *STELL* external piloting
Fremdstoff *m* *FLÜSS* s. Schmutz *m*
Frischöl *n* *Ergänzungsflüssigkeit im geschlossenen Kreislauf* make-up (replenishing) fluid *to the closed circuit*; *Neuöl* new (fresh) oil
Frostschutzadditiv *n* *FLÜSS* antifreeze [additive]
Fühler *m* s. Sensor *m*
Führungsbuchse *f* *der Kolbenstange ZYL* rod bearing (bushing)
Führungslänge *f* *ZYL* bearing length
Führungsring *m* *ZYL* wear (bearing) ring; *Leitring der Flügelzellenmaschine* cam (track, contour) ring *of a sliding-vane unit*
Füllbehälter *m* prefill tank
Fülldruck *m* *SPEICH* precharge (charging, preload, inflation) pressure
Fülleitung *f* filling (loading, charging) line
füllen *Speicher mit Flüssigkeit* charge *v*, load *v* *accumulator with fluid*; *Speicher mit Gas* [pre]charge *v*, load *v*, pressurize *v* *accumulator with gas*; *Pumpe* prime *v*, prefill *v*, supercharge *v*, boost *v* *pump*
Füllöffnung *f* *BEHÄLT* filler [opening], refill opening
Füllpumpe *f* charge (charging, booster, supercharge, prefill) pump

Füllstandsanzeiger *m* *BEHÄLT* level indicator (gage)
Füllstandsauge *n,* **Füllstandsglas** *n* *BEHÄLT* level sight glass, sight gage, gage glass
Füllstandsschalter *m* *BEHÄLT* level[-control] switch
Füllung *f* *SPEICH* filling, loading, charging
Füllung *f*: **ungenügend gefüllt werden** *Pumpe* starve *v* *pump*
Füllventil *n* *WEGEV* prefill valve
Fünfwegeventil *n* *WEGEV* five-way (5-way) valve
Funktionsablaufplan *m* cycle [sequence] diagram (plot), functional circuit
funktionserweitertes Druckbegrenzungsventil *n* multi-function pressure control valve
Funktionsschaltplan *m* circuit diagram (drawing)
Funktionsschaltzeichen *n,* **Funktionssymbol** *n* functional symbol
Funktionsverschraubung *f* function fitting
Fuß *m* s. Zylinderfuß *m*
Fußbefestigung *f* *ZYL* foot (lug) mount
fußbetätigtes Ventil *n* *STELL* foot-operated (-actuated) valve, *wenn Drehpunkt am Pedalende, auch* pedal-operated (-actuated) valve, *wenn Drehpunkt in Pedalmitte, auch* treadle-operated (-actuated) valve
Fußventil *n* dump (bottom, foot) valve; s. fußbetätigtes Ventil *n*
Fußwippenventil *n* *STELL* treadle-operated (-actuated) valve

G

Gabelbefestigung *f* ZYL clevis mount
Gas *n* **unter Druck** pressurized (pressure) gas, gas under pressure
Gasabscheidevermögen *n* FLÜSS gas release, ability of gas separation
gasbelasteter Speicher *m* hydropneumatic (gas-loaded, gas- charged, gas-oil) accumulator
gasdicht gas-tight, bubble-tight
Gasdruckspeicher *m* s. gasbelasteter Speicher *m*
Gasdynamik *f* aerodynamics, gas dynamics
gasdynamisch aerodynamic, gasdynamic
Gasflasche *f* BEHÄLT gas bottle
Gasfüllung *f* SPEICH gas charge (loading)
Gasfüllventil *n* SPEICH gas precharge (charging, loading) valve, gas valve
Gaslösungsvermögen *n* FLÜSS gas solubility, solubility of gas
Gasmechanik *f* aeromechanics, gas mechanics
Gasraum *m* SPEICH gas chamber
Gasseite *f* SPEICH gas side
Gasspeicher *m* s. gasbelasteter Speicher *m*
Geberzylinder *m* master cylinder
geborstenes Filterelement *n* burst filter element
Gebrauchsdauer *n* **von Öl** oil life
Gebrauchtöl *n* waste (used) oil
gedichteter Schieber *m* WEGEV packed spool
gedrehter Dichtring *m* special cut lip (SCL) seal, lathe-cut ring
Gefälleleckleitung *f* gravity drain line
gefaßte Dichtung *f* cased (captive) seal
gefordertes Ölvolumen *n* SPEICH demand (required) oil volume
geförderte Luft *f* KOMPR discharge (delivered) air
Gefrierschutzadditiv *n* FLÜSS antifreeze [additive]
geführter Ventilkegel *m* DRUCKV guided poppet
Gegendruck *m* back pressure
Gegendruckerzeugung *f* back pressuring
Gegendruckfließpressen *n* ein Höchstdruckverfahren ANWEND fluid-to-fluid (pressure-to-pressure) extrusion *from superpressure to somewhat lower pressure*
Gegendruckventil *n* DRUCKV counterbalance (back-pressure) valve
Gegenhub *m* ZYL reverse stroke
Gegenkolbenflüssigkeitsfeder *f* compound liquid spring
Gegenstrahlelement *n* LOGIKEL impact modulator
Gegenstrahlelement *n* **mit axialer Steuerdüse** LOGIKEL direct impact modulator, DIM
Gegenstrahlelement *n* **mit transversaler Steuerdüse** LOGIKEL transverse impact modulator, TIM
Gegenstrahlelement *n* **ohne Steuerdüsen** LOGIKEL summing impact modulator, SIM
Gegenstromwärmeübertrager *m* counterflow heat exchanger
Gehäuseablaß *m* PU/MOT case drain
Gehäuseleckanschluß *m* s. Gehäuseablaß *m*
Gehäusering *m* Leitring der Flügel-

gekerbter 182

zellenmaschine cam (track, contour) ring *of a sliding-vane unit*
gekerbter Bremsansatz *m* ZYL slotted spear
Gelenkbefestigung *f* ZYL spherical mounting
gelöste Luft *f* FLÜSS dissolved air, air in solution
gelöstes Wasser *n* FLÜSS dissolved water, water in solution
Generator *m* *s.* Pumpe *f*
geöffnete Stellung *f* WEGEV open (passing) position
geometrisches Fördervolumen (Verdrängungsvolumen) *n* PU/MOT geometric displacement [volume]
gerade Verbindung (Verschraubung) *f* straight fitting (connection)
gerades Gewinde *n* VERBIND parallel (straight) thread
Geradrohrwärmeübertrager *m* straight-tube heat exchanger
Gerät *n* *s.* Hydraulikgerät *n*; *s.* Pneumatikgerät *n*; *s z. B.* Radialkolbenpumpe *oder* -motor
Gerätelebensdauer *f* component life
Geräteschutzfilter *m,n* last-chance (point-of-use) filter
Geräteverbindung (Geräteverschraubung) *f* port fitting (connection)
Geräuschdämpfer *m* muffler, silencer, sound attenuator
Gerotormotor *m* *trochoidenverzahnter Innenzahnradmotor mit feststehendem Zahnring* gerotor motor, *progressing-tooth gear motor with fixed internal gear*
Gerotorpumpe *f* gerotor pump *cf* Gerotormotor *m*
Gesamtabscheidegrad *m* filter efficiency

Gesamtdruck *m* *s.* Gesamtdruckhöhe *f*
Gesamtdruckhöhe *f* THEOR total [pressure] head
Gesamtstromfilter *m,n* full-flow filter
Gesamtstromfilterung *f* full-flow filtration
Gesamtwirkungsgrad *m* PU/MOT overall efficiency
geschlossen: in Ruhestellung geschlossen WEGEV normally closed, NC
geschlossener Kreislauf *m* closed[-loop] circuit
geschmierter Verdichter *m* lubricated compressor
geschweißter Behälter *m* welded reservoir (tank)
geschweißter Zylinder *m* welded cylinder
geschweißtes Stahlrohr *n* [seam-] welded steel pipe
Geschwindigkeitshöhe *f* THEOR velocity (kinetic) head
Geschwindigkeitsprofil *n* THEOR velocity distribution (profile)
Geschwindigkeitsverteilung *f* *s.* Geschwindigkeitsprofil *n*
gesperrte Stellung *f* WEGEV closed (non-passing) position
Gestängestelleinheit *f* PU/MOT lever control (controller, stroker)
gesteuerte Leitung *f* WEGEV controlled connection
gesteuertes Rückschlagventil *n* pilot [-operated] check valve, pilot check
gestreckte Rohrlänge *f* developed pipeline length, length of straight pipe
gestufter Bremsansatz *m* ZYL stepped spear
getaucht *Unteröl ...* immersed, submerged, wet [mount]

geteilte Dichtung *f* split-ring seal
geteilter Flansch *m* *VERBIND* split flange
Getriebe *n* *hydrostatisches Getriebe* hydrostatic transmission
Getriebe *n* **in aufgelöster Bauweise** *Ferngetriebe n* split (non-integral, separate pump/motor) transmission
Getriebe *n* **mit äußerer Leistungsverzweigung** hydrodifferential transmission
Getriebe *n* **mit innerer Leistungsverzweigung** hydrostatic transmission with output speed summation or differencing
Getriebe *n* **mit Leistungsverzweigung** hydromechanical (mixed hydrostatic/mechanical) transmission
Gewebefilter *m,n* [woven-]screen filter, screen; *Textilfilter* woven-cloth filter, fabric filter
gewebeverstärkte Dichtung *f* fabric-reinforced seal
gewichtsbelastete Manometerprüfvorrichtung *f* dead-weight pressure gauge tester
gewichtsbelastetes Rückschlagventil *n* gravity-held check valve
Gewichtsspeicher *m* weight-loaded (weighted) accumulator
Gewichtsspeicher *m* **mit Außenführung** externally guided weight-loaded accumulator
Gewichtsspeicher *m* **mit Innenführung** self-guided weight-loaded accumulator
Gewindeansatzbefestigung *f* *ZYL* threaded end mount
Gewindeanschluß *m* threaded port
Gewindebohrungsbefestigung *f* tapped holes mounting

Gewindefitting *m* s. Gewindeverschraubung *f*
Gewindemuffe *f* Übergangsstück *mit Innengewinde* female threaded union
Gewindestopfen *m* threaded plug
Gewindestutzen *m* Übergangsstück *mit Außengewinde* male threaded union
Gewindeverschraubung *f* threaded (screwed) fitting
Giftigkeitsprüfung *f* *FLÜSS* toxicity test
Glasfaserfilter *m,n* glass-fiber filter, fiberglass filter
Glasfiberfilter *m,n* s. Glasfaserfilter *m,n*
glattes Rohr *n* *THEOR* smooth pipe
Glattmembranzylinder *m* flat-diaphragm cylinder
Gleichdruckturbinenmotor *m* impulse-type turbine motor
Gleichgangzylinder *m* equal-stroking speed cylinder
Gleichkraft-Flüssigkeitsfeder *f* constant-force liquid spring
Gleichlauf *m* synchronized motion, synchronism
Gleichlauf *m*: **Gleichlauf herstellen** synchronize *v*
Gleichlaufschaltung *f* synchronizing circuit
Gleichlauf-Teleskopzylinder *m* constant extension speed telescopic cylinder
Gleichlaufventil *n* flow equalizer, equalizing (balancing) valve
Gleichlaufzylinder *m* equal-stroking speed cylinder
Gleichrichterschaltung *f* rectifying circuit

Gleichspannungsdruckwandler *m*
  PDC pressure transducer
Gleichstrommagnet *m*   *STELL* direct current solenoid, D. C.
Gleichstromwärmeübertrager *m*   parallel [flow] heat exchanger
Gleitdichtung *f*   reciprocating (sliding) seal
Gleitdichtungskupplung *f*   slide-seal coupling
Gleitmuffenkupplung *f*   sleeve-and-poppet coupling
Gleitrahmen *m*   einer Radialkolbenpumpe slide block of a radial piston pump
Gleitring *m*   *ZYL* wear (bearing) ring
Gleitringdichtung *f*   rotary face seal, slipper ring
Gleitschuh   *PU/MOT* piston (slide, thrust) shoe, slipper pad
Gleitschuhkolben *m*   *PU/MOT* slipper piston
Gleitschuhpumpe *f*   slipper pump
Gleitsitzkupplung *f*   slide-seal coupling
Glockenfilter *m,n*   bowl (can) filter
glycolbasische Flüssigkeit *f*   polyglycol solution, waterglycol [fluid]
Graugußbehälter *m*   cast-iron reservoir (tank)
Graugußzylinder *m*   cast-iron cylinder
grenzflächenaktiver Stoff *m*   *FLÜSS* surface-active agent, surfactant
Grenznutzungsdauer *f* des Öls   oil life
Grenzschaltventil *n*   limit valve
Grenzschicht *f*   *THEOR* boundary layer
Grenzviskosität *f*   *FLÜSS* intrinsic viscosity
Grobfilter *m,n*   coarse filter
Grobfilterung *f*   coarse filtration

grobmaschig   *FILTER* coarse-meshed
großvolumige Pumpe *f*   high-flow pump
Großwinkelpumpe *f*   eine Schrägachsen-Axkoausführung large-angle pump   bent-axis axial piston pump
Grundkreislauf *m*   basic circuit
Grundöl *n*   oil base
Grundschaltung *f*   basic circuit
Grundschaltzeichen *n*   basic symbol
Grundstellung *f*   neutrale Stellung *WEGEV* normal (neutral) position, wenn mittlere Stellung s auch Mittelstellung
Grundsymbol *n*   basic symbol
Grundverschmutzung *f*   *FLÜSS* initial contamination
Grundviskosität *f*   intrinsic viscosity
Gummifederspeicher *m*   rubber-cushion accumulator
gummigespannte Dichtung *f*   rubber-energized seal
Gußbehälter *m*   cast reservoir (tank)
Gußzylinder *m*   cast-iron *or* cast-steel cylinder

H

H   *FLÜSS* Kennzeichen für Mineralöl ohne Wirkstoffzusätze
Haarnadelwärmeübertrager *m*   U-tube heat exchanger
Haftstrahlelement *n*   *LOGIKEL* wall-attachment (Coanda effect) element
Halbleitertemperaturfühler *m*   thermistor temperature sensor
halbstarre Leitung *f*   semi-rigid line
halogenierte Kohlenwasserstoffe *mpl*   s. Flüssigkeit *f* auf Basis halogenierter Kohlenwasserstoffe

Haltekreislauf *m* [load] holding circuit
Haltestrom *m* *STELL* holding current
Halteventil *n* *WEGEV* lock (holding) valve
Haltstellung *f* *WEGEV* hold position
Handablaß *m* *FILTER* manual drain
handbetätigt *STELL* hand-operated (-actuated, -controlled), manually operated (actuated, controlled)
Handeingriff *m* *STELL* manual override
Handpumpe *f* hand pump
handradbetätigt *STELL* handwheel-operated (-actuated, -controlled)
Handradstelleinheit *f* *PU/MOT* handwheel control (controller, stroker)
Handstelleinheit *f* *PU/MOT* manual (hand) control (controller, stroker)
handtasterbetätigt *STELL* palm button-operated (-actuated, -controlled)
Harzbildung *f* *FLÜSS* resinification, gum formation
harzfrei *FLÜSS* nonresinous, resin-free
Harzgehalt *m* *FLÜSS* resin (gum) content
Hauptkolben *m* *eines zweistufigen Ventils* main (primary) spool *of a two-stage valve*
Hauptkreislauf *m* main (primary) circuit
Hauptläufer *m* *KOMPR* male rotor
Hauptleitung *f* working (operating, main) line
Hauptpumpe *f* *eines hydrostatischen Getriebes* main (primary) pump *of a hydrostatic transmission*

Hauptsteuerschieber *m* s. Hauptkolben *m*
Hauptstrahl *m* *LOGIKEL* main (power, principle) jet
Hauptstrom *m* main flow
Hauptstromfilter *m,n* full-flow filter
Hauptstromfilterung *f* full-flow filtration
Hauptstufe *f* *eines Servoventils* main (primary) spool
Hauptventil *n* power [stage] valve, main (primary) valve
Hauptzylinder *m* master cylinder
HD *Hochdruck: in der HY von etwa 100 bis etwa 600 bar; in der PN von etwa 2 bis etwa 10 bar* high pressure, H. P. *in HY from about 1500 up to about psi; in PN from about 25 up to about 150 psi*
HD-Kreislauf *m* *auch: der Kreislauf mit dem höheren Druck* high-pressure circuit, H. P. circuit
HD-Seite *f* *die Seite mit dem höheren Druck* high-pressure side, H. P. side
Heatlessregenerierung *f* *Druckwechselregenerierung von Trockenmittel* pressure-swing regeneration *of a desiccant*
hebelbetätigt lever-operated (-actuated, -controlled)
Hebelrückführung *f* *STELL* lever feedback
Hebelstelleinheit *f* *PU/MOT* lever control (controller, stroker)
Hebelventil *n* lever-operated (-actuated, -controlled) valve
Heizer *m* [pre]heater
Heizmotor *m* preheating motor
Hemmpumpe *f* meter-out pump
Hemmschaltung *f* meter-out circuit

Hemmstoff *m*   *FLÜSS* inhibitor, inhibiter
herstellerverpreßte Verbindung *f*   *Schlauchverbindung* factory-attached hose fitting
HFA-Flüssigkeit *f*   *s.* Öl-in-Wasser-Emulsion *f*
HFB-Flüssigkeit *f*   *s.* Wasser-in-Öl-Emulsion *f*
HFC-Flüssigkeit *f*   *s.* Wasser-Glycol-Lösung *f*
HFD-Flüssigkeit *f*   *s.* synthetische Flüssigkeit *f*
HFDR-Flüssigkeit *f*   *s.* Phosphatesterflüssigkeit *f*
HFDS-Flüssigkeit *f*   *s.* Flüssigkeit *f* auf Basis chlorierter Kohlenwasserstoffe
Hilfsbetätigung *f*   *STELL* auxiliary (override) control
Hilfskreislauf *m*   auxiliary circuit
Hilfspumpe *f*   auxiliary pump; *Spülpumpe im geschlossenen Kreislauf* make-up (slippage, boost, scavenger) pump
Hilfsventil *n*   auxiliary valve
hindurchströmen lassen   *z.B. Flüssigkeit* pass *v*   *eg fluid*
Hinhub *m*   *ZYL* forward stroke
hinter *einem Element*   *s.* nachgeschaltete Leitung *f*
hintereinanderschalten   install *v* (connect *v*, pipe *v*) in series, cascade *v*
HL   *FLÜSS Kennzeichen für Mineralöl mit Wirkstoffzusätzen*
HLP   *FLÜSS Kennzeichen für Mineralöl mit zusätzlichen Wirkstoffen gegen Verschleiß*
Hochbehälter *m*   overhead reservoir (tank)
Hochdruck *m*   *in der HY von etwa 100 bis etwa 600 bar; in der PN von etwa 2 bis etwa 10 bar* high pressure, H. P. *in HY from about 1500 up to about psi; in PN from about 25 up to about 150 psi*
Hochdruckfilter *m*   pressure[-line] filter, high-pressure filter
Hochdruckkreislauf *m*   *auch: der Kreislauf mit dem höheren Druck* high-pressure circuit, H. P. circuit
Hochdruckleitung *f*   high-pressure line
Hochdruckpumpe *f*   high-pressure pump
Hochdruckschlauch *m*   high-pressure hose
Hochdruckseite *f*   *die Seite mit dem höheren Druck* high- pressure side, H. P. side
Hochdruck-Sprühprüfung *f*   *der Schwerentflammbarkeit FLÜSS* high-pressure spray test   *of fire resistance*
Hochmomentmotor *m*   *auch Langsamläufermotor* low speed (high torque) motor, LSHT motor
Höchstdruckadditiv *n*   *FLÜSS* extreme-pressure (EP) additive
Höchstdruckschlauch *m*   extreme-pressure hose
Hochtemperaturflüssigkeit *f*   high-temperature fluid
hochviskos   *FLÜSS* [high] viscous, thick, high-viscosity ...
hochwasserhaltige Flüssigkeit *f*   high water content fluid, HWCF, high water base fluid, HWBF
Höhenverkettung *f*   vertical valve stacking, sandwich mounting, *s also* Modulverkettung *f*
Hohlrundring *m*   *DICHT* hollow O-ring

Hohlschieber *m*   *WEGEV*   hollow spool

Horizontalverkettung *f*   horizontal valve stacking, *s also* Modulverkettung *f*

Hub *m*   *im Sitzventil* lift, stroke *of valve poppet*; *ZYL* stroke; *Hublänge* length of stroke, stroke [length]

Hubbegrenzerdeckel *m*   *eines 2-Wege-Einbauventils* stroke limiter cover *of a logic element*

Hubbegrenzungsbuchse *f*   *ZYL* stop tube

Hubeinstelldeckel *m*   *s.* Hubbegrenzerdeckel *m*

Hubeinstellung *f*   stroke adjustment

Hubende *n*   *ZYL* stroke end, end of stroke

Hubendeanzeiger *m*   *ZYL* end-of-stroke indicator (sensor)

Hubendebremse *f*   *ZYL* cylinder cushion

Hubendebremse *f* **mit konstanter Verzögerung**   *ZYL* constant-deceleration cushion

Hubendebremsung *f*   *ZYL* cushioning

Hubendeposition *f*   *ZYL* end-of-stroke position, end of stroke

Hubgeschwindigkeit *f*   *ZYL* stroke speed, stroking rate

hubgesteuerter Proportionalmagnet *m*   *STELL* controlled-stroke proportional solenoid

hubgesteuerter Zylinder *m*   motion-controlled cylinder

Hubkolbenverdichter *m*   reciprocating compressor

Hubkreislauf *m*   *ANWEND* lift circuit

Hublänge *f*   *ZYL* length of stroke, stroke [length]

Hubraum *m*   *ZYL* stroke (swept) volume

Hubschaltung *f*   *ANWEND* lift circuit

Hubscheibe *f*   *PU/MOT* cam (angle) plate

Hubumkehr *f*   *ZYL* stroke reversal

Hubumsteuerung *f*   *s.* Hubumkehr *f*

Hubvervielfältiger *m*   stroke multiplier

Hubvolumen *n*   *ZYL* stroke (swept) volume

Hubzylinder *m*   *ANWEND* jack, ram, lifting cylinder

Huckepackbefestigung *f*   *PU/MOT* piggyback mounting

Hufeisenrohrfeder *f*   *MESS* horseshoe-shaped Bourdon tube

Hülse *f*   *Schlauchverbindung* socket, shell *hose fitting*

hülsenlose Verbindung *f*   *Schlauch* socketless fitting *hose*

Hutmanschette *f*   flanged (hat, collar) seal

Hydraulik *f*   hydraulics, fluid power technology (engineering), *compounds s also with* hydraulisch, Hydro ... , fluidtechnisch, Fluidtechnik ... , Flüssigkeits ... , Öl ... ; *Hydraulikanlage* hydraulic (fluid power) system

Hydraulikaggregat *n*   *ANWEND* [hydraulic] powerpack (power unit, power package)

Hydraulikanlage *f*   hydraulic (fluid power) system

Hydraulikanschluß *m*   hydraulic (fluid power) connection (port)

Hydraulikantrieb *m*   *ANWEND* hydraulic (fluid) drive

Hydraulikanwender *m*   user of hydraulic equipment, hydraulics user

Hydraulikbagger *m* *ANWEND* hydraulic excavator
Hydraulikbauteil *n* hydraulic (fluid power) component (element)
Hydraulikbehälter *m* fluid (hydraulic) reservoir (tank)
Hydraulikelement *n* *s.* Hydraulikbauteil *n*
Hydrauliker *m* hydraulic engineer
Hydraulikfilter *m,n* hydraulic filter
Hydraulikflüssigkeit *f* hydraulic fluid (medium), *cf* Arbeitsmedium *n*
Hydraulikgerät *n* *s.* Hydraulikbauteil *n*
Hydraulikhersteller *m* manufacturer of hydraulic equipment, hydraulics manufacturer
Hydraulik-IC *m* hydraulic integrated circuit, HIC
Hydraulikindustrie *f* hydraulics industry
Hydraulikingenieur *m* hydraulic engineer
Hydraulikkomponente *f* hydraulic (fluid power) component (element)
Hydraulikkreislauf *m* hydraulic (fluid power) circuit (system)
Hydraulik-Lehrversuchsstand *m* *ANWEND* hydraulic trainer
Hydraulikmedium *n* hydraulic fluid (medium), *cf* Arbeitsmedium *n*
Hydraulikmotor *m* hydraulic (fluid) motor
Hydrauliköl *n* hydraulic oil
Hydraulikpumpe *f* hydraulic pump
Hydraulikschaltplan *m* hydraulic (fluid power) circuit diagram
Hydraulikschaltung *f* hydraulic (fluid power) circuit (system)
Hydraulikschaltzeichen *n* hydraulic symbol

Hydraulikschlauch *m* hydraulic hose
Hydraulikschrank *m* *ANWEND* hydraulic enclosure (cabinet), enclosed hydraulic panel
Hydraulikschrauber *m* *ANWEND* hydraulic nut runner (wrench)
Hydrauliksymbol *n* hydraulic symbol
Hydraulikspeicher *m* (hydraulic) accumulator
Hydrauliksteuerung *f* *ANWEND* hydraulic (fluid) control
Hydrauliksystem *n* hydraulic (fluid power) system; *Hydraulikkreislauf* hydraulic (fluid power) circuit (system)
Hydraulikventil *n* hydraulic (fluid) valve
Hydraulikverbindung *f* hydraulic fitting
Hydraulikverschraubung *f* *s.* Hydraulikverbindung *f*
Hydraulikzubehör *n* hydraulic accessories *pl*
Hydraulikzylinder *m* hydraulic (fluid-power, fluid) cylinder
hydraulisch hydraulic; *compounds s also with* Hydraulik ... , Hydro ... , fluidtechnisch, Fluidtechnik ... , Flüssigkeits ... , Öl ...
hydraulisch angetrieben (betrieben, getrieben) hydraulically (fluid-)powered (operated)
hydraulisch verriegelt *STELL* hydraulically detented
hydraulisch vorgesteuert *STELL* hydraulically piloted
hydraulische Achse *f* *für Positionieraufgaben ANWEND* hydraulic axis *for positioning*
hydraulische Antriebsachse *f* *ANWEND* hydrostatic transaxle

hydraulische Betätigungseinrichtung *f* *STELL* hydraulic operator (actuator, control)
hydraulische Bremse *f* *ANWEND* hydraulic (oil) brake
hydraulische Leistung *f* hydraulic horsepower (power), fluid power
hydraulische Nachgiebigkeit *f* *THEOR* hydraulic compliance
hydraulische Steife *f* *THEOR* hydraulic stiffness
hydraulische Stelleinheit *f* *PU/MOT* pilot control (controller, stroker); *STELL* hydraulic operator (actuator, control)
hydraulische Steuerung *f* *ANWEND* hydraulic (fluid) control
hydraulische Welle *f* *Konstantgetriebe* fixed-displacement transmission
hydraulische Zentrierung *f* *STELL* hydraulic (pressure) centering
hydraulischer Antrieb *m* hydraulic (fluid) drive
hydraulischer Durchmesser *m* *THEOR* hydraulic (characteristic) diameter
hydraulischer Hebel *m* *ANWEND* hydraulic jack
hydraulischer Radius *m* *THEOR* hydraulic (characteristic) radius
hydraulischer Stoß *m* hydraulic (fluid) shock, water (fluid) hammer, pressure surge
hydraulischer Verstärker *m* *STELL* hydraulic amplifier
hydraulischer Widerstand *m* *THEOR* hydraulic resistor
hydraulisches Verkleben (Verklemmen) *n* *des Steuerschiebers WEGEV* hydraulic lock, pressure freeze, binding, gumming

Hydro ... : *compounds s also with* hydraulisch, Hydraulik ... , fluidtechnisch, Fluidtechnik ... , Flüssigkeits... , Öl ...
Hydroachse *f* *ANWEND* hydrostatic transaxle
Hydroarmatur *f* *umgangssprachlich s.* Hydraulikverschraubung *f*
Hydrodynamik *f* hydrodynamics
hydrodynamisch hydrodynamic
hydrodynamische Kupplung *f* *ANWEND* hydrodynamic (hydraulic, fluid) coupling (clutch)
hydrodynamische Maschine *f* hydrodynamic (hydrokinetic) machine
hydrodynamische Schmierung *f* *ANWEND* hydrodynamic lubrication
hydrodynamisches Lager *n* *ANWEND* hydrodynamic bearing
Hydrofilter *m,n* hydraulic filter
Hydroformen *n* hydrostatisches Ziehen oder Tiefziehen *ANWEND* hydroforming *hydrostatic forming of sheet metal*
Hydrologikventil *n* *s.* Zweiwege-Einbauventil *n*
Hydrolysebeständigkeit *f* *FLÜSS* hydrolytic stability
Hydromechanik *f* hydromechanics
hydromechanisch hydromechanic
Hydromotor *m* hydraulic (fluid) motor
Hydropneumatik *f* hydropneumatics, air-hydraulics, airdraulics
hydropneumatisch hydropneumatic, air-hydraulic, airdraulic
hydropneumatische Pumpe *f* hydropneumatic (air-hydraulic, air-powered, air-operated) pump
hydropneumatischer Speicher *m* hydropneumatic accumulator, *als*

gasbelasteter Speicher auch: gas-loaded (gas-charged, gas-oil) accumulator, *als luftbelasteter Speicher auch:* air-loaded (air-charged, air-oil, airdraulic) accumulator
Hydropumpe *f* hydraulic pump
Hydrospeicher *m* [hydraulic] accumulator
Hydrostat *m* *Differenzdruckregler im Stromregelventil* controlled orifice, pressure compenator, hydrostat *in a compensated flow-control valve*
Hydrostatik *f* hydrostatics
hydrostatisch hydrostatic
hydrostatische Antriebsachse *f* *ANWEND* hydrostatic transaxle
hydrostatische Schmierung *f* *ANWEND* hydrostatic lubrication
hydrostatischer Antrieb *m* *ANWEND* hydrostatic drive
hydrostatisches Fließpressen *n* *ANWEND* hydrostatic extrusion
hydrostatisches Getriebe *n* hydrostatic transmission, *compounds s also with* Getriebe *n*
hydrostatisches Getriebe *n* **in offenem Kreislauf** open-circuit hydrostatic transmission
hydrostatisches Getriebe *n* **mit geschlossenem Kreislauf** closed-circuit hydrostatic transmission
hydrostatisches Lager *n* *ANWEND* hydrostatic bearing
Hydrosteuerung *f* *ANWEND* hydraulic (fluid) control
Hydroventil *n* hydraulic (fluid) valve
Hydroversorgungseinheit *f* *ANWEND* hydraulic supply, *s also* Hydraulikaggregat *n*
Hydrozylinder *m* hydraulic (fluid) power, fluid) cylinder

I

ID *Innendurchmesser* inside (inner) diameter, ID, I. D., bore [size]
idealer Förderstrom *m* *PU/MOT* ideal flow (discharge, delivery)
idealer Schluckstrom *m* *PU/MOT* ideal input (intake) flow rate
ideales Fluid *n* *THEOR* ideal (perfect) fluid
IFPA = International Fluid Power Exhibition *Chicago, USA*
Impaktmodulator *m* *LOGIKEL* impact modulator
Impedanz *f* *Druckabfall/Volumenstrom THEOR* impedance *pressure drop/flow*
Impulsfestigkeit *f* *LEIT* impulse (pulsation) resistance
Impulsstromhydraulik *f* pulsed-flow hydraulics
Impulsverstärker *m* *LOGIKEL* momentum-exchange fluid amplifier
inaktiver Bereich *m* *WEGEV* deadband, deadzone
Indikatordiagramm *m* *KOMPR* pressure-volume diagram
indirekt gesteuert *vorgesteuert STELL* pilot-operated (-actuated), piloted
induktiver Druckwandler *m* induction (inductive, electromagnetic) pressure transducer
induktiver Durchflußmeser *m* induction (electromagnetic) flowmeter
induktiver Widerstand *m* *s.* Induktivität *f*
Induktivität *f* *THEOR* inductance
Industrieflüssigkeit *f* industrial-type fluid

Industriehydraulik *f* industrial hydraulics
Industriepneumatik *f* industrial pneumatics
Inhibitor *m* *FLÜSS* inhibitor, inhibiter
inkompressibel *THEOR* incompressible
innen beaufschlagte Radialkolbenpumpe *f* centrally ported radial-piston pump, radial-piston pump with interior admission
Innendruck *m* internal pressure
Innendurchmesser *m* inside (inner) diameter, ID, I. D., bore [size]
Innenlippenring *m* flanged (hat, collar) seal
Innenwand *f* *ZYL* inner wall, bore
Innenzahnradeinheit *f* internal gear pump *or* motor
Innenzahnradmotor *m* internal gear motor
Innenzahnradmotor *m* **mit Trennsichel** crescent-seal internal gear motor, crescent motor
Innenzahnradpumpe *f* internal gear pump
Innenzahnradpumpe *f* **mit Trochoidenverzahnung** progressing-tooth gear pump
innere Flüssigkeitsrückführung *f* *ohne Ablaufdruckentlastung* *WEGEV* internal drainage
innere Leistungsverzweigung *f* *s.* Getriebe *n* mit innerer Leistungsverzweigung
innere Reibung *f* *THEOR* viscous (fluid) friction
innere Verriegelung *f* *ZYL* internal locking device
innerer Leckstrom *m* internal leakage, *bei Luft auch* blowby, *Schlupfstrom bei Pumpen und Motoren* slip [flow], slippage; *ZYL* cross-piston leakage
innerer Strömungsweg *m* *WEGEV* *s* ED way
instationäre Strömung *f* *THEOR* unsteady flow
integrierter Hydraulikschaltkreis *m* hydraulic integrated circuit, HIC
Integritätsprüfung *f* *Nachweis der einwandfreien Fertigungsqualität FILTER* fabrication integrity test
Intervallventil *n* *STROMV* time-delay (timing) valve
Iodzahl *f* *FLÜSS* iodine value, I. V.
IOP *f* = **Internationale Fachmesse** *f* **für Ölhydraulik und Pneumatik** *Zürich, CH*
isostatisches Heißpressen *n* *ANWEND* hot isostatic pressing, HIP
isostatisches Kaltpressen *n* *ANWEND* cold isostatic pressing, CIP
isostatisches Pressen *n* *ANWEND* isostatic pressing
IZ *s.* Iodzahl *f*

J

Jodzahl *f* *FLÜSS* iodine value, I. V.
Joukowski-Stoß *m* *plötzliche Stromänderung THEOR* Joukowski impact *sudden flow change*
Joysticksteuerung *f* *STELL* joystick operator (actuator, control)
JZ *s.* Jodzahl *f*

K

K-Ring *m*   *DICHT* K-ring
k-Wert *m*   *s.* Viskositäts-Gradexponent *m*
Kabelzylinder *m*   cable cylinder
Kalibrierflüssigkeit *f*   calibration fluid
Kaltanlauf *m*   cold [temperature] start
Kältefließfähigkeit *f*   *FLÜSS* low-temperature fluidity
Kälteöl *n*   low-temperature oil
Kältetrockner *m*   refrigerated air dryer, refrigeration (refrigerant-type) dryer
Kälteverhalten *n*   *FLÜSS* low-temperature characteristics
Kälteviskosität *f*   *FLÜSS* low-temperature viscosity
Kaltregenerierung *f*   *Druckwechselregenerierung von Trockenmittel* pressure-swing regeneration *of a desiccant*
Kaltstart *m*   cold [temperature] start
Kammersystem *n*   *WEGEV* chamber configuration (system)
Kammerungsring *m*   *DICHT* anti-extrusion (back-up) ring
Kanal *m*   flow path, gallery, duct, passageway, channel; *LOGIKEL* leg
Kanalsystem *n*   *WEGEV* chamber configuration (system)
Kantenfilter *m,n*   edge[-type] filter
Kantenfilterung *f*   edge filtration
Kapazität *f*   *THEOR* capacitance
kapazitiver Druckwandler *m*   capacitance pressure transducer
Kapillare *f*   capillary [tube]
Kapillarnebelöler *m*   capillary-action lubricator
Kapillarviskosimeter *n*   efflux (capillary) viscosimeter

Kapselfederdruckwandler *m*   capsule pressure transducer
Kapselfedermanometer *n*   capsule pressure gauge
kavitieren   cavitate *v*
Kavitation *f*   cavitation
Kavitationsschutzventil *n*   *WEGEV* anticavitation valve, anticav
Kavitationsverlust *m*   cavitation loss
Kegelrückschlagventil *n*   poppet check valve
Kegelsitzelement *n*   *stumpfer Kegel, Sitzventilkolben* poppet
Kegelsitzventil *n*   poppet valve
kegliger Bremsansatz *m*   *ZYL* tapered spear
kegliges Gewinde *n*   tapered (taper, pipe) thread
Keilringverschraubung *f*   compression (grip) fitting
Kennlinie *f*   *eines Druckminderventils* flow (regulation) curve *of a pressure regulator*
Kennzeichnungszeile *f*   *auf dem Schlauchmantel* lay line   on hose outer wall
Keramikfilter *m,n*   ceramic filter
Keramikventil *n*   ceramic valve
Kerbe *f*   *Steuerkerbe WEGEV* metering notch
Kettenkolbenschwenkmotor *m*   [piston-]chain rotary actuator
Kettenrückzugzylinder *m*   chain-return cylinder
Kfz-Hydraulik *f*   *Kraftfahrzeughydraulik* automotive hydraulics
Kieselgeltrocknungsmittel *n*   silica-gel desiccant
kinematische Viskosität *f*   *FLÜSS* kinematic viscosity
Kippmoment *n*   *PU/MOT* stall torque

Klauenkupplung *f*   claw coupling
Kleben *n*   *hydraulisches Verklemmen des Steuerkolbens* hydraulic lock, pressure freeze, binding, gumming
Klebverbindung *f*   adhesive fitting
Kleinaggregat *n*   *ANWEND* mini (miniature) power pack, *s also ED* power pack
Kleinpneumatik *f*   miniature (mini, micro) pneumatics
Kleinpumpe *f*   miniature (mini, micro) pump
Kleinzylinder *m*   miniature (small, mini, micro, midget) cylinder
Klemmen *n*   s. Kleben *n*
Klemmringverschraubung *f*   compression (grip) fitting
Klemm-Schlauchverbindung *f*   clamp-type hose fitting
Knickfestigkeit *f*   *ZYL* buckling resistance (strength)
Knicklänge *f*   *ZYL* buckling (column) length
Knieverlust *m*   pressure loss over sharp pipe bend
Knüppelstelleinheit *f*   *STELL* joystick operator (actuator, control)
Koaleszenzfilter *m,n*   coalescing filter
kohlenwasserstoffbasische Flüssigkeit *f*   hydrocarbon-base fluid
Kolben *m*   *allgemein ZYL* piston, *s also ED* ram; *Scheibenkolben* [headed] piston
Kolben *m* **großer Abmessungen**   *ZYL* ram
Kolben *m* **mit Kopf**   *ZYL* headed piston
Kolbenanzahl *f*   *PU/MOT* number of pistons, piston number
Kolbenbewegung *f*   *ZYL* piston travel (traverse, displacement, movement)

Kolbendämpfer *m*   piston damper
Kolbendichtung *f*   *ZYL* piston[-head] seal
Kolbendrehschieber *m*   *WEGEV* rotary spool, valve plug
Kolbendrehschieberventil *n*   *WEGEV* rotary-spool valve
Kolbendruckschalter *m*   piston pressure switch
Kolbendurchmesser *m*   *ZYL* piston diameter
Kolbeneinheit *f*   piston pump *or* motor
Kolbenelement *n*   *LOGIKEL* piston (spool) element
Kolbenfläche *f*   *ZYL* piston area; *Kolbenseite* piston face
Kolbenflächenverhältnis *n*   *ZYL* piston area ratio
Kolbenführung *f*   *ZYL* piston bearing
Kolbengerät *n*   piston pump *or* motor
Kolbengeschwindigkeit *f*   *ZYL* piston speed
Kolbenhub *m*   *ZYL* piston stroke
Kolbenkopf *m*   *ZYL* piston head
Kolbenkraft *f*   *ZYL* piston force
Kolbenlängsschieber *m*   *WEGEV* [sliding] spool
Kolbenlängsschieberventil *n*   *WEGEV* [sliding] spool valve
Kolbenlaufbahn *f*   *PU/MOT* piston trajectory
Kolbenmanometer *n*   piston pressure gauge
Kolbenmaschine *f*   piston pump *or* motor
Kolbenmotor *m*   piston motor
Kolbenposition *f*   *ZYL* piston position
Kolbenpumpe *f*   piston pump; *Tauchkolbenpumpe* plunger pump

Kolbenraum *m*   *ZYL* cap (blind, rear, blank) end [chamber]
Kolbenring *m*   *DICHT* piston ring
Kolbenringraum *m*   *ZYL* rod (head, front) end [chamber]
Kolbenrücklauf *m*   *ZYL* s. Rücklauf *m*
Kolbenschieber *m*   *WEGEV* s. Kolbenlängsschieber *m*
*n*kolbenschieber *m*   *WEGEV* *n*-land spool
Kolbenseite *f*   *ZYL* cap (rear, blind, blank) end; *Kolbenfläche* piston face
kolbenseitiger Druck *m*   *ZYL* cap (rear, blind, blank) end pressure
kolbenseitiges Zylinderende *n*   cap (rear, blind, blank) end
Kolbensitzventil *n*   poppet valve
Kolbenspeicher *m*   piston accumulator
Kolbenstange *f*   *ZYL* piston rod
Kolbenstangenabstreifer *m*   *ZYL* rod wiper [seal]
Kolbenstangenbefestigung *f*   *ZYL* rod end, rod-end coupler (coupling)
Kolbenstangendichtung *f*   *ZYL* rod seal
Kolbenstangendurchmesser *m*   *ZYL* rod diameter
Kolbenstangenende *n*   *ZYL* rod end, rod-end coupler (coupling)
Kolbenstangenfläche *f*   *ZYL* rod area
Kolbenstangenknickung *f*   *ZYL* rod buckling
Kolbenstangenkopf *m*   *ZYL* rod end, rod-end coupler (coupling)
kolbenstangenloser Zylinder *m*   rodless cylinder
Kolbenstangenquerschnitt *m*   *ZYL* rod area
Kolbenstangenraum *m*   *ZYL* rod (head, front) end chamber
Kolbenstangenschutz *m*   *als Faltenbalg ZYL* rod boot (bellows, gaiter)
Kolbenstangenseite *f*   *ZYL* rod (front, head) end
kolbenstangenseitiger Druck *m*   *ZYL* rod (head, front) end pressure
kolbenstangenseitiges Zylinderende *n*   *ZYL* rod (front, head) end
Kolbenstange-Ritzel-Schwenkmotor *m*   rack-and-pinion (piston rack) rotary actuator
Kolbenstellung *f*   *ZYL* piston position
Kolbenträger *m*   *PU/MOT* cylinder block
Kolbentrommel *f*   s. Kolbenträger *m*
Kolbenventil *n*   *WEGEV* spool (plunger, piston) valve
Kolbenverdichter *m*   reciprocating compressor
Kolbenverschleiß *m*   *ZYL* piston wear
Kolbenvorlauf *m*   *ZYL* s. Vorlauf *m*
Kolbenvorschub *m*   *ZYL* piston travel (traverse, displacement, movement)
Kolbenweg *m*   s. Kolbenvorschub *m*
Kolbenzahl *f*   *PU/MOT* number of pistons, piston number
Kolbenzähler *m*   piston-type flowmeter
Kollaps-/Berstfestigkeitsprüfung *f*   *FILTER* collapse-burst test
Kollapsdruck *m*   *FILTER* collapse pressure
Kombinationsfilter *m,n*   composite media filter
kombinierte Dichtung *f*   composite (combined, dual-material) seal
kombinierter Schalldämpfer-Filter *m*   muffler-reclassifyer

Kommutatorventil *n* *PU/MOT* commutator [valve]
Kompaktgetriebe *n* integral (packaged) transmission
Kompensationsdüse *f* *im Druckminderventil* aspirator *of a pressure regulator*
kompensieren *Druck* pressure-balance (-compensate) *v*; *Leckverluste* compensate *v*, make-up *v* *for leakage*
Komponente *f* *s.* Hydraulikkomponente *f*; *s.* Pneumatikkomponente *f*
kompressibel compressible
Kompressiblität *f* compressibility
Kompressiblitätsausgleich *m* *SPEICH* pressure-volume compensation
Kompressibilitätsstrom *m* compressibility flow
Kompressibilitätszahl *f* *FLÜSS* compressiblity [value]
kompressible Strömung *f* *THEOR* compressible flow
Kompressionsarbeit *f* *THEOR* compression work
Kompressionsmodul *m* *FLÜSS* bulk modulus [of elasticity]; *s also* Sekantenkompressionsmodul *m*
Kompressionstrom *m* *THEOR* compressibility flow
Kompressor *m* compressor, *compounds s also with* Verdichter
Kompressorraum *m* compressor room
Kompressorstufe *f* compressor stage
komprimieren compress *v*
Kondensatablaß *m* *FILTER* condensate drain (discharge)
Kondensatleitung *f* drain leg
Konduktanz *f* *reeller Leitwert* *THEOR* conductance
konisches Gewinde *n* tapered (taper, pipe) thread

Konsolbefestigung *f* *PU/MOT* bracket mounting
Konstantbremse *f* *ZYL* constant-deceleration cushion
Konstantdämpfung *f* *s.* Konstantbremse *f*
Konstantdrossel *f* *STROMV* fixed (non-adjustable) restrictor (restriction)
Konstantgetriebe *n* fixed-displacement transmission
Konstantleistungsgetriebe *n* constant power transmission
Konstantmomentgetriebe *n* constant torque transmission
Konstantmotor *m* fixed-displacement (constant-volume) motor
Konstantpumpe *f* fixed-displacement (constant-volume, constant-delivery) pump
Kontaktmanometer *n* contact pressure gauge
Kontinuitätsgleichung *f* *THEOR* continuity equation
Kontraktionsverlust *m* contraction loss
Kontrollfläche *f* *Fluiddynamik* control surface *fluid dynamics*
Kontrollgebiet *n* *Fluiddynamik* control area *fluid dynamics*
Kontrollvolumen *n* *Fluiddynamik* control volume *fluid dynamics*
Kopf *m* *Kolbenkopf ZYL* piston head; *Zylinderkopf* cylinder head
Kopf *m*: **mit Kopf** *Kolben ZYL* headed *piston*
kopieren *nachformen* *ANWEND* copy *v*
Kopierventil *n* *ANWEND* copying (tracer) valve
Korrekturbetätigung *f* *STELL* override control

korrodierende Wirkung *f* *FLÜSS* corrosivity, corrosiveness
Korrosionsanfälligkeit *f* *FLÜSS* corrodibility, corrosiveness
Korrosionsempfindlichkeit *f* *s.* Korrosionsanfälligkeit *f*
Korrosionsinhibitor *m* *FLÜSS* corrosion inhibitor, anti-corrosion additive
Korossionsschutzadditiv *n* *s.* Korrosionsinhibitor *m*
Korrosionsschutzfähigkeit *f* *FLÜSS* anti-corrosive power
Korrosivität *f* *FLÜSS* corrosiveness, corrosivity
Kraftfahrzeughydraulik *f* automotive hydraulics
kraftgesteuerter Proportionalmagnet *m* *STELL* controlled-force proportional solenoid
Kraftmeßdose *f* *hydraulisch* load cell, fluid gauge transducer
kraftstoffgekühlter Wärmeübertrager *m* fluid-to-fuel heat exchanger
Kraftzylinder *m* *ANWEND* power cylinder
Kratzer *m* *FILTER* scraper, knife
Kreiselpumpe *f* centrifugal (rotodynamic) pump
Kreisformrohrfeder *f* C-type Bourdon tube
Kreiskolbenverdichter *m* lobed-rotor compressor
Kreislauf *m* circuit, system; *Funktionsschaltplan* circuit diagram
Kreislauf *m* **mit ablaufseitiger Stromsteuerung** meter-out circuit
Kreislauf *m* **mit Sperrstellung** closed-center circuit
Kreislauf *m* **mit Spülung** scavenger circuit
Kreislauf *m* **mit Umlaufstellung** open-center circuit
Kreislauf *m* **mit Volumenstromquelle** constant flow circuit
Kreislauf *m* **mit zulaufseitiger Stromsteuerung** *f* meter-in circuit
Kreislauf *m* **ohne Volumenstrombeeinflussung** constant flow circuit
Kreislaufdruck *m* circuit pressure
Kreislauf-Druckbegrenzungsventil *n* crossport relief [valve]
Kreislaufentwurf *m* circuit design
kreislaufgebundene Unterplatte *f* custom (circuit, cross drilled) manifold, manifold block, drilled plate
Kreislaufsicherheitsventil *n* closed-circuit relief [valve]
Kreislaufsimulator *m* flowboard, fluid breadboard, patchboard
Kreislaufspülung *f* *Leckölergänzung im geschlossenen Kreislauf* fluid make-up (replenishment) *to the closed circuit*
Kreislaufzweig *m* circuit branch
Kreisschieber *m* *WEGEV* rotary plate (disk)
Kreisschieberventil *n* *WEGEV* rotary-plate (-disk) valve
Kreuzverschraubung *f* cross [fitting]
Kreuzverschraubung *f* **mit vierseitigem Rohranschluß** union cross
Kreuzstück *n* cross fitting
Kriechdrehzahl *f* *PU/MOT* inching (creeping) speed
kritische Geschwindigkeit *f* *Luftstrom durch Drosselstellen* critical (choked) flow *airflow through restrictions*
kritische Reynolds-Zahl (Re-Zahl) *f* critical Reynolds' number (R. N.)
Krümmerverbindung *f* *Schlauch-*

leitung elbow connector (end fitting) *of a hose assembly*
Krümmerverlust *m* bend loss
Kufenbefestigung *f* skid mount
Kugeldrehschieberventil *n* *s.* **Kugelhahn** *m*
Kugelelement *n* *LOGIKEL* [moving] ball (sphere) element
Kugelfallviskosimeter *n* [falling] ball viscometer, dropping viscometer
Kugelhahn *m* *WEGEV* ball (spherical plug) valve
Kugelkolben *m* *kugelförmiger Kolben PU/MOT* ball piston; *mit kugeligem Ende* ball-end piston
Kugelkolbenmotor *m* ball-piston motor
Kugelkolbenpumpe *f* ball-piston pump
Kugelrohrgelenk *n* spherical swivel joint
Kugelrückschlagventil *n* ball check [valve]
Kugelspeicher *m* spherical accumulator
Kugelumlaufschwenkmotor *m* ball screw rotary actuator
Kugelventil *n* ball valve
kühlen cool *v*
Kühlmittel *m* coolant, cooling medium
Kühlmitteleinspritzung *f* *KOMPR* coolant injection
Kühlung *f* cooling
Kulissenschwenkmotor *m* [scotch-]yoke rotary actuator
kundenspezifische Dichtung *f* custom seal
Kunststoffrohr *n* *als Halbzeug* plastic tubing
Kupferrohr *n* copper tubing

Kupferstreifenprüfung *f* *der korrodierenden Wirkung FLÜSS* copper-strip test *corrosivity*
Kupplung *f* *s.* Schlauchkupplung *f*
Kupplungsdose *f* *VERBIND* coupling socket, female coupling half, coupler
Kupplungshälfte *f* *VERBIND* coupling half
Kupplungsnippel *m* *s.* Kupplungsstecker *m*
Kupplungsstecker *m* *VERBIND* coupling plug (nipple), male coupling half
Kupplungsträger *m* *VERBIND* multi-tube connection, multi-connector, cluster fitting
Kurbelschwenkmotor *m* [piston-]crank rotary actuator
kurvenscheibenbetätigt *STELL* cam plate-operated (- actuated, -controlled)
Kurzhubventil *n* short-stroke valve
Kurzhubzylinder *m* short-stroke cylinder, *s also ED* pancake cylinder
Kurzschlußventil *n* bypass valve

L

l/min *auch L/min; Liter je Minute* unit of flow rate: 1 L/min = 0.2199 gpm
L-Gehäuse-Filter *m,n* L-type (L-ported) filter
L-Ring *DICHT* L-ring
Labyrinthdichtung *f* labyrinth seal
laden *Speicher mit Flüssigkeit* charge *v*, load *v* *accumulator with fluid*; *Speicher mit Gas* [pre]charge *v*, load *v*, pressurize *v* *accumulator*

Ladepumpe

with gas; *Pumpe* prime *v*, prefill *v*, supercharge *v*, boost *v* pump
Ladepumpe *f* charge (charging, booster, supercharge, prefill) pump
Lagegeber *m* *ZYL* position (displacement, piston, cylinder) sensor
Lamelle *f* *im Druckluftmotor* vane *in an air motor*; *FILTER* disk, plate, washer
Lamellenfilter *m,n* disk (plate) filter
Lamellenmotor *m* *PN* vane motor
Lamellenpaket *n* *FILTER* disk (plate, washer) pack
Lamellenverdichter *m* rotary vane (sliding-vane, vane) compressor
laminar *THEOR* laminar
laminare Strömung *f* laminar (viscous) flow
Laminar-Turbulenz-Umschlag *m* *THEOR* laminar-to-turbulent transition
Laminarität *f* *THEOR* laminar regime (flow mode)
Laminarwiderstand *m* laminar-type (viscous) restriction, choke
Landmaschinenhydraulik *f* agricultural hydraulics
Langhubflüssigkeitsfeder *f* long-stroke type liquid spring
Langhubzylinder *m* long-stroke cylinder
Langsamläufermotor *m* *auch Hochmomentmotor* low speed (high torque) motor, LSHT motor
Längsdrehschieber *m* combination sliding and rotary spool valve
Längsfestigkeitsprüfung *f* *FILTER* end load test
Längskerbe *f* *STROMV* axial notch
Längsverkettung *f* horizontal valve stacking, *s also* Modulverkettung *f*

Lässigkeit *f* *DICHT* permeability
Lastdiagramm *n* *s.* Lastgrößen-Zyklusdiagramm *n*; *s.* Last-Weg-Zyklusdiagramm *n*
Lastdruck *m* load pressure
lastdruckgesteuerte Pumpe *f* load-sensing (load-and-flow- sensing, load-compensated) pump
Lastdruckrückführung *f* *STELL* load-pressure feedback
Lastdrucksteuerung *f* *einer Pumpe* load-sensing (power-matching) control *of a pump*; *s also ED* feather *v*
Lastdrucksteuerung *f* **mit Druckabschneidung** load-sensing/pressure-limiting control
Lastempfindlichkeit *f* load sensitivity
Lastgrößen-Zyklusdiagramm *n* load-magnitude cycle plot, load plot
Lastmoment *n* load torque
Lastrückführung *f* *ZYL* load return
Lastschwankung *f* load variation
Lasttragfähigkeit *f* *FLÜSS* load-carrying capacity
Lastumkehr-Vorspannventil *n* *DRUCKV* overcenter valve
Last-Volumenstrom-Verhalten *n* load-flow characteristics
Last-Weg-Zyklusdiagramm *n* load-displacement cycle plot, load plot
Laufrad *n* *KOMPR* impeller
Laufspindel *einer Schraubenpumpe* driven screw, idler rotor *of a screw pump*
Laval-Geschwindigkeit *f* *kritische Geschwindigkeit des Luftstroms durch Drosselstellen THEOR* critical (choked) flow *airflow through restrictions*
leck leaky
leck sein leak *vi*

Leckage *f* leakage, seepage
Leckageausgleich *m* leakage compensation
leckagefrei leak-free
Leckagefreiheit *f* zero-leakage
Leckageinhibitor *FLÜSS* anti-leak additive
Leckageleitung *f* drain (leakage) line
Leckagestrom *m* leakage flow (current)
Leckdruck *m* leakage exhaust (drain line) pressure
lecken leak *vi*
leckend leaky
Leckflüssigkeit *f* leakage [fluid]
Leckflüssigkeitsabführung *f* drainage
  Leckflüssigkeitsabführung *f*: mit einer Leckflüssigkeitsabführung ausstatten *z. B. ein Gehäuse* drain *vt eg a case*
Leckflüssigkeitsanschluß *m* drainage (drain) port
Leckflüssigkeitsausgleich *m* leakage compensation
Leckflüssigkeitsfreiheit *f* zero-leakage
Leckflüssigkeitsgefälleleitung *f* gravity drain line
Leckflüssigkeitskanal *m* drain (leakage) passage
Leckflüssigkeitsrücklauf *m* drain return, drainback
Leckflüssigkeitsstrom *m* leakage flow (current)
Leckfreiheit *f* zero-leakage
Leckleitung *f* drain (leakage) line
Leckluft *f* leakage air
Lecköl *n* leakage oil
Leckölergänzung *f* in einem geschlossenen Kreislauf fluid make-up (replenishment) *to a closed circuit*
Leckölergänzungsleitung *f* im geschlossenen Kreislauf make-up (replenishing, slippage compensation) line *in a closed circuit* .
Leckquerschnitt *m* leakage area
Leckspalt *m* leakage gap (clearance)
Leckstelle *f* leak [point]; *z. B. Öl durch eine Leckstelle austreten lassen* leak *vt eg joint leaks oil*; *z. B. Luft dringt durch eine Leckstelle ein* leak *vi eg air leaks into the system*
Leckstrom *m* leakage flow (current); *Volumenstrom* leakage flow [rate], leakage rate
Leckstrom *m* **im Kippunkt** *PU/MOT* stalled leakage flow
Leckstrom *m* **zwischen den Anschlüssen** *PU/MOT* cross-port leakage, *in a pump also* outlet-to-inlet leakage
Leckverlust *m* leakage, seepage
Leckverluste *mpl* **bei Nenndrehzahl** *PU/MOT* running leakage
Leckverluststrom *m* leakage flow (current)
Leckvolumenstrom *m* leakage flow [rate], leakage rate
Leckwasser *n* leakage water
Leckweg *m* leakage path
Lederdichtung *f* leather packing
Leerhub *m* *ZYL* no-load (idle) stroke
Leerlaufdruck *m* *PU/MOT* idling pressure
Leerlaufregelung *f* *KOMPR* on-line/ off-line control
legieren *Hydraulikflüssigkeit* dope *v*, inhibit *v* *hydraulic fluid*
legiertes Öl *n* doped (inhibited) oil
Leicht[bau]zylinder *m* lightweight cylinder
Leistungs-Abmessungs-Verhältnis *n* capacity-to-size ratio

Leistungsbewegung *f*   *ZYL* power motion
Leistungsdiagramm *n*   *s.* Leistung-Weg-Diagramm *n*
Leistungsregelung *f*   *einer Pumpe* constant-horsepower (torque limiter, horsepower limiter) control *of a pump*
Leistungsregler *m*   *einer Pumpe* power compensator *of a pump*
Leistungsstrahl *m*   *LOGIKEL* main (power, principle) jet
Leistungsverhältnis *n*   *LOGIKEL* index of performance, IP
Leistungsverlust *m*   power loss
leistungsverzweigtes Getriebe *n*   hydromechanical (mixed hydrostatic/mechanical) transmission
Leistungszylinder *m*   *ANWEND* power cylinder
Leistungszyklusprofil *n*   *s.* Leistung-Weg-Diagramm *n*
Leistung-Weg-Diagramm *n*   power-displacement (-demand pattern) plot, power plot
Leitapparat *m*   *KOMPR* distributor
Leitblech *n*   *BEHÄLT* baffle [plate]
Leitblech *n*: mit Leitblechen ausstatten   *BEHÄLT* baffle *v*
leiten   *z. B. Flüssigkeit zum Verbraucher* port *v*, direct *v*, route *v*, pipe *v eg fluid to the actuator*
Leitplatte *f*   *BEHÄLT* baffle [plate]
Leitplatte *f*: mit Leitplatten ausstatten   baffle *v*
Leitrad *n*   *KOMPR* distributor
Leitring *m*   *der Flügelzellenmaschine* cam (track, contour) ring *of a sliding-vane unit*
Leitung *f*   flow line, pipeline, line, conduit, conductor; *Verbindungsleitung* connecting line, connector, junction
Leitungsabschnitt *m*   pipe run
Leitungsbruchventil *n*   *STROMV* flow fuse, automatic shut-off valve, maximum flow control valve
Leitungsdruck *m*   line pressure
Leitungseinbau *m*   [in-]line mounting
Leitungsfilter *m,n*   *für Rohrleitungseinbau* [in-]line filter
Leitungsnetz *f*   piping, pipework
Leitungsquerschnitt *m*   line [cross-sectional] area, line size
Leitungsventil *n*   *für Rohrleitungseinbau* in-line (direct) mounted valve, [in-]line valve
Leitungsverlust *m*   line loss *frequently pl*
Leitungsverteiler *m*   header
Leitungsverzweigung *f*   pipe (line) branching
Leitwert *m*   *reell*, Konduktanz *f THEOR* conductance
Lenkblech *n*   *BEHÄLT* baffle [plate]
Lenkblech *n*: mit Lenkblechen ausstatten   baffle *v*
Lenkkraftverstärker *m*   *ANWEND* steering booster (cylinder)
Liefermenge *f*   *KOMPR* delivery (discharge, flow) rate
liefern   *z. B. Flüssigkeit an einen Motor* deliver *v*, supply *v*, provide *v eg fluid to an actuator*
Lieferung *f*   supply, admission, delivery
Liniendichtung *f*   band (line) seal
Linsensteuerplatte *f*   *PU/MOT* lens-shaped valve (port) plate
Lippendichtung *f*   lip seal
Lippenscheibe *f*   *DICHT* bevel washer seal

Liter je Minute, l/min, L/min *unit of flow rate: 1 L/min = 0.2199 gpm*
Load-sensing-Pumpe *f* *lastdruckgesteuerte Pumpe* load-sensing (load-and-flow-sensing, load-compensated) pump
Load-sensing-Steuereinheit *f* load-sensing controller (flow compensator)
Load-sensing-Steuerung *f* *Lastdrucksteuerung einer Pumpe* load-sensing (power-matching) control *of a pump*; *s also ED* feather *v*
Load-sensing-Steuerung *f* **mit Druckabschneidung** load-sensing/pressure-limiting control
Load-sensing-Wegeventil *n* *WEGEV* load-sensing directional control valve
4-Loch-Flansch *m* *VERBIND* four-bolt flange
Logikelement *n* *LOGIKEL* fluid logic device
Logikventil *n* cartridge (hydraulic) logic valve, logic element (cartridge valve)
lösbare Verbindung (Verschraubung) *f* detachable (separable) fitting
Losbrechkraft *f* *PU/MOT* breakaway force
Losbrechmoment *n* *PU/MOT* breakaway torque
Lösekraft *f* *Steuerkolben WEGEV* unlocking force *valve spool*
lösen *z.B. Schlauchkupplung* disconnect *v* *eg hose coupling*; *Steuerkolben losreißen* unlock *v* *valve spool*
Loshälfte *f* *einer Schlauchkupplung* coupling half attached to the hose
losreißen *Steuerkolben* unlock *v* *valve spool*

Lösungsvermögen *n* *FLÜSS s. Gas- oder Luftlösungsvermögen n*
lötlose Verbindung (Verschraubung) *f* brazeless (solderless) fitting
Lötverbindung *f* brazed (soldered) fitting
Lötverschraubung *f* *s. Lötverbindung f*
LS-Ventil *n* *s. Load-sensing-Wegeventil n*
Luft *f* air; *Druckluft* compressed air, *compounds s also with* pneumatisch, Pneumatik..., Pneumo..., Druckluft...
Luft *f*: aus der Luft stammend airborne
Luft *f*: in der Luft enthalten airborne
Luft *f*: Luft aufnehmen *Flüssigkeit nimmt Luft auf* be aerated *fluid is aerated*
Luft *f*: Luft ausscheiden release *v* air, outgas *v*
Luft *f*: Luft entfernen dearate *v*, bleed *v*, vent *v* *entrained air*
Luft *f* im Ansaugzustand atmospheric (free) air
Luft *f* im Normzustand standard air
Luft *f* unter Druck pressurized (pressure) air, air under pressure
Luft *f* unter Norm[al]bedingungen standard air
Luftabscheidevermögen *n* *FLÜSS* air release, ability of air separation
Luftanschluß *m* pneumatic (compressed air, air) connection, air port
Luftaufbereiter *m* air conditioner [unit], filter-regulator-lubricator, FRL (*pl* FRLs)

Luftaufbereitung *f* air conditioning, compressed air preparation
Luftaufnahme *f* *durch das Öl* aeration *of the fluid*
Luftausscheidevermögen *n* *FLÜSS* air release, ability of air separation
Luftbedarf *m* [compressed] air demand
luftbelasteter Speicher *m* hydropneumatic (air-loaded, air-charged, air-oil, airdraulic) accumulator
Luftblase *f* *FLÜSS* air bubble
luftdicht airtight
Luftdruckspeicher *m* *s.* luftbelasteter Speicher *m*
Luftdurchsatz *m* *s.* Luftvolumenstrom *m*
Lufteindringstelle *f* *in einer Hydraulikanlage* air leak *in a hydraulic system*
Lufteinschluß *m* air trap (inclusion)
Luftfahrtflüssigkeit *f* aircraft fluid
Luftfahrthydraulik *f* *ANWEND* aircraft hydraulics
Luftfeder *f* air spring
Luftfilmförderung *f* *ANWEND* air film handling
Luftfilter *m,n* *HY* air filter; *PN* air [line] filter, pneumatic filter; *BEHÄLT* air breather filter
Luftfilter-Prüffeinstaub *m* air cleaner fine test dust, AC fine test dust, ACFTD
Luftfilter-Prüfgrobstaub *m* air cleaner coarse test dust, AC coarse test dust, ACCTD
Luftflasche *f* *BEHÄLT* air bottle, air cylinder
Luftgehalt *m* *FLÜSS* air content
luftgekühlt air-cooled, *also as a verb:* air-cool *v*

luftgekühlter Ölkühler *m* air-cooled (oil-to-air) oil cooler
luftgetrieben pneumatically (air-) powered, air-operated
lufthaltige Flüssigkeit *f* aerated fluid
Luftkompressor *m* air compressor
Luftkühlung *f* air cooling
Luftlager *n* *ANWEND* pneumatic (air) bearing
Luftleckage *f* *s.* Luftleckverlust *m*
Luftleckverlust *m* air leakage
Luftleitung *f* airline, [compressed-]air line, pneumatic line
Luftleitungsablaß *m* airline drain
Luftlösungsvermögen *n* *FLÜSS* air solubility, solubility of air
Luftmotor *m* [compressed-]air motor
Luftnetz *n* [compressed-]air mains
Luft-Öl-Kühler *m* air-cooled (oil-to-air) oil cooler
Luftprobenahme *f* air sampling
Luftsack *m* air trap (inclusion)
luftschaltender Magnet *m* *Trockenmagnet STELL* air-gap solenoid
Luftschlauch *m* air (compressed air, pneumatic) hose
Luftspalt *m* *in einem Magneten* air gap *in a solenoid*
Luftspeicher *m* *s.* luftbelasteter Speicher *m*
Luftstrahl *m* *LOGIKEL* air jet (beam, stream)
Luftstrom *m* air current (stream, flow); *Volumenstrom* air flow [rate]
Luftsystem *n* pneumatic (compressed air, air) circuit (system)
Luftumlauf *m* air circulation
Lüftung *f* *BEHÄLT* breathing, ventilation, aeration
Lüftungsöffnung *f* *BEHÄLT* breather

hole, ventilating eyelet, aeration opening
Lüftungsorgan *n*   BEHÄLT air breather
Luftventil *n*   pneumatic (compressed-air, air) valve
Luftverbrauch *m*   [compressed-]air consumption
Luftverdichter *m*   air compressor
Luftverteilung *f*   [compressed-] air distribution system
Luftvolumenstrom *m*   air flow [rate]
Luftzirkulation   air circulation
Luftzufuhr *f*   [compressed-] air supply
Luftzylinder *m*   pneumatic linear actuator, pneumatic (compressed-air, air) cylinder

M

m-Wert *m*   *s.* Viskositäts-Richtungskonstante *f*
Machzahl *f*   *Verhältnis zwischen lokaler Strömungs- und Schallgeschwindigkeit* THEOR Mach number, M ratio of local flow to sound velocity
Magnet *m*   VENTILE solenoid
Magnetablaßventil *n*   FILTER solenoid[-operated] drain valve
Magnetabscheider *m*   magnetic separator
Magnetanker *m*   solenoid armature
magnetbetätigtes Ventil *n*   solenoid-operated (-actuated) valve, solenoid valve
Magnetelement *n*   FILTER magnetic element
Magnetfilter *m,n*   magnetic filter
Magnetkolben *m*   ZYL magnet[ic] piston

Magnetkolbenzylinder *m*   magnet rodless (magnetic piston) cylinder
Magnetpatrone *f*   FILTER magnetic cartridge
Magnetsäule *f*   *s.* Magnetpatrone *f*
Magnetspule *f*   STELL solenoid coil
Magnetstelleinheit   VENTILE solenoid operator (actuator)
Magnetstopfen *m*   FILTER magnetic plug
Magnetstößel *m*   STELL solenoid push[pin]
Magnetventil *n*   solenoid-operated (-actuated) valve, solenoid valve
Magnetvorabscheidung *f*   FILTER magnetic pre-separation
Magnetzylinder *m*   *s.* Magnetkolbenzylinder *m*
Manometer *n*   [pressure] gauge
Manometerabsperrventil *n*   gauge isolating valve, gauge cock
Manometerdämpfer *m*   gauge snubber (pulsation damper)
Manometerprüfvorrichtung *f* **mit Gewichten**   dead-weight pressure gauge tester
Manschette *f umgangssprachlich noch für* Dichtung *f*; *s.* Dachmanschette *f*; *s.* Topfmanschette *f*
Manschettenrücken *m*   heel *of a lip ring*
Mantel *m*   *s.* Schlauchmantel *m*; *s.* Zylindermantel *m*
manuelle Stelleinheit *f*   PU/MOT manual (hand) control (controller, stroker)
Maschenweite *f*   FILTER mesh [size]
Maschine *f s z. B.* Radialkolbenmaschine *f*
maschinengebundene Unterplatte *f*

Massestrom

custom (circuit, cross drilled) manifold, manifold block, drilled plate
Massestrom *m*   *THEOR* mass flow [rate] ; *Massetransport* mass flow
Materialverträglichkeitsprüfung *f*   *FILTER* material compatibility test
Maximaldruckventil *n*   *s.* Druckbegrenzungsventil *n*
maximaler Drehwinkel *m*   *eines Schwenkmotors* maximum rotation [angle]
Mechanik *f* **flüssiger und gasförmiger Körper**   fluid mechanics
mechanisch-hydraulische Stelleinheit *f*   *PU/MOT* hydromechanical control, mechanical servo control
mechanische Betätigungseinrichtung *f*   *VENTILE* mechanical operator (actuator, control)
mechanische Stelleinheit *f*   *PU/MOT* mechanical control (controller, stroker); *VENTILE*   *s.* mechanische Betätigungseinrichtung *f*
mechanischer Verlust *m*   mechanical loss
mechanischer Wirkungsgrad *m*   *PU/MOT* mechanical efficieny
Medium *n*   hydraulic fluid (medium), *cf* Arbeitsmedium *n*
Mehrbereichsnebelöler *m*   proportional flow lubricator
Mehrbereichsöl *n*   high viscosity index oil
Mehrdrucksystem *n*   multi-pressure system
Mehrebenen-Rohrgelenk *n*   multiplane swivel joint
Mehrfachdrossel-Bremsansatz *m*   *ZYL* multiple-orifice (piccolo) spear
Mehrfachunterplatte *f*   multiple[-valve] manifold (subplate, subbase), multistation subplate
Mehrfachverteiler *m*   *VERBIND* header
Mehrfluidsystem *n*   multi-fluid system
Mehrkantenschieber *m*   *WEGEV* valve spool with multiple metering edges
Mehrkreispumpe *f*   multiple-flow (-section) pump, multiple pump
Mehrmembranenlogikventil *n*   *LOGIKEL* stacked-diaphragm logic valve
Mehrpositionszylinder *m*   positional (digital) cylinder
Mehrradpumpe *f*   multiple-gear pump
Mehrstellungszylinder *m*   *s.* Mehrpositionszylinder *m*
Mehrstrompumpe *f*   *s.* Mehrkreispumpe *f*
mehrstufiger Verdichter *m*   multi-stage compressor
mehrstufiges Ventil *n*   multi-stage valve
Mehrwege-Drehverbindung *f*   rotating manifold
membranbetätigtes Ventil *n*   diaphragm-operated (-actuated) valve, diaphragm valve
Membrandruckschalter *m*   diaphragm pressure switch
Membrandruckwandler *m*   diaphragm pressure transducer
Membranelement *n*   *LOGIKEL* diaphragm element
Membranenpaket *n*   diaphragm stack
Membranensatz *m*   *s.* Membranenpaket *n*
Membranfilter *m,n*   *z. B. aus Celluloseacetat* membrane filter   *eg of cellulose acetate*
Membranhub *m*   diaphragm stroke

Membrankammer *f*   *STELL* diaphragm chamber
Membrankolben *m*   *ZYL* diaphragm piston
Membran-Kugel-Element *n*   *LOGIKEL* diaphragm-ball (-sphere) element
Membranmanometer *n*   diaphragm pressure gauge
Membranmotor *m*   *ein Druckluftmotor* diaphragm motor   *type of air motor*
Membranpaketlogikventil *n*   stacked-diaphragm logic valve
Membranschutzvorrichtung *f*   *SPEICH* transfer barrier
Membransicherheitsventil *n*   hydraulic (pressure) fuse
Membranspeicher *m*   diaphragm accumulator
Membranventil *n*   *WEGEV* diaphragm valve; *membranbetätigtes Ventil* diaphragm-operated (-actuated) valve, diaphragm valve
Membranverdichter *m*   diaphragm compressor
Membranzylinder *m*   diaphragm cylinder
Meßblende *f*   orifice meter
Meßdüse *f*   flow nozzle
Meßfühler *m*   *s.* Sensor *m*
Meßpumpe *f*   *Dosierpumpe* metering pump
Meßturbine *f*   turbine (propeller) flowmeter
Meßwandler *m*   *s.* Sensor *m*
Metallabstreifer *m*   *ZYL* metal scraper
Metallfilter *m,n*   metal filter
metallgefaßte Dichtung *f*   metal-cased seal

Metallschlauch *m*   flexible metallic hose, metal flexible hose; *Wellrohrschlauch* corrugated metal hose
Mikroaggregat *n*   *ANWEND* micro power pack, *s also* Hydraulikaggregat *n*
Mikroemulsion *f*   *FLÜSS* microemulsion
Mikrofilter *m,n*   micronic filter
Mikrofilterung *f*   micronic filtration, superfiltration
Mikroglasfaserfilter *m,n*   microfiberglass filter
Mikronebelöler *m*   micro-mist (-fog) lubricator, extra-fine-fog lubricator
Mikropneumatik *f*   miniature (mini, micro) pneumatics
Mikropumpe *f*   miniature (mini, micro) pump
Mikrozylinder *m*   miniature (small, mini, micro, midget) cylinder
Mindestimpulszahl *f*   *LEIT* minimum total pulsation
mineralölbasische Flüssigkeit *f*   petroleum[-based] fluid, mineral-oil base fluid
Miniaggregat *n*   *ANWEND* mini (miniature) power pack, *s also* power pack
Miniaturpumpe *f*   miniature (mini, micro) pump
Miniaturzylinder *m*   miniature (small, mini, micro, midget) cylinder
Minipneumatik *f*   miniature (mini, micro) pneumatics
Mischbarkeit *f*   *FLÜSS* miscibility
mitreißen   *z.B. Luft FLÜSS* entrain *v   eg air*
Mitteldruckpumpe *f*   medium-pressure pump

Mitteldruckschlauch *m* medium-pressure hose
Mittelläufermotor *m* medium-speed motor
Mittelstellung *f* WEGEV center (middle, crossover) position, midposition
Mittelstellungsverhalten *n* WEGEV crossover characteristics
mittelviskos FLÜSS medium-viscosity
Mittelzapfen *m* PU/MOT pintle valve, valve spindle, porting pintle
Mittenzapfenbefestigung *f* ZYL intermediate trunnion mount
mittlerer Kompressionsmodul *m* FLÜSS secant (mean, average) bulk modulus
Mitschleppströmung *f* drag flow
MIZ *f* Mindestimpulszahl LEIT minimum total pulsation
Mobilhydraulik *f* ANWEND mobile hydraulics
Modellunterplatte *f* flowboard, fluid breadboard, patchboard
modulationsgesteuertes Ventil *n* modulating control valve, MCV
Modulventil *n* sectional (stack, stackable, modular, gang, sandwich) valve
Modulverkettung *f* sectional (stack, modular, gang, sandwich) mounting, valve stacking
Molekularsieb-Trockenmittel *n* molecular sieve desiccant
Monoblockventil *n* monoblock valve unit
Montageverschmutzung *f* FLÜSS built-in contamination
Montagewand *f* s. Ventilmontageplatte *f*
Motor *m* Hydraulikmotor hydraulic (fluid) motor; Druckstromverbraucher actuator
Motor *m* **für begrenzte Drehzahl** *s*. Drehwinkelmotor *m*
Motor *m* **mit konstantem Verdrängungsvolumen** fixed-displacement (constant-volume) motor
Motor *m* **mit unbegrenztem Drehwinkel** rotary motor
Motor *m* **mit veränderbarem Verdrängungsvolumen** variable-displacement (-volume) motor
Motoranschluß *m* WEGEV motor port (connection)
Motorbefestigung *f* motor mount
Motorbetrieb *m*: **im Motorbetrieb** functioning as rotary motor
Motordrehzahl *f* motor speed (rpm)
Motorgehäuse *n* motor case (housing, casing)
Motorgeräusch *n* motor noise
Motorkennlinie *f* motor characteristics
Motorkennlinienfeld *n* family of motor characteristic curves, motor curves plot
Motorkörper *m* motor body
Motor-Pumpen-Aggregat *n* ANWEND [hydraulic] powerpack (power unit, power package)
Motorsteuerkreislauf *m* motor (actuator) control circuit (system)
Motorstromteiler *m* STROMV rotary flow divider
Motorverdrängungsvolumen *n* motor displacement [volume], displacement
Motorverhalten *n* motor characteristics
Motorverstellung *f* des hydrostatischen Getriebes motor displacement control of a hydrostatic transmission

Motorwirkungsgrad *m*   motor efficiency
Motorzylinder *m*   motor cylinder
Muffe *f*   *s.* Gewindemuffe *f*
Multikupplung *f*   *VERBIND* multitube connection, multi-connector, cluster fitting
Multipass-Prüfung *f*   *FILTER* multipass test
Multischlauch *m*   hose bundle
muskelkraftbetätigt   *STELL* manually operated (actuated, controlled)

N

nach *einem Element*   *s.* nachgeschaltete Leitung *f*
nachformen   *kopieren ANWEND* copy *v*
Nachfüllflüssigkeit *f*   make-up fluid
Nachfülleitung *f*   *im geschlossenen Kreislauf* make-up (replenishing, slippage compensation) line   *in a closed circuit*
Nachfüllstelle *f*   add-fluid point
nachgeschaltete Leitung *f*   downstream line
nachgiebige Dichtung *f*   flexible (elastic, resilient) seal
nachkühlen   aftercool *v*
Nachkühler *m*   aftercooler
Nachkühlung *f*   aftercooling
Nachlaufsteuerung *f*   *ANWEND* follower control
Nachsaugbehälter *m*   *BEHÄLT* surge tank
Nachsaugeventil *n*   make-up (replenishing) check
Nachschaltung *f*   *von Gasflaschen SPEICH* adding   *of gas bottles*

nachtropffreie Kupplung *f*   no-spill coupling
Nachwärmer *m*   after-warmer
Nachweis *m* **der einwandfreien Fertigungsqualität**   *FILTER s.* Integritätsprüfung *f*
Nadel[drossel]ventil *n*   *STROMV* needle valve
Näherungssensor *m*   *LOGIKEL* proximity sensor
nahtloses Rohr *n*   seamless tubing, solid drawn tubing
naphthenbasisches (naphthenisches) Öl *n*   naphthene-base (naphthenic) oil
nasser Magnet *m*   öldruckdicht *STELL* [oil-]immersed solenoid
nasser Torquemotor *m*   öldruckdicht *STELL* [oil-]immersed torque motor
Naßfilter *m,n*   wet filter
natives Öl *n*   *s.* Naturöl *n*
Naturöl *n*   native oil
ND   Nenndruck nominal (rated) pressure, pressure rating; *Nenndurchmesser, Nennweite = Innendurchmesser* [inside] nominal diameter, nominal ID (bore, size); *Niederdruck in der HY unter etwa 100 bar; in der PN unter etwa 0,1 bar* low pressure, L. P. *in HY up to about 1500 psi; in PN up to about 1.5 psi*
ND-Kreislauf *m*   *Niederdruckkreislauf, auch der Kreislauf mit dem niederen Druck* low-pressure circuit, L. P. circuit
ND-Seite *f*   *Niederdruckseite = die Seite mit dem niederen Druck* low-pressure side, L. P. side
Nebel *m*   *Ölnebel* oil mist (fog)
Nebelöler *m*   oil-mist (-fog) lubricator
Nebenkreislauf *m*   circuit branch

Nebenschluß *m* bypass
Nebenschluß *m*: **im Nebenschluß ableiten** bypass *v*, divert *v*, bleed[-off] *v*, vent *v*
Nebenschluß *m*: **im Nebenschluß fließen** bypass *v*
Nebenschlußkreislauf *m* *s*. Nebenstromkreislauf *m*
Nebenstrom *m* bypass [flow]
Nebenstromfilter *m,n* bypass[-flow] filter
Nebenstromfilterung *f* bypass[-flow] filtration
Nebenstromkreislauf *m* bleed-off (bypass) circuit
Nebenstromsteuerung *f* bleed-off (bypass) flow control
negative Überdeckung *f* WEGEV underlap, negative overlap
Nennbedingungen *fpl* nominal (rated) conditions
Nenndrehmoment *n* nominal (rated) torque, torque rating
Nenndrehzahl *f* nominal (rated) speed (rpm), speed (rpm) rating
Nenndruck *m* nominal (rated) pressure, pressure rating
Nenndurchfluß[strom] *m* nominal flow [rate], rated flow, capacity rating
Nenndurchmesser *m* Innendurchmesser [inside] nominal diameter, nominal ID (bore, size)
Nennfilterfeinheit *f* nominal filtration (filter) fineness (rating)
Nennförderstrom *m* nominal delivery (discharge) rate, delivery (discharge, output flow, flow) rating, nominal output flow
Nenngröße *f* bei Geräten Nennstrom und Nenndruck bzw. Nennweite nominal size *flow and pressure ratings or nominal ID of components*
Nennhub *m* Nennhublänge ZYL nominal (rated) stroke [length]
Nennkraft *f* nominal (rated) force, force rating
Nennleistung *f* abgegebene Leistung nominal (rated) horsepower (output, capacity), horsepower (output, capacity) rating, rating
Nennmoment *n* *s*. Nenndrehmoment *n*
Nennluftverbrauch *m* nominal (rated) air consumption, air consumption rating
Nennquerschnitt *m* nominal (rated) area (size, cross-section)
Nennstrom *m* *s*. Nennvolumenstrom *m*
Nenntemperatur *f* nominal (rated) temperature, temperature rating
Nennverdrängungsvolumen *n* nominal (rated) displacement [volume], capacity rating
Nennviskosität *f* FLÜSS nominal (rated) viscosity, viscosity rating
Nennvolumenstrom *m* nominal flow [rate], rated flow, capacity rating
Nennweite *f* Innendurchmesser [inside] nominal diameter, nominal ID (bore, size)
Netzanschluß *m* mains tap
Netzdruck *m* mains pressure
Netzleitung *f* distribution line
neubefüllen *z.B. mit Öl* refill *v* *eg with oil*
Neuöl *n* new (fresh) oil
neutrale Stellung *f* WEGEV normal (neutral) position, *wenn mittlere Stellung s. auch* Mittelstellung *f*

Neutralisationszahl *f* *FLÜSS* neutralization value, acid value, A. V.
newtonsche Flüssigkeit *f* *FLÜSS* newtonian fluid
Newtonsches Gesetz *n* *Zähigkeit THEOR* Newton's law *of viscosity*
NG *Nenngröße f bei Geräten Nennstrom und Nenndruck bzw. Nennweite* nominal size *flow and pressure ratings or nominal ID of components*
nicht ablaufdruckentlastet *WEGEV* internally drained
nicht armierte Dichtung *f* unreinforced (homogeneous) seal
nicht ausgeglichen *s.* nicht druckausgeglichen
nicht brennbar *FLÜSS s.* nicht entflammbar
nicht entlastet *s.* nicht druckausgeglichen
nicht korrodierend wirkend *FLÜSS* non-corrosive
nicht nachgiebige Befestigung *f* fixed (rigid) mount
nicht-newtonsche Flüssigkeit *f* *THEOR* non-newtonian fluid
nicht-rotierende Strömung *f* irrotational (non-rotational) flow
nicht-stationäre Strömung *f* unsteady flow
nicht stellbare Pumpe *f* fixed-displacement (constant-volume, constant-delivery) pump
nicht stellbarer Motor *m* fixed-displacement (constant-volume) motor
nicht umsteuerbarer Motor *m* non-reversible (nonreversing) motor
nicht unter Druck nonpressurized, unpressurized, atmospheric, pressureless

nicht verstellbare Drossel *f* *STROMV* fixed (non-adjustable) restrictor (restriction)
nicht-viskose Flüssigkeit *f* *THEOR* nonviscous (inviscid) fluid
nicht-viskositätserhöhte Flüssigkeit *f* unthickened fluid
nicht vorgespannt, drucklos nonpressurized, unpressurized, atmospheric, pressureless
nicht-vorgesteuertes Druckbegrenzungsventil *n* direct-acting relief valve
nicht-vorgesteuertes Ventil *n* direct-acting (one-stage) valve, direct valve
nicht-wasserhaltige Flüssigkeit *f* non-aqueous (non-water, waterfree, anhydrous) fluid
nicht wiederverwendbare Kupplung *VERBIND* permanent coupling
nicht zusammendrückbar incompressible
Nichtentflammbarkeit *f* *FLÜSS* non-[in]flammability, uninflammability, flameproofness
Niederdruck *m in der HY unter etwa 100 bar; in der PN unter etwa 0,1 bar* low pressure, L. P. *in HY up to about 1500 psi; in PN up to about 1.5 psi*
Niederdruckbegrenzungsventil *n* low-pressure relief valve
Niederdruckfilter *m, n* low-pressure filter
Niederdruckkreislauf *m auch: der Kreislauf mit dem niederen Druck* low pressure circuit, L. P. circuit
Niederdruckleitung *f* low-pressure line
Niederdruckluft *f* low-pressure air

Niederdruckpumpe *f* low-pressure pump
Niederdruckschlauch *m* low-pressure hose
Niederdruckseite *f* *die Seite mit dem niederen Druck* low-pressure side, L. P. side
niedrigviskos, dünnflüssig *FLÜSS* low viscous, thin, low-viscosity ...
Nippel *m* *Schlauchverbindung* tailpiece, nipple, insert *hose fitting*
Niveauanzeiger *m* *BEHÄLT* level indicator (gage)
niveaugesteuerter Ablaß *m* level-controlled drain
Nockenventil *n* cam-operated (-actuated) valve, cam valve
Normalbedingungen *f pl* *von Temperatur und Druck* standard (normal) conditions *of temperature and pressure; s. also* Luft *f* unter Normalbedingungen
Normbehälter *m* standard reservoir (tank)
Normdruck *m* standard (normal) pressure
Normtemperatur *f* standard (normal) temperature
Normverschraubung *f* standardized fitting
Normvolumen *n* standard (normal) volume
Normzustand *m* *s.* Normbedingungen *f pl*
Normzylinder *m* standard cylinder
Notbetätigung *f* *STELL* emergency control, override
Notpumpe *f* emergency (stand-by) pump
Notrückzugsschaltung *f* *ZYL* failsafe-retract circuit

Notventil *n* emergency valve
NPT = **National Pipe Taper** *thread USA,* National Standard Taper
NPT-Anschluß *m* NPT port
NPT-Rohrgewinde *n* NPT thread, National Standard Taper thread
NPTF = **National Pipe Taper Fuel and Oil** *thread USA,* National Taper Pipe
NPTF-Rohrgewinde *n* NPTF thread, Dryseal pipe thread, National Taper Pipe thread
Nullastdrehzahl *f* *PU/MOT* runaway (free) speed
Nullastvolumenstrom *m* *PU/MOT* no-load (zero-load) flow
Nulldurchgangsverhalten *n* *WEGEV* crossover characteristics
Nullförderdruck *m* *PU/MOT* deadhead pressure, *s also ED* deadhead *v*
Nullförderstellung *f* *PU/MOT* deadhead position
Nullförderstrom *m* zero delivery
Nullförderstrom *m*: mit Nullförderstrom laufen *Nullhubpumpe* deadhead *v* *pressure-compensated pump*
Nullhubpumpe *f* pressure-compensated (-controlled) pump
Nullhubregelung *f* pressure-compensator (-compensation) control
Nullhubregler *m* pressure compensator
Nullstellung *f* *WEGEV* normal (neutral) position, *wenn mittlere Stellung s. auch* Mittelstellung *f*
Nullüberdeckung *f* *WEGEV* zero (line-to-line) lap
Nullverdrängungsvolumen *n* *PU/MOT* zero displacement
Nutring *m* U-seal, U-cup, double lip seal

Nutzhub *m* *ZYL* power (working, operating) stroke
Nutzvolumen *n* s. Speichernutzvolumen *n*
NW *Nennweite = Innendurchmesser* [inside] nominal diameter, nominal ID (bore, size)
Nylonrohr *n* nylon tubing
NZ *Neutralisationszahl FLÜSS* neutralization value, acid value, A. V.

O

O-Ring *m* *DICHT* O-ring
O-Ringstopfen *m* O-ring plug
O-Ring-Verschraubung *f* face-seal (flat-faced) fitting
Oberflächenfilter *m,n* surface filter
Oberflächenfilterung *f* surface filtration
Oberflächenspannung *f* *FLÜSS* surface tension
Oberflächentemperaturfühler *m* surface temperature sensor
oberhalb *eines Elements* s. vorgeschaltete Leitung *f*
offen: in Ruhestellung *f* **offen** *WEGEV* normally open, NO
offener Kreislauf *m* open[-loop] circuit
Offendruck *m* *bei dem auf der gesamten Elementoberfläche Blasen austreten FILTER* open bubble point *when bubbles appear over the entire surface of the element*
offenporig *FILTER* open-pore
Offenstellung *f* *WEGEV* open (passing) position
öffnen *VENTILE* open *v, of a seated valve also* unseat *v, of a pressure control valve also* crack *v*, respond *v*, blow[-off] *v*
Öffnungscharakteristik *f* *einer Drossel* restriction (throttle, metering, flow, area) characteristics
Öffnungsdruck *m* **des Umgehungsventils** bypass cracking pressure
Öffnungsdruckstoß *m* opening pressure surge, opening shock
Öffnungsschlag *m* s. Öffnungsdruckstoß *m*
Öffnungsweg *m* *DRUCKV* opening stroke
Öffnungsventil *n* *WEGEV* normally closed (NC) valve
Öl *n* oil, *compounds s also with* hydraulisch, Hydraulik..., Hydro..., fluidtechnisch, Fluidtechnik..., Flüssigkeits...
Öl *n* **unter Druck** pressurized (pressure) oil, oil under pressure
Ölabnutzung *f* oil wear
Ölabscheider *m* oil remover (scrubber)
Ölanschluß *m* hydraulic (fluid power) connection (port)
Ölaufbereitungsgerät *n* oil conditioner (filtration unit, reconditioner), filter cart, portable filter unit, fluid transfer cart, kidney machine
Ölbadmagnet *m* *nicht druckdicht STELL* wet pin solenoid
ölbasische Flüssigkeit *f* oil-base fluid
Ölbehälter *m* fluid (hydraulic) reservoir (tank)
ölbeständig oil-resistant, resistant to oil
Öldampf *m* oil vapour
öldicht oiltight
Öldruckbremse *f* *ANWEND* hydraulic (oil) brake

öldruckdichter Magnet *m* STELL [oil-]immersed solenoid
öldruckdichter Torquemotor *m* STELL [oil-]immersed torque motor
öleingespritzter Kompressor *m* oil-injected compressor
Öleinspritzelement *m* oil injector
Öleinspritzung *f* oil injection
Öleintrag *m* oil carry-over
Öler *m* lubricator
ölfest oil-resistant, resistant to oil
Ölfilter *m,n* PN oil filter
Ölfiltrationswagen *m* *s.* Ölaufbereitungsgerät *n*
Ölförderstrom *m* *Druckluftschmierung* oil-feed (lubrication, drip) rate
ölfrei non-lubricated, non-lube, unlubricated, oil-free, oilless, dry
ölfreier Verdichter *m* non-lubricated (non-lube, oil-free, oilless) compressor
Ölgebrauchsdauer *f* oil life
Ölgehalt *m* oil content
Ölgerät *n* lubricator
ölgeschmiert oil-lubricated
Ölgrenznutzungsdauer *f* oil life
Ölhydraulik *f* hydraulics, fluid power technology; oil hydraulics; *s.* Ölhydraulikanlage *f*
Ölhydraulikanlage *f* oil hydraulic system
Öl-in-Wasser-Emulsion *f* *HSA-Flüssigkeit* FLÜSS oil-in-water emulsion, *wrongly* water-soluble oil
Ölkühler *m* oil cooler
Ölkühlung *f* oil cooling
öllos *s.* ölfrei
öllöslich FLÜSS soluble in oil, oil-soluble
Öl-Luft-Wärmeübertrager *m* air-cooled (oil-to-air) heat exchanger
Ölnebel *m* oil mist (fog)

Ölnebelgerät *n* oil-mist (-fog) lubricator
Ölnebelschmierung *f* oil-mist (-fog) lubrication
Ölprobe *f* oil sample
Ölregeneration *f* oil reclamation (purification)
Ölregenerator *m* oil regenerator
Ölreinigungsgerät *n* oil conditioner (filtration unit, reconditioner), filter cart, portable filter unit, fluid transfer cart, kidney machine
Ölrückgewinnung *f* oil reclamation (purification)
Ölrückgewinnungsgerät *n* oil reclaimer (purifier)
Ölsäule *f* oil column
Ölschaum *m* oil foam (froth)
Öl-Service-Aggregat *n* *s.* Ölreinigungsgerät *n*
Ölstand *m* BEHÄLT oil level
Ölsumpf *m* oil sump
Öltasche *f* oil trap
Ölumlauf *m* oil recirculating (circulation)
Ölumwälzung *f* *s.* Ölumlauf *m*
Ölversagen *n* oil breakdown
Ölverschleiß *m* **oil wear**
Öl-Wasser-Wärmeübertrager *m* water-cooled (oil-to-water) heat exchanger
Ölwechsel *m* oil change
Ölwechselfrist *f* oil change period
Ölwechselgerät *n* oil conditioner (filtration unit, reconditioner), filter cart, portable filter unit, fluid transfer cart, kidney machine
Ölwiederaufbereitung *f* oil reclamation (purification)
Ölwiederaufbereitungsgerät *n* oil reclaimer (purifier)
Ölzuführstrom *m* *Druckluft-*

schmierung oil feed (lubrication, drip) rate
Orbitmotor *m* orbit motor
örtlicher Druckverlust *m* local pressure loss
Oszillierantrieb *m zur Umgehung der Haftreibung STELL* dither drive *to eliminate breakaway friction*
oszillieren *mit geringer Amplitude* dither *v*
Ovalring *m DICHT* oval (elliptical) ring
Ovalring-Flügelzellenpumpe *f* elliptical cam ring-type vane pump
Ovalzylinder *m* oval cylinder
Oxidationsbeständigkeit *f FLÜSS* oxidation (oxidative) stability, resistance to oxidation
Oxidationshemmer *m s.* Oxidationsinhibitor *m*
Oxidationsinhibitor *m FLÜSS* oxidation inhibitor, antioxidant

P

P-Anschluß *m Druckanschluß WEGEV* pressure port
Packung *f* packing [seal], seal assembly
Papierbandfilter *m,n* wound-ribbon paper filter, paper ribbon filter
Papierfilter *m,n* paper filter
Papierscheibenfilter *m,n* paper-disk (-plate, -washer) filter
Papiersternfilter *m,n* pleated paper filter
parabolischer Bremsansatz *m ZYL* parabolic spear
paraffinbasisches (paraffinisches) Öl *n* paraffin-base (paraffinic) oil

parallelschalten install (connect, pipe) *v* in parallel, parallel *v to, with*
Parallelschaltung *f* parallel circuit
Parallelspalt *m* parallel-walled gap
Partikel *n* particle, *s also ED* particulate
Partikelgrobzählung *f FLÜSS* raw particle count
Partikelzähler *m FLÜSS* particle counter
Partikelzählung *f FLÜSS* particle count
Pascalsches Gesetz *der hydrostatischen Druckfortpflanzung* Pascal's law *of hydrostatic pressure transmission*
passieren lassen pass *v eg valve passes fluid*
passive Vorsteuerung *f STELL* internal piloting
Passivdrossel *f Stelldrossel im Stromregelventil* variable orifice *in the compensated flow-control valve*
Patrone *f FILTER* filter cartridge
pedalbetätigtes Ventil *n* foot-operated (-actuated) valve, *wenn Drehpunkt am Pedalende, auch* pedal-operated (-actuated) valve, *wenn Drehpunkt nicht am Pedalende, auch* treadle-operated (-actuated) valve
Pfeifenreinigerprüfung *der Schwerentflammbarkeit FLÜSS* pipe-cleaner test *of fire resistance*
pfeilverzahnte Zahnradpumpe *f* herringbone-gear pump
pflanzenölbasische Flüssigkeit *f* vegetable-oil base fluid
Phasenkompensationsschalldämpfer *m* phase-shift muffler
phosphatesterbasische Flüssigkeit *f* phosphate ester base fluid

Phosphatesterflüssigkeit *f* phosphate ester fluid
piezoelektrischer Druckwandler *m* piezoelectric pressure transducer
piezoresistiver Druckwandler *m* piezo-resistive pressure transducer
Pitotrohr *n* *zur Druckmessung in Luftströmen* pitot tube *for pressure measurement in air currents*
Plansitzventil *n* globe valve
Plastikrohr *n* *als Halbzeug* plastic tubing
Platte *n* *einer Zahnradpumpe* plate of a gear pump; *FILTER* disk, plate, washer
Plattenfedermanometer *n* diaphragm pressure gauge
Plattenfilter *m,n* disk (plate) filter
Plattenpaket *n* *FILTER* disk (plate, washer) pack
Plattensitzventil *n* *s.* Plansitzventi *n*
Plattenspalt *m* *THEOR* flat-plate clearance
Plattenventil *n* *WEGEV* flat-slide (plate) valve
Plattenwärmeübertrager *m* plate heat exchanger
Plattenzahnradpumpe *f* sandwiched gear pump
Platzdruck *m* burst pressure
plötzliche Erweiterung *f* *THEOR* sudden (abrupt) enlargement
plötzliche Verengung *f* *THEOR* sudden (abrupt) contraction
Plunger *m* *Tauchkolben ZYL* ram, plunger
Pneumatik *f* pneumatics, *compounds s also with* **pneumatisch, Pneumo ... , Druckluft ..., Luft ...;** *s.* Pneumatikanlage *f*

Pneumatikanlage *f* pneumatic (compressed air, air) system
Pneumatikanschluß *m* pneumatic (compressed air, air) connection (port)
Pneumatikantrieb *m* *ANWEND* pneumatic (compressed air, air) drive
Pneumatikanwender *m* user of pneumatic equipment, pneumatics user
Pneumatikbauteil *n* *s.* Pneumatikelement *n*
Pneumatikelement *n* pneumatic component (element)
Pneumatiker *m* pneumatic engineer
Pneumatikfilter *m,n* air [line] filter, pneumatic filter
Pneumatikgerät *n* *s.* Pneumatikbauteil *n*
Pneumatikhersteller *m* manufacturer of pneumatic equipment, pneumatics manufacturer
Pneumatikindustrie *f* pneumatics industry
Pneumatikingenieur *m* *s.* Pneumatiker *m*
Pneumatikkomponente *f* *s.* Pneumatikbauteil *n*
Pneumatikleitung *f* airline, air (compressed air, pneumatic) line
Pneumatikmotor *m* [compressed] air motor
Pneumatiknetz *n* [compressed] air mains
Pneumatikschaltung *f* pneumatic (compressed air, air) circuit (system)
Pneumatikschaltzeichen *n* pneumatic symbol
Pneumatikschlauch *m* [compressed] air (pneumatic) hose
Pneumatiksteuerung *f* *ANWEND* pneumatic (compressed air, air) control

Pneumatiksymbol *n* pneumatic symbol
Pneumatiksystem *n* *s.* Pneumatikanlage *f*; *s.* Pneumatikschaltung *f*
Pneumatikventil *n* pneumatic (compressed air, air) valve
Pneumatikzubehör *n* pneumatic accessories
Pneumatikzylinder *m* pneumatic linear actuator, pneumatic (compressed air, air) cylinder
pneumatisch pneumatic; *compounds s also with* Pneumatik ... , Pneumo ... , Druckluft ... , Luft ...
pneumatisch-akustischer Sensor *m* *LOGIKEL* acoustic wave sensor, fluidic ear
pneumatisch angetrieben pneumatically powered, air-powered (-operated)
pneumatisch vorgesteuert pneumatically (air-)piloted
pneumatische Achse *f* *für Positionieraufgaben* pneumatic axis *for positioning*
pneumatische Betätigungseinrichtung *f* *STELL* pneumatic (air) operator (actuator, control)
pneumatische Digitaltechnik *f* pneumatic (air) logic control, ALC
pneumatische Kupplung *f* *ANWEND* pneumatic (air) clutch
pneumatische Leistung *f* pneumatic (fluid) power
pneumatische Logik *f* *s.* pneumatische Digitaltechnik *f*
pneumatische Stelleinheit *f* *s.* pneumatische Betätigungseinrichtung *f*
pneumatische Steuerung *f* pneumatic (compressed air, air) control

pneumatischer Antrieb *m* pneumatic (compressed air, air) drive
pneumatischer Drehantrieb *m* *s.* Pneumatikmotor *m*
pneumatisches Fördern *n* *ANWEND* pneumatic (air) conveying
pneumatisches Relais *n* *LOGIKEL* pneumatic relay
Pneumo ... : *compounds s also with* pneumatisch, Pneumatik ... , Druckluft ... , Luft ...
Pneumohydraulik *f* hydropneumatics, air-hydraulics
pneumohydraulisch hydro-pneumatic, air-hydraulic
pneumohydraulische Stelleinheit *f* *STELL* air-controlled hydraulic pilot operator
pneumohydraulischer Druckübersetzer *m* air-oil (air-hydraulic) intensifier (booster)
pneumonisches Element *n* *Strahlelement LOGIKEL* fluidic
pneumopneumatische Stelleinheit *f* air-controlled air-pilot operator
pneumostatisch pneumostatic
Polhöhe *f* *FLÜSS* viscosity pole height
Polviskosität *f* *FLÜSS* pole viscosity
Polyblockpumpe *f* *Mehrkreispumpe* multiple-flow (-section) pump, multiple pump
Polyethylenrohr *n* polyethylene tubing
polyglycolesterbasische Flüssigkeit *f* polyglycol ester fluid
Polymerlösung *f* *Wasser-Glycol-Lösung* polyglycol solution, water-glycol fluid, water-glycol
Polyolesterflüssigkeit *f* polyol ester fluid

Polyurethanrohr *n* polyurethane tubing
Porenfilter *m,n* porous filter
Porenkörper *m* *LOGIKEL* porous plug
Positioniereinrichtung *f* *ZYL* stroke positioner
Positionsanzeigezylinder *m* position indicating cylinder
Positionserfassung *f* *ZYL* position sensing
Positionsgeber *m* *ZYL* position (displacement, piston, cylinder) sensor
positive Schaltüberdeckung *f* *WEGEV* crossover lap
positive Überdeckung *f* *WEGEV* overlap, lap
Potentiometerdruckwandler *m* potentiometric pressure transducer
Pourpoint *m* *FLÜSS* pour (flow) temperature
Pourpointerniedriger *m* *FLÜSS* pourpoint depressant
Prallfänger *m* shock absorber (suppressor), impact absorber, damper, decelerator
Prallplatte *f* *des Düse-Prallplatte-Ventils STELL* flapper *of the flapper-and-nozzle valve*
Prandtlsches Staurohr *n* *zur Messung des Geschwindigkeitsdrucks* pitot-static tube *for measurement of the dynamic pressure*
Prandtl-Zahl *f* *Wärmeleitung in strömenden Medien THEOR* Prandtl number, Pr *heat conduction of fluid flow*
Pressenkolben *m* *ANWEND* press ram (plunger)
Pressenwasser *n* *ANWEND* press water
Pressenzylinder *m* *ANWEND* press ram (cylinder)
Preßhub *m* *ANWEND* press[ing] stroke
Preßluft *f* compressed air
Preßpumpe *z. B. einer Umformmaschine* power pump *of a hydraulic press*; *Wasserhydraulik* [compressed] water pump
Preßverbindung *f* crimped-on (crimp, swaged-on) fitting
Preßwasser *n* compressed water
Preßwasseranlage *f* water [hydraulic] system, water hydraulic
Preßwasserpumpe *f* [compressed] water pump
Preßwassersystem *n* *s.* Preßwasseranlage *f*
Preßzahl *f* *FLÜSS* compressiblity [value]
Preßzylinder *m* *ANWEND* press ram (cylinder)
Primärsteuerung *f* *STROMV s.* Zulaufsteuerung *f*
Primär- und Sekundärverstellung *f* *eines hydrostatischen Getriebes* pump and motor displacement control *of a hydrostatic transmission*
Primärverstellung *f* *eines hydrostatischen Getriebes* pump displacement control *of a hydrostatic transmission*
Probehahn *m* *s.* Probenahmeventil *n*
Probenahme *f* *FLÜSS* sampling
Probenahmeventil *n* sampling valve
Profildichtung *f* moulded (preformed) seal
Profilweichdichtung *f* squeeze-type moulded seal
Proportional-Druckventil *n* proportional pressure control valve

Proportionalhydraulik *f* proportional valve technology *application of hydraulic proportional valves*
Proportionalmagnet *m* *STELL* force motor, proportional solenoid
Proportionalnebelöler *m* proportional flow lubricator
Proportional-Stromventil *n* proportional flow control valve
Proportionalventil *n* proportional valve
Proportionalwegeventil *n* proportional directional control valve
Prozeßluft *f* *für höchste Anforderungen* process air *of very high purity*
Prüfanschluß *m* test port (connection)
Prüfdruck *m* proof pressure
Prüffeinstaub *m* fine test dust
Prüfgrobstaub *m* coarse test dust
Prüfmanometer *n* test pressure gauge
Prüfstand *m* test bench (stand)
Prüfstaub *m* test dust
Prüfung *f* der Schwerentflammbarkeit *FLÜSS* fire resistance test
Prüfverschmutzung *f* test contaminant
Pulsation *f* *des Förderstroms PU/MOT* output pulsation (ripple, fluctuation), pump pulsation (ripple)
Pulsationsdämpfung *f* pulsation damping (attenuation)
pulsationsfrei *PU/MOT* pulsation-(ripple-)free, smooth
Pulsationsgeräusch *m* pulsation (ripple) noise
Pulsationsglättung *f* pulsation damping (attenuation)
pulsieren *PU/MOT* fluctuate *v*, pulsate *v*

Pumpbetrieb *m*: im Pumpbetrieb functioning as pump
Pumpe *f* pump
Pumpe *f* für eine Drehrichtung unidirectional pump
Pumpe *f* für umkehrbare Drehrichtung reversible (birotational, bi-directional) pump
Pumpe *f* in Kleinstbauweise miniature (mini, micro) pump
Pumpe *f* mit drehendem Förderteil rotary pump
Pumpe *f* mit drehrichtungsabhängiger Förderrichtung uni-flow pump
Pumpe *f* mit einem Förderstrom single[-flow] pump
Pumpe *f* mit Förderstromregler flow-compensated pump
Pumpe *f* mit großem Förderstrom high-flow pump
Pumpe *f* mit konstantem Verdrängungsvolumen fixed-displacement (constant-volume, constant-delivery) pump
Pumpe *f* mit mehreren Förderströmen multiple[-flow] pump, multi-section pump
Pumpe *f* mit niedrigem Förderstrom low-flow (low-volume) pump
Pumpe *f* mit Übernullsteuerung (mit umkehrbarer Förderrichtung) over-center (reversible) pump
Pumpe *f* mit veränderbarem Verdrängungsvolumen variable-displacement (-volume, -delivery) pump
Pumpelement *n* pumping element
Pumpenaggregat *n* *ANWEND* [hydraulic] powerpack (power unit, power package)
Pumpenanschluß *m* *WEGEV* pump port (connection)

Pumpenausfall *m* pump failure
Pumpenbefestigung *f* pump mount
Pumpendrehzahl *f* pump speed
Pumpenfrequenz *f* *Wellendrehfrequenz mal Anzahl der Pumpelemente* pump frequency *shaft frequency multiplied by number of pumping elements*
Pumpengehäuse *n* pump case (housing, casing)
Pumpengeräusch *n* pump noise
Pumpenkennlinie *f* pump characteristics
Pumpenkennlinienfeld *n* family of pump characteristic curves, pump curves plot
Pumpenkörper *m* pump body
Pumpen-Motor-Einheit *f* *arbeitet als Pumpe oder Motor* pump/motor unit *functions either as pump or as rotary motor*
Pumpenrad *n* *der Strömungskupplung* input impeller *of a hydrodynamic coupling*
Pumpensektion *f* *einer Mehrkreispumpe* multiple pump section
Pumpensteuerkreislauf *m* pump control circuit (system)
Pumpensteuerung *f* *s.* Pumpensteuerkreislauf *m*
Pumpenstrom *m* pump delivery (discharge rate, discharge, output, flow rate, flow)
Pumpenträger *m* pump bracket
Pumpenverdrängungsvolumen *n* [pump] displacement, output [volume]
Pumpenverhalten *n* pump characteristics
Pumpenversagen *n* pump failure
Pumpenverstellung *f* *eines hydrostatischen Getriebes* pump displacement control *of a hydrostatic transmission*
Pumpenwirkungsgrad *m* pump efficiency
Pumpenzylinder *m* pump cylinder

Q

Quadrat[schnur]ring *m* *DICHT* square ring
Quarzdruckwandler *m* piezoelectric pressure transducer
quasi-drehungsfreie (quasi-nichtrotierende) Strömung *f* *THEOR* quasi-irrotational (-nonrotational) flow
Quelle *f* *THEOR* source
Quellverhalten *n* *DICHT* swell characteristics
Querschieberventil *n* *WEGEV* gate valve
Querschnittsänderung *f* change of sectional area, section change
Querstrahlelement *n* *LOGIKEL* transverse impact modulator, TIM
Quetschöl *n* *PU/MOT* trapped (entrapped, pocketed) oil
Quetschölnut *f* *PU/MOT* trapping relief groove
Quetschverbindung *f* crimped-on (crimp, swaged-on) fitting

R

radiale Dichtung *f* radial seal
Radialkolbeneinheit *f* radial piston pump *or* motor
Radialkolbengetriebe *n* radial-piston transmission

Radialkolbenmaschine *f*   s. Radialkolbeneinheit *f*
Radialkolbenmotor *m*   radial piston motor, *compounds s with* Radialkolbenpumpe *f*
Radialkolbenpumpe *f*   radial piston pump
Radialkolbenpumpe *f* **mit Außenbeaufschlagung**   peripherally ported radial piston pump, radial piston pump with exterior admission
Radialkolbenpumpe *f* **mit außenliegender Hubkurve (mit äußerer Kolbenabstützung)**   externally-guided radial pump
Radialkolbenpumpe *f* **mit feststehendem Kolbenträger**   fixed-block radial pump
Radialkolbenpumpe *f* **mit Innenbeaufschlagung**   centrally ported radial piston pump, radial piston pump with interior admission
Radialkolbenpumpe *f* **mit innenliegender Hubkurve (mit innerer Kolbenabstützung)**   centrally guided radial pump
Radialkolbenpumpe *f* **mit rollengeführten Kolben**   rolling-piston radial pump
Radialkolbenpumpe *f* **mit rotierendem Kolbenträger**   rotary-block radial pump
Radialkolbenpumpe *f* **mit Steuerzapfen**   pintle-valve (pintle-ported, valve-spindle) radial-piston pump
Radialverdichter *m*   centrifugal compressor
Rad[naben]motor *m*   ANWEND wheel motor
Raketenflüssigkeit *f*   missile fluid

rastgesichertes Ventil *n*   detent-positioned valve
Raststellung *f*   STELL detent position
rattern   z. B. *Ventile* chatter *v*, flutter *v*   *eg valves*
Raum *m* **unter Druck**   pressurized volume
Reaktionsturbinenmotor *m*   reaction-type turbine motor
reales Fluid *n*   THEOR real (true) fluid
Rechteckkolben *m*   ZYL rectangular piston
Rechteckring *m*   DICHT rectangular section ring
rechtwinklig getrenntes Rohrende *n*   square-cut pipe end
Reduzierstutzen *m*   reducing nipple
Reduzierung *f*   s. Reduzierverschraubung *f*
Reduzierventil *n*   [pressure-]reducing valve, reducer; *für Druckluft* pressure regulator   *for compressed air*
Reduzierverschraubung *f*   reducer [fitting]
Reduzierverschraubung *f* **mit zweiseitigem Rohranschluß**   reducer (reducing) union
Referenzbedingungen *fpl*   reference conditions
Referenzdruck *m*   reference pressure
Referenzzustand *m*   reference conditions *pl*
Regeldrossel *f*   *Differenzdruckregler im Stromregelventil* controlled orifice, pressure compenator, hydrostat   *in a compensated flow-control valve*
Regeleinrichtung *f*   PU/MOT s. Stelleinrichtung *f*

Regelkolben *m* *im Stromregelventil* compensator spool *in a flow regulator*
Regelkurve *f* *eines Druckminderventils* flow (regulation) curve *of a pressure regulator*
Regeneration *f* **des Öls** oil reclamation (purification)
Regenerativtrockner *m* regenerative-type dryer
regenierbares Filterelement *n* cleanable (re-usable, recleanable) filter element, cleanable
regenerierbares Trockenmittel *n* regenerative desiccant
Regler *m* *s.* Druckluftregler *m*
Reibmoment *n* *THEOR* friction torque
Reibung *f* friction, *s also ED* viscous drag
reibungsarmer Zylinder *m* low-friction cylinder
Reibungsdämpfer *m* friction damper
reibungsfrei frictionless
reibungsfreie Flüssigkeit *f* *THEOR* nonviscous (inviscid) fluid
Reibungskraft *f* frictional force
Reibungsverlust *m* friction loss
Reihe *f*: **in Reihe schalten** install *v* (connect *v*, pipe *v*) in series, cascade *v*
Reihenkolbenmotor *m* in-line piston motor
Reihenkolbenpumpe *f* in-line piston pump
Reihenschaltung *f* series circuit
reinfluidisches Element *n* *ohne bewegte Teile LOGIKEL* pure fluid (non-moving part) element
Reinheitsgraduierung *f* *FLÜSS* cleanliness (contamination) code
reinhydraulisch all-hydraulic, purely hydraulic
reinhydraulischer Druckübersetzer *m* oil-to-oil intensifier (booster)
reinigungsfähiges Filterelement *n* cleanable (re-usable, recleanable) filter element, cleanable
Reinigungsflüssigkeit *f* cleaning fluid
Reinigungsöffnung *f* *BEHÄLT* clean-out opening
Reinluft *f* clean air
reinpneumatisch all-pneumatic, purely pneumatic
reinpneumatischer Druckübersetzer *m* air-to-air intensifier (booster)
Reinseite *f* *eines Filters* clean side *of a filter*
Reißscheibe *f* *DRUCKV* rupture (blow-out) disk
Restdruckablaßventil *n* *WEGEV* residual-pressure exhaust valve
Restentlüfter *m* *s.* Restdruckablaßventil *n*
Reversierpumpe *f* *mit umkehrbarer Förderrichtung* over-center (reversible) pump
Reynolds-Zahl *f* *THEOR* Reynolds' number, R. N.
richtungseinstellbare Verschraubung *f* adjustable fitting, banjo
richtungseinstellbare Winkelverschraubung *f* adjustable (swivel nut) elbow, single banjo
Richtungskonstante *f* *Viskositäts-Temperatur-Verhalten FLÜSS* [viscosity] slope coefficient *viscosity- temperature characteristics*
Richtungssteuerdeckel *m* *eines 2-Wege-Einbauventils* directional control cover *of a logic element*

Richtungssteuereinrichtung *f*   directional control
Richtungsventil *n*   routing valve, *but s* Wegeventil *n*
Ringdichtung *f*   seal ring, ring seal
Ringdrossel *f*   STROMV annular orifice
Ringkanal *m*   WEGEV circumferential (annular) groove
Ringkolben *m*   ZYL annular piston
Ringleitung *f*   ring main [system]
Ringnut *f*   WEGEV *s.* Ringkanal *m*
Ringraum *m*   annulus; ZYL rod-end (head-end, front-end) chamber
Ringraumfläche *f*   *s.* Ringraumquerschnitt *m*
Ringraumquerschnitt *m*   ZYL annulus [area]
Ringraumseite *f*   ZYL annulus side; *Ausfahrseite* rod (front, head) end
ringraumseitiger Druck *m*   ZYL rod-end (head-end, front-end) pressure
Ringspalt *m*   annulus, annular gap
Ringstrahlverstärker *m*   LOGIKEL focused-jet amplifier
Ringverschlußkupplung *f*   VERBIND ring-lock coupling
Ringwulstdämpfer *m*   *für Druckpulsation* toroidal damper   *to attenuate pressure pulsation*
Rippenrohr *n*   finned tube, fintube
rizinusbasische Flüssigkeit *f*   caster-base fluid
Rohr *n*   *gröberer Toleranz, größeren Innendurchmessers, größerer Wanddicke* pipe, *s also* ED piping, pipe *v*; *Präzisionsrohr* tube, *s also* ED tubing, pipe *v*; *Zylinderrohr* cylinder barrel
Rohr *n* **aus einem nichtrostenden Stahl**   stainless steel tubing

Rohrabschneider *m*   pipe (*or* tube) cutter, *s.* Rohr *n*
Rohrbiegevorrichtung *f*   pipe (*or* tube) bender, *s.* Rohr *n*
Rohrbogenverlust *m*   THEOR pipe bend loss
Rohrbruch *m*   pipe fracture
Rohrbruchventil *n*   flow fuse, automatic shut-off valve, maximum flow control valve
Rohrbündel *n*   tube (tubing) bundle
Rohrbündelwärmeübertrager *m*   tube-bundle heat exchanger
Rohreintrittsreibung *f*   THEOR pipe entrance friction
Röhrenwärmeübertrager *m* **mit Mantel**   shell-and-tube heat exchanger
Rohrentgrater *m*   pipe deburrer
Rohrfeder *f*   Bourdon (bourdon) tube (spring)
Rohrfederdruckschalter *m*   bourdon-tube pressure switch
Rohrfederdruckwandler *m*   bourdon-tube pressure transducer
Rohrfedermanometer *n*   bourdon-[-tube] pressure gauge
rohrförmiger Drosselwiderstand *m*   THEOR long hole restriction
Rohrgelenk *m*   swivel joint (connection), swivel
Rohrgruppenverbindung *f*   multi-tube connection, multi- connector, cluster fitting
Rohrhalter *m*   tube clamp, pipe support
Rohrkabel *n*   tube (tubing) bundle
Rohrkrümmer *m*   pipe (*or* tube) bend, *s.* Rohr *n*
Rohrkrümmerverlust *m*   THEOR pipe bend loss
Rohrleitung *f*   pipeline, tube line;

Rohrleitungsabschnitt

Gesamtheit von Rohren piping, pipework
Rohrleitungsabschnitt *m* pipe run
Rohrleitungseinbau *m* *z. B. von Filtern* in-line (line) mounting
Rohrleitungssystem *n* *s.* Rohrnetz *n*
Rohrnetz *n* piping, pipework
Rohrrauheit *f* pipe[wall] roughness
Rohrrauheitsverlust *m* *THEOR* pressure loss due to pipewall roughness
Rohrrauhigkeit *f* *s.* Rohrrauheit *f*
Rohrreibung *f* *THEOR* pipe friction
Rohrreibungsverlust *m* *THEOR* pipe friction loss
Rohrschelle *f* tube clamp, pipe support; *elastisch und dämpfend* pipe (*or* tube) cushion, *s.* Rohr *n*
Rohrschneidegerät *n* pipe (*or* tube) cutter, *s.* Rohr *n*
Rohrventil *n* *für Rohrleitungseinbau* in-line [mounted] valve, line (direct mounted) valve
Rohrverbindung *f* *s.* Rohrverschraubung *f*
Rohrverschluß *m* pipe (*or* tube) plug, *s.* Rohr *n*
Rohrverschraubung *f* pipe (*or* tube) fitting, *s.* Rohr *n*, *s also ED* union
Rohrverteiler *m* header
Rohrverzweigung *f* pipe (line) branching
Rohrwand[ung] *f* pipe (*or* tube) wall, *s.* Rohr *n*
Rohwasser *f* raw water
Rollenhebelventil *n* roller lever-operated (-actuated) valve, roller valve
Rollenstößelventil *n* roller plunger-operated (-actuated) valve, roller valve
Rollenverschlußkupplung *f* *VERBIND* roller-lock coupling

Rollflügel *m* *PU/MOT* roller vane
Rollflügelmotor *m* roller-vane motor
Rollflügelpumpe *f* roller-vane pump
Rollmembran *f* *SPEICH, ZYL* rolling diaphragm
Rollmembranzylinder *m* rolling-diaphragm cylinder
Rootsgebläse *n* Roots blower
Rootsmotor *m* Drehkolbenmotor lobed-element (lobe) motor
Rootspumpe *f* Drehkolbenpumpe lobed-element (lobe) pump
Rost- und Oxidationsinhibitor *m* *FLÜSS* rust and oxidation inhibitor, R&O inhibitor
Rostinhibitor *m* *FLÜSS* rust inhibitor, anti-rust additive
Rostschutzadditiv *n* *s.* Rostinhibitor *m*
Rotameter *n* Schwebekörper-Durchflußmesser [tapered-tube] rotameter
Rotationsdichtung *f* rotary seal
Rotationsmotor *m* *Motor mit unbegrenztem Drehwinkel* rotary motor
Rotationsviskosimeter *n* rotational (rotating cylinder) viscometer
rotierende Verbindung *f* rotary joint (connection, union), rotating distributor
rotierender Zylinder *m* rotating cylinder
Rotor *m* rotor
Rückbördelverbindung *f* inverted flare fitting
Rückdruck *m* return [line] pressure
Rückflüssigkeit *f* return (exhaust) fluid
Rückfluß *m* return [flow]
Rückflußleitung *f* return line
Rückführkanal *m* *STELL* feedback duct

Rückhalterate *f* *FILTER* filtration (filter) rating (fineness)
Rückhaltevermögen *n* *für Schmutz FILTER* dirt (retention, contaminant-holding, filter) capacity
Rückhub *m* *ZYL* return (in, inward, retract, retraction, withdrawal, draw-back) stroke
Rücklauf *m* return [flow]; *ZYL s.* Rückhub *m*
Rücklauf *m* **mit freiem Durchfluß** free[-flow] return
Rücklaufanschluß *m* *WEGEV* return port (connection)
Rücklaufdruck *m* return [line] pressure
Rücklauf-Druckbegrenzungsventil *n* port relief [valve]
Rücklauffilter *m,n* return[-line] filter
Rücklaufgeschwindigkeit *f* *ZYL* retraction (withdrawal, draw-back, return) speed
Rück[lauf]leitung *f* return line
Rücköl *n* return (exhaust) oil
Rückschlagventil *n* check [valve], non-return valve
Rückschlagventil *n* **für Bohrungseinbau** cartridge check [valve]
Rückspeiseschaltung *f* regenerative (differential) circuit
rückspülen *durch Stromumkehr säubern* backwash *v*, back-flush *v*
Rückstelldruck *m* reset pressure
Rückstrom *m* backflow
Rück[volumen]strom *m* return flow [rate]
Rückstromverteiler *m* *in einem Behälter* diffuser *in a reservoir*
Rückwasser *n* return (exhaust) water
Rückzug *m* *ZYL* return (in, inward, retract, retraction, withdrawal, draw-back) stroke
Rückzugfeder *f* *ZYL* return (retract) spring
Rückzuggeschwindigkeit *f* *ZYL* retraction (withdrawal, draw-back, return) speed
Rückzugzylinder *m* retract (return, draw-back, kicker) cylinder
ruhende Dichtung *f* static seal
Ruhestellung *f* *WEGEV* normal (neutral) position, *wenn mittlere Stellung s auch* Mittelstellung *f*
Ruhestellung *f*: **in Ruhestellung durchlässig** *LOGIKEL* normally passing
Ruhestellung *f*: **in Ruhestellung geschlossen** *WEGEV* normally closed, NC
Ruhestellung *f*: **in Ruhestellung nicht durchlässig** *LOGIKEL* normally nonpassing
Ruhestellung *f*: **in Ruhestellung offen** *WEGEV* normally open, NO
Rund[dicht]ring *m* O-ring
Rundringstopfen *m* O-ring plug
Rundschnur *f* O-section sealing (packing) strip

S

SAE = Society of Automotive Engineers USA
SAE-Anschluß *m* *gerades Gewinde, mit O-Ring* SAE port
SAE-Flansch *m* *geteilter 4-Loch-Flansch* SAE flange *split 4-hole flange*
SAE-Grad *m* *FLÜSS s.* Viskositätsklasse *f*

SAE-Klasse *f*   *FLÜSS* s. Viskositätsklasse *f*
SAE-Rohrgewinde *n*   gerade SAE thread
Sammelanschlußplatte *f*   custom (circuit, cross drilled) manifold, manifold block, drilled plat
Sammelleitung *f*   header [line]
Sattelring *m*   eines Dichtungssatzes female (back) support ring, female adaptor *of a V-ring assembly*
Sättigungsdruck *m*   saturated vapour pressure, saturation pressure
säubern: durch Stromumkehr säubern backwash *v*, back-flush *v*
Sauganschluß *m*   suction (input, inlet, intake, supply) port
Saugdrosselung *f*   *KOMPR* inlet (intake) throttling
Saugdruck *m*   suction (input, inlet) pressure
saugen   aus dem Behälter *PU/MOT* draw *v* *from the reservoir*
Saugen *n*   *PU/MOT* suction, sucking
Saugfähigkeit *f*   *PU/MOT* suction (intake) capacity
Saugfilter *m,n*   suction (intake, inlet) filter
Saughöhe *f*   *PU/MOT* suction head
Saughub *m*   *PU/MOT* suction stroke
Saugkanal *m*   suction (input, inlet, intake) channel
Saugleistung *f*   *PU/MOT* suction power (horsepower)
Saugleitung *f*   suction (input, inlet, intake) line
Saugnapf *m*   *ANWEND* suction cap
Saugniere *f*   *PU/MOT* suction (inlet) kidney
Saugraum *m*   suction (input, inlet, intake) chamber
Saugrohr *n*   im Druckminderventil aspirator *of a pressure regulator*
Saugseite *f*   suction (input, inlet, intake) side
Saugstrom *m*   s. Saugvolumenstrom *m*
Saugung *f*   *PU/MOT* suction, sucking
Saugventil *n*   *PU/MOT* suction (inlet) valve
Saugverhalten *n*   *PU/MOT* suction (intake) characteristics
Saugvermögen *n*   *PU/MOT* suction (intake) capacity
Saugvolumenstrom *m*   suction (input, inlet, intake) flow [rate]
Säurezahl *f*   *FLÜSS* neutralization value, acid value, A. V.
schädlicher Raum *m*   *ZYL* clearance (dead) volume
schälen   Schlauchdecke skive *v* hose cover
Schalldämpfer *m*   muffler, silencer, sound attenuator
Schalldämpfer-Filter *m*   muffler-reclassifyer
Schalldämpfer *m* **mit schallschluckendem Material**   absorptive muffler
schalldichtes Gehäuse *n*   sound-proof enclosure, cocoon
Schallgeschwindigkeit *f*   sound (acoustic) velocity, sonic speed, speed of sound, celerity
schallnahe Strömung *f*   transonic flow
schällose Schlauchverbindung *f*   no-skive type hose fitting
Schallschutzhaube *f*   sound-proof enclosure, cocoon
Schaltdruck *m*   *STELL* switching pressure
Schaltdruckstoß *m*   switching shock (pressure surge)

Schaltebene *f* *WEGEV* control plane
schalten *Ventil* shift *vi* and *vt*   valve
Schaltentlüftung *f* *WEGEV* crossover bleed, cross-bleed
Schaltfilter *m,n* switch-over filter
Schaltfolge *f* switching (stepping) sequence
Schaltgeschwindigkeit *f* *WEGEV* shifting velocity, shifting (switching, stroking) speed
Schaltkraft *f* *STELL* operating (actuating, control, stroke) force
Schaltleckstrom *m* *WEGEV* crossover leakage
Schaltmagnet *m* *im Gegensatz zum Proportionalmagneten STELL* on-off solenoid
Schaltplan *m* circuit diagram (drawing), symbolic (graphical) diagram
Schaltpumpe *f* dual-pressure (high-low) pump
Schaltschlag *m* switching shock (pressure surge)
Schaltstellung *f* *WEGEV* control (spool) position
Schaltstoß *m* *s.* Schaltschlag *m*
Schaltüberdeckung *f* *WEGEV* crossover lap
Schaltung *f* *Grundschaltung* basic circuit; *Kreislauf, System* circuit, system
Schaltung *f* **für Volumenstrombeeinflussung** demand flow circuit
Schaltung *f* **mit Spülung** scavenger circuit
Schaltungssimulator *m* flowboard, fluid breadboard, patchboard
Schaltventil *n* *SPEICH* unloading valve
Schälverbindung *f* Schlauchverbindung skive-type (cover-removed) fitting *hose fitting*
Schaltweg *m* *WEGEV* spool stroke
Schaltzeichen *n* symbol
Schaltzeit *f* *WEGEV* shifting (switching, stroking) time
schärfen *Schlauchende, abschälen Schlauchdecke* skive *v* hose cover
scharfkantige Blende *f* sharp-edged orifice
schärflose Schlauchverbindung *f* no-skive type hose fitting
Schärfverbindung *f* Schlauchverbindung *s*. Schälverbindung *f*
Schaum *m* foam, froth
Schaumbildung *f* *FLÜSS* foaming, frothing, foam (froth) formation
schäumen *FLÜSS* foam *v*, froth *v*
Schaumhemmer *m* *FLÜSS* anti-foaming additive, defoamer, foam inhibitor (depressant)
Schaumneigung *f* *FLÜSS* foaming tendency
Schäumwiderstand *m* *FLÜSS* foaming resistance, resistance to foaming
Scheibe *f* *FILTER* filter disk (plate, washer); *DICHT* seal washer, washer seal
Scheibendichtung *f* seal washer, washer seal
Scheibenfilter *m,n* disk (plate) filter
Scheibenkolben *m* *ZYL* [headed] piston
Scheibenkolbenzylinder *m* piston-type cylinder
Scheibenpaket *n* *FILTER* disk (plate, washer) pack
Scheibenzähler *m* wobble-plate (nutating disk) flowmeter

Scherfestigkeit *f*   *FLÜSS* shear stability (strength)
Schergefälle *n*   s. Schergeschwindigkeit *f*
Schergeschwindigkeit *f*   shear[ing] rate, rate of shear
Scherschlußventil *n*   *WEGEV* shear valve
Scherströmung *f*   s. Schleppströmung *f*
Scherung *f*   shear
Scherungsversagen *n*   Bruch der langen Kettenmoleküle *FLÜSS* shear breakdown
Schieber *m*   *WEGEV* valve spool
Schieber *m* **mit Dichtelementen**   *WEGEV* packed spool
Schieberbund *m*   *WEGEV* spool land
schiebergedichtetes Ventil *n*   *WEGEV* pack[ing]less-spool valve
schiebergesteuerte Pumpe *f*   spool-valve pump
Schieberhub *m*   *WEGEV* spool stroke
Schieberkammer *f*   *WEGEV* spool (control) chamber
Schieberkante *f*   *WEGEV* land edge
Schieberstellung *f*   *WEGEV* control (spool) position
Schieberventil *n*   *WEGEV* spool (plunger, piston) valve
Schieberweg *m*   maximaler *WEGEV* spool stroke; Verschiebung *WEGEV* spool travel (traverse, displacement)
Schiebung *f*   shear
Schiefscheibe *f*   *PU/MOT* swashplate
Schiffshydraulik *f*   marine hydraulics
Schlaffmembran *f*   *STELL* slack diaphragm
Schlamm *m*   aus Schmutzpartikeln kleiner etwa 5 µm *FLÜSS* silt, sludge   contaminant particles less than about 5 µm

Schlammbildung *f*   *FLÜSS* silting, sludge formation
Schlammbildungswiderstand *m*   *Schlammtragvermögen FLÜSS* resistance to sludge (deposit) formation
Schlauch *m*   [flexible] hose
Schlaucharmierung *f*   hose reinforcement (support)
Schlauchaufroller *m*   hose reel
Schlauchbruchventil *n*   hose-break valve
Schlauchbrücke *f*   hose carrier
Schlauchbündel *n*   hose bundle
Schlauchdecke *f*   hose cover
Schlaucheinbindung *f*   hose assembly
Schlauchklemme *f*   hose clamp
Schlauchkupplung *f*   quick-connect (-disconnect, -acting) coupling, quick disconnect
Schlauchkupplung *f* **mit Abreißsicherung**   breakaway-type (pull-break type) quick-disconnect coupling
Schlauchkupplung *f* **mit Drahtgeflechteinlage[n]**   wire braid hose
Schlauchkupplung *f* **mit einfacher Drahtgeflechteinlage**   single-wire (1-wire) braid hose
Schlauchkupplung *f* **mit mehrfacher Drahtgeflechteinlage**   multiple-wire braid hose
Schlauchkupplung *f* **mit Spiralarmierung**   spiral wire wrap hose
Schlauchkupplung *f* **mit Textilgeflechteinlage[n]**   fabric braid hose
Schlauchkupplung *f* **mit vierfacher Spiralarmierung**   4-spiral wire wrap hose
Schlauchkupplung *f* **mit zwei Draht-**

geflechteinlagen double wire (2-wire) braid hose
Schlauchleitung *f* hose (flexible) line; *einbaufertig montiert* hose assembly; *Schlauch* [flexible] hose
Schlauchmantel *m* hose cover
Schlauchpresse *f* crimping device, crimper
Schlauchpumpe *f* flexible-tube pump
Schlauchschelle *f* hose clamp
Schlauchschutzhülle *f* hose protector (guard)
Schlauchseele *f* hose inner tube
Schlauchspeicher *m* flexible hose accumulator
Schlauchsteg *m* s. Schlauchstütze *f*
Schlauchstütze *f* hose carrier
Schlauchumhüllung *f* hose cover
Schlauchverbindung *f* hose fitting, hose [end] coupling
Schlauchverbindung *f* **mit zweiseitigem Schlauchanschluß** union hose connector
Schlauchverschlingung *f* hose kinking
Schlauchverschraubung *f* s. Schlauchverbindung *f*
Schlauchverstärkung *f* hose reinforcement (support)
schleichend öffnendes Ventil *n* slow-opening valve
schleichend schließendes Ventil *n* slow-closure valve
Schleifring *m* hydraulischer s. Drehverbindung *f*
Schleppströmung *f* *THEOR* drag flow
schlichte Strömung *f* s. laminare Strömung *f*
Schlierenlinie *f* *THEOR* streak line
Schließdruck *m* *DRUCKV* closing (shutoff, reseat) pressure

Schließdruckstoß *m* closing (shutoff) pressure surge, closing (shutoff) shock
Schließelement *n* *eines Sperrventils* seating member *of a check valve*
schließen/sich *VENTILE* close *v*, *of a seat valve also* reseat *v*
Schließschlag *m* s. Schließdruckstoß *m*
Schließventil *n* *WEGEV* normally open (NO) valve
Schließweg *m* *DRUCKV* closing (shutoff) stroke
Schließzeit *f* *VENTILE* closing time
Schlitzdichtung *f* *ZYL* slot seal
schlitzgesteuerte Pumpe *f* port (valve) plate controlled pump
Schlitzzylinder *m* slotted cylinder
schlucken *Motor* absorb *v*, displace *v* *motor*
Schluckstrom *m* input (inlet, intake) flow [rate]
Schluckverhalten *m* motor characteristics
Schluckvolumen *n* motor displacement [volume]
Schlupfdrehzahl *f* *PU/MOT* speed loss
Schlupfstrom *m* *PU/MOT* slip [flow], slippage
schmieren, beölen *Druckluft* lubricate *v* *compressed air*
Schmierfähigkeit *f* *FLÜSS* lubricity, oiliness
Schmierfähigkeitsverbesserer *m* *FLÜSS* lubricity additive, oiliness agent
Schmutz *m* contaminant, contamination, dirt
Schmutzabscheidevermögen *n* *FLÜSS* contaminant release, ability of contaminant separation

Schmutzabstreifring *m* DICHT wiper [seal]; *gering nachgiebig:* scraper [seal]
Schmutzaufnahmevermögen *n* FILTER dirt (retention, contaminant-holding, filter) capacity
Schmutzbart *m* particulate accumulation
schmutzempfindlich contaminant-(dirt-)sensitive, sensitive to contamination (dirt)
Schmutzempfindlichkeit *f* contaminant (dirt) sensitivity
Schmutzempfindlichkeitsklasse *f eines Geräts* contaminant- (dirt-)sensitivity class (grade) *of a component*
schmutzfrei FLÜSS free of contamination, contaminant- free
Schmutzgehalt *m* FLÜSS contaminant (dirt) content
schmutzhaltig FLÜSS contamination-loaded, dirty
Schmutzkappe *f* dust (protector) cap
Schmutzkennwert *m* FLÜSS contamination (dirt) characteristic
Schmutzklasse *f* FLÜSS contamination (dirt) class
Schmutzkonzentration *f* contaminant (dirt) concentration
Schmutzniveau *n* FLÜSS contamination (contaminant, dirt) level
Schmutzpartikel *n* contaminant (dirt) particle
Schmutzquelle *f* contamination source
Schmutzseite *f eines Filters* contaminated (dirt, dirty) face *of a filter*
Schmutzteilchen *n* contaminant (dirt) particle
Schmutztragvermögen *n* FILTER dirt (retention, contaminant-holding, filter) capacity; FLÜSS contaminant- (dirt-)carrying capacity
Schnappbefestigung *f* ZYL snap mount[ing]
schnarren *z. B. Ventile* chatter *v,* flutter *v eg valves*
schnattern *s.* schnarren
Schneckenrohrfeder *f* spiral Bourdon tube
Schneidentonverstärker *m* LOGIKEL edgetone amplifier
Schneidkante *f* VERBIND cutting edge
Schneidlippe *f s.* Schneidkante *f*
Schneidring *m* VERBIND cutting sleeve (ferrule)
Schneidringverschraubung *f* bite (ferrule) fitting
schnell schließendes Ventil *n* fast (quick) closure valve
Schnelläufermotor *m* high-speed [low-torque] motor
Schnellentlüftungsventil *n* WEGEV quick-exhaust (- release) valve, rapid-escape valve
Schnellkupplung *f* quick-connect (-disconnect, -acting) coupling, quick disconnect
schnellösbare Kupplung *f s.* Schnellkupplung *f*
Schnellschaltventil *n* fast-acting valve
Schnelltrennkupplung *f s.* Schnellkupplung *f*
Schnittbild *n* cutaway diagram
Schottverschraubung *f* bulkhead fitting (connector)
Schottverschraubung *f* **mit zweiseitigem Rohranschluß** bulkhead union
Schrägachsenmotor *m* bent-axis (angled) axial piston motor

Schrägachsenpumpe *f* bent-axis (angled) axial piston pump
Schrägkolbenpumpe *f* inclined-piston pump
Schrägscheibe *f* *PU/MOT* swashplate
Schrägscheibenmotor *m* swashplate axial piston motor
Schrägscheibenpumpe *f* swashplate axial piston pump
Schrägtrommelmotor *m* bent-axis (angled) axial piston motor
Schrägtrommelpumpe *f* bent-axis (angled) axial piston motor
Schraube *f* *der Schraubenpumpe* rotor, screw *of the screw pump*
Schraubeneinheit *f* screw pump *or* motor
Schraubenmotor *m* screw motor
Schraubenpumpe *f* screw pump
Schraubenrohrfeder *f* helical Bourdon tube
Schraubenverdichter *m* helical (rotary screw) compressor
Schraubkolbenschwenkmotor *m* piston-and-helix (helical spline, helix) rotary actuator
Schrittmotor *m* *STELL* stepper (stepping, step) motor, stepper
Schrittzylinder *m* stepper-positioned (stepping) cylinder
Schub *m* *THEOR* shear
Schubkolbentrieb *m* linear actuator
Schutzdichtung *f* exclusion (external, protective) seal
Schutzkappe *f* dust (protector) cap
Schwallöl *n* *BEHÄLT* churning oil
schwanken *z. B. Förderstrom* fluctuate *v*, pulsate *v* *eg pump output*
schwappen *Öloberfläche im Behälter* slosh *vi* *oil surface in the reservoir*
Schwebekörper-Durchflußmesser *m* free-float (tapered-tube) flowmeter, rotameter
Schwefelgehalt *m* *FLÜSS* sulfur *USA* (sulphur *GB*) content
Schweißbehälter *m* welded reservoir (tank)
Schweißkegelverschraubung *f* socket welt fitting with cone extension, conical seal union fitting
Schweißkugelverschraubung *f* socket weld fitting with spherical seating surface, spherical seal union fitting
schweißlose Verschraubung *f* weldless fitting
Schweißnippelverschraubung *f* s. Schweißkegelverschraubung *f*
Schweißverschraubung *f* weld[ed] fitting
Schweißzylinder *m* welded cylinder
Schwellendruck *m* threshold pressure
Schwenkaugenbefestigung *f* *ZYL* eye (ear, single-ear) mount
Schwenkbefestigung *f* *ZYL* pivot (swivel) mount
Schwenkgabelbefestigung *f* *ZYL* clevis mount
Schwenkgehäuse *n* *einer Schrägachsenmaschine* swivel housing, tilting cylinder block *of a bent-axis unit*
Schwenkkopfmotor *m* bent-axis (angled) axial piston motor
Schwenkkopfpumpe *f* bent-axis (angled) axial piston pump
Schwenkkörper *m* s. Schwenkgehäuse *n*; s. Schwenkrahmen *m*
Schwenkmotor *m* rotary actuator
Schwenkmotor *m* **mit Drehkeilwelle** piston-and-helix (helical spline, helix) rotary actuator

Schwenkmotor *m* **mit einer Zahnstange** single-rack rotary actuator
Schwenkmotor *m* **mit Kugelumlaufmutter** ball screw rotary actuator
Schwenkmotor *m* **mit zwei Zahnstangen** double-rack rotary actuator
Schwenkrahmen *m* *einer Radialkolbenpumpe* tilting block *of a radial piston pump*
Schwenkscheibe *f* *PU/MOT* adjustable-angle (variable-angle) swashplate (cam plate)
Schwenkverschraubung *f* adjustable fitting, banjo
Schwenkwinkel *m* *der Schrägscheibe PU/MOT* swashplate (tilt) angle
Schwenkzapfenbefestigung *f* *ZYL* trunnion (stud) mount
Schwenkzylinder *m* *STELL* tilting cylinder
Schweredruck *m* gravity (gravitational) head
schwerentflammbare Flüssigkeit *f* fire resistant (FR) fluid
Schwerentflammbarkeit *f* *FLÜSS* fire resistance
schwerkraftrückgezogener Tauchkolben *m* gravity-return (weight-returned) ram
Schwerkraftrückzug *m* *ZYL* gravity (weight) return
Schwerkraftvorfüllung *f* *einer Pumpe* gravity flooding (feeding)
schwimmerbetätigt *STELL* float-operated (-actuated, -controlled)
schwimmerbetätigter Ablaß *m* *FILTER* float drain
Schwimmerschalter *m* *BEHÄLT* float level switch
Sechskantenschieber *m* *WEGEV* valve spool with six metering edges

Sechswegeventil *n* six-way (6-way) valve
Seitenplatte *f* *PU/MOT* wear (seal, pressure) plate
seitlich befestigter Zylinder *m* side mount[ed] cylinder
Sekantenkompressionsmodul *m* *FLÜSS* secant (mean, average) bulk modulus
Sekundärsteuerung *f* *s.* abflußseitige Stromsteuerung *f*
Sekundärverstellung *f* *eines hydrostatischen Getriebes* motor displacement control *of a hydrostatic transmission*
Sekundärviskosität *f* *FLÜSS* dilatational (second) viscosity
selbstansaugend self-priming
selbstausrichtende Verbindung *f* self-align fitting
selbstbördelnde Verbindung *f* self-flare fitting
selbstdichtende Kupplung *f* self-sealing coupling
Selbstentzündungstemperatur *f* *FLÜSS* auto- (self-, spontaneous, autogenous) ignition temperature, SIT, AIT
selbstgesteuertes Ventil *n* *STELL* internally piloted valve
selbstreinigendes Filterelement *n* self-cleaning filter element
selbstschmierend *FLÜSS* self-lubricating, self-lube
Selbststeuerdruck *m* *STELL* internal pilot pressure
Selbststeuerung *f* *STELL* internal piloting
selbsttätiger Ablaß *m* *FILTER* automatic drain
selbstwirkende Dichtung *f* pressure-

energized (-actuated, - activated) seal, self-energized (automatic) seal
Senkbremsventil *n*   lowering valve
Senke *f*   *THEOR* sink
senkrechtes Rohrstück *n*   standpipe
Senkventil *n*   lowering valve
Sensorverschraubung *f*   sensor fitting
Serie *f*: **in Serie schalten**   install *v* (connect *v*, pipe *v*) in series, cascade *v*
Serienschaltung *f*   series circuit
Servoantrieb *m*   servodrive
Servobremse *f*   *ANWEND* servo (power) brake
servogesteuert   *STELL* servocontrolled
Servohydraulik *f*   use of hydraulic equipment in closed-loop control systems
Servolenkung *f*   *ANWEND* power[-assisted] steering
Servomotor *m*   servomotor
Servopumpe *f*   servopump
Servoregler *m*   mit mechanischer Ausgangsgröße servomechanism, servo
Servostelleinheit *f*   *PU/MOT* servo control[ler] (stroker), [stroker] servo; *VENTILE* servoactuator
Servosteuerung *f*   *STELL* servo[control]
Servoventil *n*   servovalve
Servoverstärker *m*   servoamplifier
Servozylinder *m*   *STELL* servo cylinder
Sichel *f*   *PU/MOT* s. Trennsichel *f*
Sicherheitsschaltung *f*   *KREISL* safety circuit
Sicherheitsventil *n*   relief [valve], safety valve
Sickerweg *m*   leakage path
Siebfilter *m,n*   [woven-]screen filter, screen

Siebkorb *m*   bag-type strainer
Siebscheibenfilter *m,n*   wire-screen disk filter
Siebsternfilter *m,n*   pleated-screen filter
Siedepunkt *m*   *FLÜSS* boiling point (temperature), b. p.
Signalanschluß *m*   signal port
Silicageltrocknungsmittel *n*   silica-gel desiccant
Silicatesterflüssigkeit *f*   silicate ester fluid
siliconölbasische Flüssigkeit *f*   silicone fluid
sinken   z. B. Druck decay *v*, drop *v* eg pressure
Sinterbronzefilter *m,n*   sintered bronze filter
Sinterdrahtgewebefilter *m,n*   sintered wire-cloth filter
Sinterfaserfilter *m,n*   sintered fiber filter
Sintermetallfilter *m,n*   sintered metal-powder filter
Sinterplastikfilter *m,n*   sintered plastic filter
Sitz *m*   *VENTILE* valve seat
Sitzventil *n*   seated (seating, seat) valve
sitzventilgesteuerte Pumpe *f*   check-valve (seated-valve) pump
Sitzventilkolben *m*   poppet
Sommeröl *n*   summer-grade oil
Sonderkreislauffilterung *f*   off-line filtration
Sonderventil *n*   special-function valve
Sonderzylinder *m*   specialty (special) cylinder
Spalt *m*   Leckspalt leakage gap (clearance); Filterspalt filter[ing] gap
Spaltausgleich *m*   z. B. in der Zahn-

Spaltdichtung

radpumpe clearance compensation eg in a gear pump
Spaltdichtung *f* clearance seal
Spaltdrossel *f* gap-type restriction
Spaltfilter *m,n* edge[-type] filter
Spaltfilterung *f* edge filtration
Spaltformel *f* *THEOR* gap (clearance) formula
Spalthöhe *f* gap height
Spaltlänge *f* gap length
Spalträumer *m* *FILTER* scraper, knife
Spaltverlust *m* gap (clearance) loss
Spaltweite *f* gap height
Spannkreislauf *m* *ANWEND* clamping circuit
Spannmembran *f* *STELL* prestressed diaphragm
Spannzylinder *m* *ANWEND* clamping cylinder
Speicher *m* *HY* [hydraulic] accumulator; *PN* receiver; *LOGIKEL* memory (maintain postion) valve
Speicher *m* **mit Trennwand** separated (separator) accumulator
Speicher *m* **ohne Trennwand** non-separated (non-separator, free-surface) accumulator
Speicherblase *f* accumulator bag (bladder)
Speicherdruck storage pressure
Speicherkreislauf *m* accumulator circuit
speichern *Energie, Druck, Flüssigkeit* store *v* *energy, pressure, fluid*
Speichernutzvolumen *n* accumulator capacity
Speicherventil *n* *LOGIKEL* memory (maintain postion) valve
Speisedruck supply (feed, input, inlet) pressure

Speisedruckbegrenzungsventil *n* *im geschlossenen Kreislauf* charge pressure relief valve *in a closed circuit*
Speisedüse *f* *LOGIKEL* supply nozzle (tube)
Speiseleitung *f* input (inlet, intake) line, *also, to a pump* suction line, *also, to a motor* supply (feed, pressure) line
Speiseluft *f* *LOGIKEL* supply air
speisen deliver *v*, supply *v*, provide *v*
Speisepumpe *f* charge (charging, booster, supercharge, prefill) pump
Speisestrahl *m* *LOGIKEL* supply jet
Speisung *f* supply, admission, delivery
sperren *z. B. Strom, Leitung* block [off] *v*, isolate *v* *eg flow, line*
Sperrflügelpumpe *f* *s.* Sperrschieberpumpe *f*
Sperrplatte *f* *einer Steuersäule* port (top) blanking plate, end plate (cover) *of a valve stack*
Sperrschiebermotor *m* vanes-in-stator (stationary vanes) motor
Sperrschieberpumpe *f* vanes-in-stator (stationary vanes) pump
Sperrstellung *f* *WEGEV* closed-center (non-passing) position
Sperrtrommelmotor *m* [rotary] abutment motor
Sperrtrommelpumpe *f* [rotary] abutment pump
Sperrventil *n* check [valve]
Spezialdichtung *f* *kundenspezifische Dichtung* custom seal
Spezialventil *n* special-function valve
sphärischer Steuerspiegel *m* *PU/MOT* spherical valve plate
Spielausgleich *m* *z. B. in der Zahn-*

radpumpe clearance compensation eg in a gear pump
Spielausgleichsventil *n* *DRUCKV* backlash valve
Spindel *f* *der Schraubenpumpe* rotor, screw *of a screw pump*; *Ventilspindel* valve stem (spindle)
Spindelventil *n* stem-operated (screwdown) valve
spiralarmierter Schlauch *m* spiral wire wrap hose
Spiralrohr *n* coil[ed] tube
Spitzendruck *m* peak pressure
Splitflansch *m* *geteilt VERBIND* split flange
Splitflanschverbindung *f* split-flange fitting
Sprühprüfung *f* *der Schwerentflammbarkeit FLÜSS* spray test *of fire resistance*
Spülaggregat *n* s. Ölwechselgerät *n*
spülen flush *v*, rinse *v*; *Teil des Öls in hydrostatischen Getrieben mit geschlossenem Kreislauf austauschen* scavenge *v* exchange part of oil in closed-circuit hydrostatic transmissions
Spülflüssigkeit *f* flushing (rinsing) fluid
Spülpumpe *f* flushing pump; *im geschlossenen Kreislauf* make-up (slippage, boost, scavenger) pump
Spülung *f* *Außenfilterung* external (bulk) filtration
Stahlgußzylinder *m* cast steel cylinder
Stahlrohr *n* steel tubing
Stahlwollefilter *m,n* steel-wool filter
Standanzeiger *m* *BEHÄLT* level indicator (gage)
Standardbedingungen *fpl* *von Temperatur und Druck* standard (normal) conditions *of temperature and pressure*; *s also* Luft *f* unter Standardbedingungen
Standarddruck *m* standard (normal) pressure
Standardmanometer *n* industrial pressure gauge
Standardtemperatur *f* standard (normal) temperature
Standardvolumen *n* standard (normal) volume
Standrohr *n* standpipe
Stange *f* *Kolbenstange ZYL* piston rod
stangengezogenes Rohr *n* *DOM* tubing *drawn over mandrel*
stangenloser Zylinder *m* rodless cylinder
stark quellende Flüssigkeit *f* high-swell fluid
starre Befestigung *f* fixed (rigid) mount
starre Leitung *f* rigid line
Startdruck *m* *PU/MOT* starting pressure
Startviskosität *PU/MOT* starting (cold-start) viscosity
Statik *f* *flüssiger und gasförmiger Körper* fluid statics
stationärer Druck *m* steady[-state] pressure
stationäre Strömung *f* steady flow
Stationärhydraulik *f* industrial hydraulics
Stationärpneumatik *f* industrial pneumatics
statische Dichtung *f* static seal
statische Flüssigkeitsprobenahme *f* static fluid sampling
statischer Druck *m* static pressure
Stator *m* *PU/MOT* stator

Staubkappe *f* dust (protector) cap
Staudruck *m* *THEOR* dynamic pressure; *Gegendruck* back pressure
Stecknippelverbindung *f* insert fitting; *mit Wulstnippel* barb fitting
Steckpumpe *f* cartridge (plug-in) pump
Steckverbindung *f* plug-in fitting
Steg *m* *am Steuerschieber WEGEV* land
stehenbleiben infolge Überlastung *z. B. Hydromotor* stall *vi eg hydromotor*
Steilgewindeschwenkmotor *m* piston-and-helix (helical spline, helix) rotary actuator
stellbare Pumpe *f* variable-displacement (-volume, -delivery) pump
stellbarer Motor *m* variable-displacement (-volume) motor
Stelldrossel *f im Stromregelventil* variable orifice *in the compensated flow-control valve*
Stelldruck *m* *STELL* operating (actuating, control) pressure
Stelleinheit *f* *PU/MOT* control[ler], stroker; *VENTILE* operator, actuator, control [mechanism]
stellen *VENTILE* operate *v*, actuate *v*, control *v*; *z. B. Volumenstrom einstellen* meter *v eg flow*
Stellkolben *m* *PU/MOT* stroking (control) piston; *VENTILE* operating (actuating, control) piston
Stellkopf *m* *PU/MOT* control[ler], stroker; *VENTILE* operator, actuator, control [mechanism]
Stellkraft *f* operating (actuating, control, stroke) force
Stellmagnet *m* *VENTILE* solenoid
Stellmagnet *m* **mit einseitiger Wirkungsrichtung** single-acting solenoid
Stellmagnet *m* **mit zweiseitiger Wirkungsrichtung und Nullstellung** two-way solenoid
Stellmoment *n* operating (actuating, control) torque
Stellmotor *m* *VENTILE* servomotor
Stellungsgeber *m* *ZYL* position (displacement, piston, cylinder) sensor
n-**Stellungsventil** *n* *n*-position valve
Stellwegbegrenzer *m* *einer Pumpenstelleinheit* stroke limiter *of a pump control*
Stellzylinder *m* slave cylinder
Sternfilterelement *n* pleated filter element
Stetigventil *n* *Ventil mit Zwischenstellungen* infinite-position valve; *Proportionalventil* proportional valve
Steueranschluß *m* *VENTILE* control port (connection)
Steuerblende *f* *STROMV* metering orifice
Steuerblock *n* valve unit (block)
Steuerbund *m* *WEGEV* land
Steuerdruck *m* control pressure; *VENTILE* operating (actuating, control) pressure
Steuerdruckverhältnis *n* *LOGIKEL* control pressure ratio, CPR
Steuerdruckflußstrom *m* control flow [rate]
Steuerdüse *f* *LOGIKEL* control tube (nozzle)
Steuerebene *f* *WEGEV* control plane
Steuereingang *m* control input
Steuerelement *n* *VENTILE* valving (valve control) element
Steuerflüssigkeit *f* *STELL* control fluid

Steuerfläche *f*   *PU/MOT* valving surface
Steuerkammer *f*   *WEGEV* spool (control) chamber
Steuerkanal *m*   *LOGIKEL* control duct
Steuerkante *f*   *WEGEV* metering (control) edge
Steuerkerbe *f*   *STROMV* metering notch (groove)
Steuerknüppelbetätigung *f*   *STELL* joystick operator (actuator, control)
Steuerkolben *m*   *STROMV* metering piston
Steuerkolbenkante *f*   *WEGEV* land edge
Steuerkolbenverschiebung *f*   *WEGEV* spool travel (traverse, displacement)
Steuerkolbenverstärker *m*   *STELL* hydraulic-follower amplifier
Steuerkonus *m*   *PU/MOT* valving cone
Steuerkraft *f*   *STELL* operating (actuating, control, stroke) force
Steuerkreis[lauf] *m*   control circuit; *Vorsteuerkreislauf* pilot circuit
Steuerläufer *m*   *KOMPR* female rotor
Steuerleitung *f*   pilot control (pressure) line, pilot line
Steuerluft *f*   *für Meß- und Regelzwecke* instrument [quality] air
Steuermagnet *m*   solenoid
Steuermoment *n*   *STELL* operating (actuating, control) torque
Steuermotor *m* **mit linearer Bewegung**   *Proportionalmagnet STELL* force motor, proportional solenoid
steuern   *STELL* operate *v*, actuate *v*, control *v*; *Volumenstrom* meter *v* flow rate
steuern: im Abfluß steuern   meter-out *v*

steuern: im Zufluß steuern   meter-in *v*
Steuerniere *f*   *PU/MOT* kidney
Steueröffnung *f*   *WEGEV* metering orifice
Steuerplatte *f*   *PU/MOT* port (valve) plate
Steuerrohr *n*   *LOGIKEL* control tube (nozzle)
Steuerrückflüssigkeit *f*   *STELL* control return fluid
Steuersäule *f*   valve bank (stack), ganged valve
Steuerschaltung *f*   control circuit
Steuerscheibe *f*   *Hubscheibe PU/MOT* cam (angle) plate; *Ventilplatte* port (valve) plate
Steuerschieber *m*   *WEGEV* valve spool, *wenn Kolbenlängsschieber, auch:* [sliding] spool
Steuerschieberbund *m*   s. Steuerschieberkolben *m*
Steuerschieberkolben *m*   *WEGEV* spool land
Steuerschiebersteg *m*   s. Steuerschieberkolben *m*
Steuerspalt *m*   *WEGEV* metering orifice
Steuerspiegel *m*   *PU/MOT* valving surface
Steuerstrahl *m*   *LOGIKEL* control jet
Steuerstrom *m*   *STELL* control flow [rate]
Steuertechnik *f* **mit bewegten Teilen**   *LOGIKEL* moving part logic, MPL
Steuervolumenstrom *m*   *STELL* control flow [rate]
Steuerwalze *f*   *PU/MOT* valving cone
Steuerzapfen *m*   *PU/MOT* pintle valve, valve spindle, porting pintle
Steuerzylinder *m*   master cylinder

Stiftbefestigung *f*   ZYL pin mount[ing]
Stillsetzregelung *f*   KOMPR start-stop control
Stirnflächenbefestigung *f*   boden- oder ausfahrseitig ZYL face mount[ing] cap or front end
Stockpunkt *m*   FLÜSS solidification (setting) point; *als Pourpoint angegeben* FLÜSS pour (flow) temperature
Stockpunkterniedriger *m*   FLÜSS pour-point depressant
Stoppventil *n*   zum Anhalten des Kolbens in einer Zwischenstellung ZYL stop valve
Stoßdämpfer *m*   shock absorber (suppressor), impact absorber, damper, decelerator; *Druckstoßdämpfer* desurger, water-hammer (shock-pressure) absorber, surge suppressor, pressure snubber
Stößel *m*   des Elektromagneten STELL solenoid push[pin]; VENTILE valve plunger
stößelbetätigtes Ventil *n*   plunger-operated (-actuated) valve
Stößelventilkupplung *f*   VERBIND stem-valve coupling
Stoßfänger *m*   s. Stoßdämpfer *m*
Stoßmagnet *m*   STELL push-type solenoid
Stoßverlust *m*   THEOR impact loss
Stoßverschraubung *f*   butt-joint fitting
Stoßwelle *f*   shock wave
Strahlablenkelement *n*   LOGIKEL beam (jet) deflection element
Strahlablösung *f*   THEOR jet separation
Strahldruck *m*   jet impact pressure

Strahldüse *f*   des Strahlrohrventils impact nozzle   of the jet-pipe valve
Strahlelement *n*   LOGIKEL fluidic
strahlelementbetätigt   STELL fluidic-operated (-actuated, -controlled)
strahlgesteuert   LOGIKEL jet-controlled
Strahlkraft *f*   THEOR jet [impact] force, jet thrust
Strahlrohr *n*   des Strahlrohrventils jet (input) pipe (tube)   of the jet-pipe valve
Strahlrohrventil *n*   jet-pipe (-action) valve
Strahlunterbrechersensor *m*   LOGIKEL interruptible jet sensor
Strahlverdichter *m*   entrainment (jet) compressor
Strahlwinkel *m*   jet angle
Strom *m*   flow, flow stream, stream, current; *Förderstrom einer Pumpe* delivery (discharge) rate, discharge flow, output flow [rate], delivery, discharge, output; *frequently used for* Volumenstrom flow [rate], *more precisely* volumetric flow rate, volume rate of flow
Strombahn *f*   THEOR streamline, path line
Strombegrenzungsventil *n*   flow regulator, pressure-compensated flow-control valve; *als Stromsicherheitsventil* excess-flow (flow-limiting) valve
strömen   z. B. durch eine Öffnung pass *v*   eg a restriction, through a restriction, flow *v*   eg through a restriction
Stromfaden *m*   THEOR stream filament
Stromkompensation *f*   STROMV flow compensation

Stromlinie *f*   *THEOR* flow line, streamline; *Strombahn* streamline, path line
stromlos   *Elektromagnet STELL* de-energized *solenoid*
stromlos machen *den Elektromagneten* de-energize *v solenoid*
Strommesser *m*   flowmeter
Strommessung *f*   flow [rate] measurement
Strom-Proportionalventil *n*   proportional flow control valve
Stromquelle *f*   constant-flow supply (source)
stromregelndes Servoventil *n*   *STELL* flow control servovalve
Stromregelventil *n*   flow regulator, pressure-compensated flow-control valve
Stromregler *m*   s. Stromregelventil *n*
Stromrichtung *f*   s. Strömungsrichtung *f*
Stromröhre *f*   *THEOR* streamtube
Stromschreiber *m*   flow recorder
Stromschwankungsdauerfestigkeit *f*   *FILTER* flow fatigue strength
Stromschwankungsdauerfestigkeitsprüfung *f*   *FILTER* flow fatigue test
Stromsicherheitsventil *n*   excess-flow (flow-limiting) valve
Stromsensor *m*   flow sensor
Stromspitze *f*   *THEOR* flow surge (peak)
Stromstärke-Druck-Wandler *m*   current-to-pressure transducer, I/P-transducer
Stromsteuerverschraubung *f*   flow control fitting
Stromteiler *m*   s. Stromteilventil *n*
Stromteilventil *n*   flow-dividing valve, flow divider

Stromüberschuß *m*   excess (surplus) flow, overflow
Stromübertragungsfaktor *m*   *STELL* flow gain
Stromumkehr *f*: durch **Stromumkehr säubern** *FILTER* backwash *v*, backflush *v*
Strömung *f*   [fluid] flow, fluid motion, [flow] stream
Strömung *f* **mit Schallgeschwindigkeit**   sonic flow
Strömungsablösung *f*   flow separation
Strömungsabriß *m*   s. Strömungsablösung *f*
Strömungsanzeiger *m*   flow indicator
Strömungsausbildung *f*   development of flow, flow development
Strömungselement *n*   *LOGIKEL* fluidic
Strömungsform *f*   *THEOR* flow mode (pattern)
Strömungsgeschwindigkeit *f*   flow velocity
Strömungsgleichung *f*   flow equation
Strömungskraft *f*   flow[-produced] force
Strömungskupplung *f*   *ANWEND* hydrodynamic (hydraulic, fluid) coupling (clutch)
Strömungslehre *f*   fluid dynamics
Strömungsmaschine *f*   hydrodynamic (hydrokinetic) machine
strömungsmechanisches Element *n*   s. Strömungselement *n*
Strömungspumpe *f*   centrifugal (rotodynamic) pump
Strömungsrichtung *f*   flow direction, direction of flow
Strömungsrichtung *f*: gegen die **Strömungsrichtung**   s. vorgeschaltete Leitung *f*

Strömungsrichtung *f*: in Strömungsrichtung *s.* nachgeschaltete Leitung *f*
Strömungsschalter *m*   flow switch
Strömungsverdichter *m*   turbocompressor
Strömungswächter *m*   fluid indicator
Strömungsweg *m*   flow path, gallery, duct, passageway, channel
Strömungswiderstand *m*   *Eigenschaft* resistance to flow, flow (fluid) resistance; *Drosselstelle* orifice, throttle, choke, restrictor, restriction
Stromventil *n*   flow-control valve, control valve, flow valve
Stromventil *n* **für Bohrungseinbau** cartridge flow valve, flow (restrictor) cartridge
Stromvereinigungsventil *n*   flow integrator, flow-combiner valve
Stromverstärker *m*   flow amplifier
Stromverstärkung *f*   flow amplification (gain)
strukturintegrierter Behälter *m*   integral reservoir (tank)
Stufendruckverhältnis *n*   *KOMPR* stage pressure ratio
Stufenflügel *m*   *PU/MOT* step vane
Stufenkolben *m*   *STROMV* differential spool
Stufenpumpe *f*   staged pump
Stufenschaltmotor *m*   *s.* Schrittmotor *m*
Stulpmembran *f*   *ZYL* rolling diaphragm
Stützdruck *m*   *LOGIKEL* supporting pressure
Stützring *m*   *DICHT* anti-extrusion (back-up) ring; *eines Dichtungssatzes* male (inside) support ring, male adaptor *of a V-ring assembly*

Summenleistungsregelung *f*   *einer Pumpe* summation (horsepower) torque limiter control *of a pump*
summierender Impaktmodulator *m*   *LOGIKEL* summing impact modulator, SIM
Sumpffilter *m,n*   in-tank (in-reservoir, tank, reservoir, sump, submerged, immersion) filter
Superhochdruckadditiv *n*   *FLÜSS* extreme-pressure (EP) additive
Superhochdruckschlauch *m*   extreme-pressure hose
Symbol *n*   symbol
synchronisieren   *Gleichlauf herstellen* synchronize *v*
Synchronisierung *f*   synchronization
Synchronisierventil *n*   flow equalizer, equalizing (balancing) valve
Synchronlauf *m*   synchronized motion, synchronism
Synchronlaufschaltung *f*   synchronizing circuit
synthetische Flüssigkeit *f*   synthetic [fluid]
System *n*   *Anlage* system, *s eg* Hydrauliksystem *n*; *Kreislauf, Schaltung* circuit, system
Systemdruck *m*   system pressure
Systemluft *f*   *Luft im Druckluftsystem* system air   *the air in a compressed air system*
SZ   *Säurezahl FLÜSS* neutralization value, acid value, A. V.

T

T-Anschluß *m*   *s.* Behälteranschluß *m*
T-Gehäuse-Filter *m,n*   T-type (T-ported) filter

T-Ring *m*   *DICHT* T-ring, T-seal
T-Verschraubung *f*   tee [fitting]
T-Verschraubung *f* **mit Aufschraubabzweig**   female branch (side) tee
T-Verschraubung *f* **mit Aufschraubkappe im durchgehenden Teil**   female run tee
T-Verschraubung *f* **mit dreiseitigem Rohranschluß**   union tee
T-Verschraubung *f* **mit Einschraubabzweig**   male branch (side) tee
T-Verschraubung *f* **mit Einschraubzapfen im durchgehenden Teil**   male run tee, street tee
T-Verschraubung *f* **mit Schottabzweig**   bulkhead branch (side) tee
Tandemzylinder *m*   tandem cylinder
Tangentenkompressionsmodul *m*   *FLÜSS* tangent bulk modulus
Tank *m*   fluid (hydraulic) reservoir (tank)
tatsächlicher Förderstrom *m*   *PU/MOT* actual pump output (delivery, discharge rate, flow rate)
tatsächlicher Schluckstrom *m*   *PU/MOT* actual motor input (inlet, intake) flow rate
Tauchkolben *m*   *ZYL* ram, plunger
Tauchkolbenzylinder *m*   ram (plunger) cylinder
Tauchkolbenpumpe *f*   plunger pump
Tauchthermometer   immersion thermometer
Taumelscheibe *f*   *PU/MOT* wobble plate
Taumelscheibenmotor *m*   wobble-plate axial piston motor
Taumelscheibenpumpe *f*   wobble-plate axial piston pump
Taupunkt *m*   *FLÜSS* dewpoint
Teerzahl *f*   *FLÜSS* tar number

Teilchen *n*   particle, *s also ED* particulate
Teilchenzähler *m*   particle counter
Teilchenzählung *f*   particle count
Teilförderung *f*   *PU/MOT* partial delivery (discharge, output)
Teilkreislauf *m*   partial circuit
Teillastwirkungsgrad *m*   *KOMPR* partload efficiency
Teilstromfilter *m,n*   partial-flow filter
Teilstromfilterung *f*   partial-flow filtration
Teleskopdämpfer *m*   telescopic damper
Teleskopkolben *m*   *als Tauchkolben ZYL* telescopic plunger
Teleskoprohr *n*   *LEIT* telescopic line, expansion fitting; *ZYL* telescopic tube
Teleskopzylinder *m*   telescopic (telescoping) cylinder
Tellerventil *n*   *WEGEV* globe valve
temperaturbetätigtes Umgehungsventil *n*   thermal bypass valve
Temperaturkompensation *f*   *STROMV* temperature compensation
temperaturkompensiertes Stromventil *n*   temperature-compensated flow-control valve
Temperaturmessung *f*   temperature measurement
Temperaturschalter *m*   temperature (thermal) switch
Temperaturschreiber *m*   temperature recorder, thermograph
Temperatur-Viskositäts-Verhalten *n*   *FLÜSS* viscosity-temperature (VT) characteristics
Tensid *n*   *FLÜSS* surface-active agent, surfactant
Tensionsthermometer *n*   vapour-pressure (-tension) thermometer

textilarmierte (textilbewehrte) Dichtung *f* fabric-reinforced seal
Textilfilter *m,n* woven-cloth filter, fabric filter
textilgeflechtarmierter Schlauch *m* fabric braid hose
Textilmantel *m* *s.* Textilumflechtung *f*
Textilumflechtung *f* *LEIT* fabric cover
thermisch ausgelöstes Umgehungsventil *n* thermal bypass valve
thermische Beständigkeit *f* *FLÜSS* thermal stability
thermischer Durchflußmesser *m* thermal flowmeter
Thermistortemperaturfühler *m* thermistor temperature sensor
Tiefenfilter *m,n* [in-]depth filter
Tiefenfilterung *f* [in-]depth filtration
Tieftemperaturverhalten *n* *FLÜSS* low-temperature characteristics
Tieftemperaturviskosität *f* *FLÜSS* low-temperature viscosity
Topfmanschette *f* *DICHT* cup seal (ring)
Torque-Motor *m* *Drehmomentmotor STELL* torque motor
Totflüssigkeitsraum *m* fluid trap
Totölraum *m* oil trap
Totpunkt *m* *ZYL* dead point (center)
Totraum *m* *ZYL* clearance (dead) volume
Totwasser *n* dead water
Totzone *f* *WEGEV* deadband, deadzone
Toxizitätsprüfung *f* *FLÜSS* toxicity test
Trägheitslast *f* *KREISL* inertial load
Transportbewegung *f* *ZYL* transport motion

transsonische Strömung *f* transonic flow
Trapezring *m* *DICHT* trapezoidal ring
Treibscheibenzähler *m* wobble-plate (nutating disk) flowmeter
Treibspindel *f* *der Schraubenpumpe* driving screw, drive rotor *of the screw pump*
trennen *z. B. Schlauchkupplung* disconnect *eg* quick-disconnect coupling
Trennglied *n* *SPEICH* separator, diaphragm
Trennkolben *m* *SPEICH* separator piston, piston separator
Trennkörper *m* *PU/MOT* crescent-shape separator, crescent
Trennschlauch *m* separator tube
Trennsichel *f* *s.* Trennkörper *m*
Trennwand *f* *SPEICH s.* Trennglied *n*
trochoidenverzahnte Innenzahnradpumpe *f* progressing-tooth gear pump
trocken non-lubricated, non-lube, unlubricated, oil-free, oilless, dry
trockener Drehmomentmotor (Torque-Motor) *m* *STELL* air-gap torque motor
Trockenlaufverdichter *m* non-lubricated (non-lube, oil-free, oilless) compressor
Trockenluft *f* dry air
Trockenluftpneumatik *f* non-lube (dry) pneumatics
Trockenmagnet *m* *STELL* air-gap solenoid
Trockenmittel *n* *AUFBER* drying compound, desiccant
Trockner *m* *AUFBER* dryer, *also* drier, dehumidifier, dehydrator

Trocknungsbelüfter *m*   *BEHÄLT* desiccant-style breather
Trocknungsfilter *m,n*   water-removing filter
Tropfenanzahl *f*   *Nebelöler* number of drips *air-line lubricator*
Tropföler *m*   gravity-feed oiler
Trübungspunkt *m*   *FLÜSS* cloud point
Tülle *f*   *Schlauchverbindung* tailpiece, nipple, insert *hose fitting*
Turbine *f*   turbine
Turbinen-Durchflußmesser *m*   turbine (propeller) flowmeter
Turbinenmotor *m*   *Luftmotor* turbine motor *air motor*
Turbinenrad *n*   *der Strömungskupplung* runner, output impeller *of a hydrodynamic coupling*
Turbinen-Volumenstromsensor *m*   turbine flow sensor
Turbomotor *m*   *s.* Turbinenmotor *m*
Turbopumpe *f*   centrifugal (rotodynamic) pump
Turboverdichter *m*   turbo-compressor
turbulent   *THEOR* turbulent
turbulente Strömung *f*   turbulent flow
Turbulenz *f*   turbulence
Turbulenzeinbauten *mpl*   turbulator, turbulence inducer
Turbulenzprobenehmer *m*   *FLÜSS* turbulent sampler
Turbulenzverstärker *m*   *LOGIKEL* turbulence amplifier
Turbulenzwiderstand *m*   *THEOR* non-viscous restriction, orifice
turbulisieren   *z. B. einen Strahl LOGIKEL* make *v* turbulent *eg a jet*
Turmverkettung *f*   vertical valve stacking, sandwich mounting, *s also* Modulverkettung *f*

U

U-Rohrwärmeübertrager *m*   U-tube heat exchanger
Überdeckung *f*   *positive Überdeckung WEGEV* overlap, lap
Überdeckungswinkel *m*   *WEGEV* angle of overlap
Überdrehzahl *f*   *PU/MOT* overspeed
Überdruck *m*   pressure, gauge pressure; *unerwünschter Überdruck* overpressure
Überdruck *m***: unter Überdruck setzen**   pressurize *v*
Überdruckturbinenmotor *m*   reaction-type turbine motor
Überdruckventil *n*   relief [valve]
Übergangsstück *n* **mit Außengewinde**   *Gewindestutzen* male threaded union
Übergangsstück *n* **mit Innengewinde**   *Gewindemuffe* female threaded union
Überlastschutz *m*   overload protection
Überlastsicherung *f*   *s.* Überlastschutz
Übernullsteuerung *f*   *PU/MOT* over-center control
Überschallströmung *f*   supersonic flow
Überschneidungsentlüftung *f*   *WEGEV* crossover bleed, cross-bleed
Überstrom *m*   excess (surplus) flow, overflow
Überströmverlust *m*   *DRUCKV* overflow (excess-flow, surplus-flow) loss
Überströmventil *n*   relief [valve]
Ultraschall-Durchflußmesser *m*   ultrasonic flowmeter
Umfangskerbe *f*   *STROMV* circumferential groove

Umfangsleckverlust *m* in einer Zahn-
radpumpe peripheral leakage
in a gear pump
Umgebungsdruck *m* ambient pressure
Umgebungsluftdruck *m* atmospheric
pressure
Umgebungsverschmutzung *f* environmental contamination
umgehen bypass *v*
Umgehung *f* bypass
Umgehungsfilter *m,n* bypass filter
Umgehungsleitung *f* bypass line
Umgehungsventil *n* bypass valve
umgewälzte Ölmenge *f* recirculated
oil volume
Umkehrfilter *m,n* automatischer
Elementewechsel bei Stromumkehr
bidirectional filter
Umkehrmagnet *m* STELL double-
acting (dual-operation, push-pull
type, two-way) solenoid
Umkehrpunkt *m* ZYL dead center
(point)
umlaufen Flüssigkeit circulate *vi*
fluid
  umlaufen lassen Flüssigkeit circulate *vt* fluid
umlaufender Zylinder *m* ZYL
rotating cylinder
Umlaufkolbenverdichter *m* rotary
compressor
Umlaufstellung *f* WEGEV tandem
(open center) position
umleiten z. B. den Flüssigkeitsstrom bypass *vt*, divert *vt*,
bleed[-off] *vt*, vent *vt*
eg fluid flow
Umlenkblech *n* BEHÄLT baffle
[plate]
Umlenkblech *n*: **mit Umlenkblech**
ausstatten baffle *v*

Umlenkplatte *f* einer Steuersäule
[top] crossover plate
Umlenkverlust *m* bend loss
Umschaltdruck STELL switching
pressure
Umschaltfilter *m,n* switch-over filter
Umsteueransprechempfindlichkeit *f*
PU/MOT reversibility response
Umsteuerbereich *m* zwischen Saug-
und Druckzone PU/MOT cross-over
zone from suction to pressure side
Umsteuerentlastungsventil *n*
DRUCKV crossover (crossport) relief
valve
Umsteuer-Gegendruckventil *n*
DRUCKV overcenter valve
Umsteuermotor *m* reversible (birotational, bidirectional) motor
umsteuern z. B. Zylinder reverse *vi*
and *vt* eg cylinder
Umsteuerpumpe *f* für umkehrbare
Drehrichtung reversible (birotational,
bidirectional) pump
Umsteuerzeit *f* PU/MOT reversing
time
Umströmungsschaltung *f* regenerative
(differential) circuit
umwälzen Flüssigkeit recirculate *v*
fluid
Unausgeglichenheit *f* unbalance, imbalance
unbeölt non-lubricated, non-lube, unlubricated, oil-free, oilless, dry
unbewehrte Dichtung *f* unreinforced
(homogeneous) seal
Unbrennbarkeit *f* s. Unentflammbarkeit *f*
undicht leaky
unentflammbar FLÜSS non-[in]flam-
mable, non-flam, uninflammable,
flame-proof

Unentflammbarkeit *f* *FLÜSS* non-[in]flammability, uninflammability, flameproofness
ungefettetes Wasser *n* non-lubricated water
ungelöste Luft *f* *FLÜSS* entrained (free, undissolved) air
ungenügend Flüssigkeit *f* **erhalten Pumpe** starve *v* pump
Ungleichförmigkeit *f* **des Förderstroms** output pulsation (ripple, fluctuation), pump pulsation (ripple)
Universalrohrgelenk *n* spherical swivel joint; *Mehrebenenrohrgelenk* multi-plane swivel joint
unlegiertes Öl *n* uninhibited (undoped) oil
Unterdeckung *f* *WEGEV* negative overlap, underlap
Unterdeckungswinkel *WEGEV* angle of underlap
Unterdruck *m* vacuum, reduced pressure
Unterdruckbegrenzungsventil *n* vacuum-relief valve
Unterdruckwelle *f* depression wave
unterflächenmontiert *VENTILE* base- (foot-)mounted, *cf* flächenmontiert
unterhalb *eines Elements* *s.* nachgeschaltete Leitung *f*
Unterölmotor *m* immersed (submerged, wet mount, wet) motor
Unterplatte *f* subplate, subbase, manifold, mounting plate
Unterplatte *f*: **auf Unterplatte montieren** manifold *v*, manifold-mount *v*
Unterplatte *f*: **mittels Unterplatte verketten** manifold *v*

Unterplattenanbau *m* *VENTILE* manifold (subplate, subbase) mount[ing]
Unterplattenbefestigung *f* *PU/MOT* subplate (subbase) mount[ing]
Unterplattenventil *n* manifold (subplate, subbase) valve
Unterschallströmung *f* subsonic flow
unterteilen *BEHÄLT* baffle *v*
unverstärkte Dichtung *f* unreinforced (homogeneous) seal
unvollständige Füllung *f* **der Pumpe** pump [intake] starvation
Urverschmutzung *f* *FLÜSS* initial contamination

V

V-Kerbe *f* *STROMV* vee notch, V-notch
Vakuum *n* *Unterdruck* vacuum, reduced pressure
vakuumdicht vacuum-tight
Vakuumpumpe *f* vacuum pump
Vakuumsaugnapf *m* *ANWEND* vacuum suction cup
VDMA *m* = **Verein** *m* **Deutscher Maschinenbau-Anstalten**
Ventil *n* valve, control valve, *s also* ED valving
Ventil *n*: **mit Ventilen ausstatten** valve *v* *eg* single-valved coupling
Ventil *n* **für Batterieverkettung** sectional (stack, stackable, modular, gang, sandwich) valve
Ventil *n* **für Bohrungseinbau** cartridge [insert] valve
Ventil *n* **für Bohrungseinbau mit Flanschbefestigung** flanged cartridge valve

Ventil *n* **für Modulverkettung**
s. Ventil *n* für Batterieverkettung
Ventil *n* **für Rohrleitungseinbau** in-line [mounted] valve, line (direct mounted) valve
Ventil *n* **für Unterplattenanbau** subplate (subbase, manifold) valve
Ventil *n* **mit** *n* **Anschlüssen** *n*-port valve
Ventil *n* **mit Federabhub** spring-offset valve
Ventil *n* **mit Federrückzug** spring-return valve
Ventil *n* **mit positiver Schaltüberdeckung** *WEGEV* closed-crossover valve
Ventil *n* **mit scharfkantigen Schieberbunden** *WEGEV* square-land valve
Ventil *n* **mit Sperrstellung** *WEGEV* closed-center valve
Ventil *n* **mit Umlaufstellung** *WEGEV* open-center (tandem-center) valve
Ventil *n* **mit Zwischenstellungen** *Stetigventil* infinite-position valve
Ventilanschluß *m* valve port (connection)
Ventilaufnahmebohrung *f* cartridge valve (mounting) cavity
Ventilaufnahmeplatte *f* valve (mounting) panel
Ventilbatterie *f* valve bank (stack), ganged valve
Ventilbaugruppenumrandung *f* *in Schaltplänen* valve envelope *in system diagrams*
Ventilbetätigungseinrichtung *f* valve operator (actuator, control mechanism, control)
Ventilblock *n* valve unit (block)

Ventilbüchse *f* *WEGEV* valve sleeve (liner)
Ventilcharakteristik *f* valve response (characteristics)
Ventildruckverlust *m* valve [pressure] loss
Ventileinbauplatte *f* cartridge valve manifold
Ventil-Einschraubpatrone *f* screw-in cartridge valve
Ventilelement *n* valving (valve control) element
Ventil-Endschalter *m* limit valve
Ventilflattern *n* *WEGEV* valve chatter (flutter)
Ventilgehäuse *n* valve body (envelope)
Ventilgeräusch *n* valve[-generated] noise
ventilgesteuerte Pumpe *f* valve-controlled pump
Ventil-Grenzschalter *m* *WEGEV* limit valve
Ventilinsel *f* *Kompakteinheit mit integrierter Elektronik und Sensorik* [pre]packaged control unit *with integral sensing elements*
Ventilkarte *f* *steckbare Mikroprozessor-Ventil-Kombination* valve card *combined microprocessor-air valves plug-in unit*
Ventilkegel *m* *stumpf* poppet
Ventilkombination *f* **in Monoblockbauweise** monoblock valve unit
Ventilkörper *m* valve body (envelope)
Ventilkugel *f* *SPERRV* ball type poppet, valve ball; *VENTILE* valve ball
ventillose Axialkolbenpumpe *f*
s. wegegesteuerte Axialkolbenpumpe *f*
ventillose Kupplung *f* *VERBIND* unvalved coupling
ventillose Radialkolbenpumpe *f*

s. wegegesteuerte Radialkolbenpumpe *f*
Ventilmontageplatte *f*   valve (mounting) panel
Ventilplatte *f*   *PU/MOT* port (valve) plate
Ventilrattern *n*   valve chatter (flutter)
Ventilschieber *m*   *WEGEV* valve spool
Ventilschnarren *n*   valve chatter (flutter)
Ventilsitz *m*   valve seat
Ventilstelleinheit *f*   valve operator (actuator, control mechanism, control)
Ventilsteuerkreislauf *m*   valve control circuit (system), valve control
Ventilsteuerung *f*   *s.* Ventilsteuerkreislauf *m*
Ventilstößel *m*   valve plunger
Ventilumrandung *f*   *in Schaltplänen* valve envelope   *in symbolic diagrams*
Ventilverhalten *n*   valve response (characteristics)
Ventilverlust *m*   valve [pressure] loss
Venturidüse *f*   Venturi [tube]
Verbesserer *m*   *FLÜSS* improving additive, improver
verbinden   connect *v*, port *v*
verbinden: mit dem Behälter verbinden   vent *v* [to tank], exhaust *v*, dump *v*, release *v*, relieve *v*, unload *v*
Verbindung *f*   *s.* Verbindungsleitung *f*; *s.* Rohrverschraubung *f*; *s.* Schlauchverbindung *f*
Verbindung *f* **mit geteiltem Flansch**   split-flange fitting
Verbindungsleitung *f*   connecting line, connector, junction
Verbindungsplatte *f*   *VENTILE* connector plate
Verbraucher *m*   actuator

Verbraucheranschluß *m*   *WEGEV* actuator (working) port
Verbrauchersteuerkreislauf *m*   motor (actuator) control circuit (system)
Verbundstoffdichtung *f*   bonded seal (washer)
Verbundverstellung *f*   *Primär- und Sekundärverstellung eines hydrostatischen Getriebes* pump and motor displacement control *of a hydrostatic transmission*
Verdampfungsrückstand *m*   *FLÜSS* evaporation deposit (residue)
verdichten   compress *v*
Verdichter *m*   compressor
Verdichter *m* **in Boxerbauart**   opposed[-cylinders] compressor
Verdichterrad *n*   impeller
Verdichterraum *m*   compressor room
Verdicker *m*   *FLÜSS* viscosity improver, thickener
verdrängen   displace *v*
Verdrängerelement *n*   *PU/MOT* displacing member (element)
Verdrängerhydraulik *f*   *s.* Hydraulik *f*
Verdrängereinheit *f*   positive-displacement pump *or* motor
Verdrängermotor *m*   positive-displacement motor
Verdrängerpumpe *f*   positive-displacement pump
Verdrängung *f*   *PU/MOT* displacement
Verdrängungsraum *m*   *PU/MOT* displacing (displacement) chamber
Verdrängungsverdichter *m*   positive-displacement compressor
Verdrängungsvolumen *n*   *PU/MOT* displacement [volume]
Verdrängungsvolumen *n* **bei Nullförderstrom**   *PU/MOT* zero displacement

verdrehgesicherter Zylinder *m*   non-rotating cylinder
Verdrehsicherung *f*   *ZYL* anti-rotation device
Verdunstungswiderstand *m*   *unterhalb des Siedepunkts FLÜSS* evaporation resistance   *below boiling temperature*
vereinfachtes Schaltzeichen *n* **(Symbol** *n***)**   simplified symbol
Vereisungsschutzschmierstoff *m*   *AUFBER* anti-freeze lubricant
Verengungsverlust *m*   *THEOR* contraction loss
β-Verhältnis *n*   *ein Maß für den Filterwirkungsgrad* beta ratio, β ratio *(β for X μm particle size = particles > X μm upstream / particles > X μm downstream)*
verharztes Öl *n*   resinous oil
Verharzung *f*   *FLÜSS* resinification, gum formation
verkettbar   *VENTILE* banking, stacking, stackable, ganged, modular
verkettbare Pumpe *f*   stack pump
verkettbare Unterplatte *f*   *s*. Verkettungsunterplatte *f*
verketten   *Ventile* bank *v*, stack *v*, gang *v*   *valves*
Verkettungsart *f*   valve mounting type
verkettungsfähig   *VENTILE* banking, stacking, stackable, ganged, modular
Verkettungsunterplatte *f*   stacking subplate, modular manifold, ganged subplate (subbase)
verkleben   *s*. verklemmen
verklemmen   *Ventilkolben* lock *v*, jam *v*, stick *v*, gum *v*   *valve spool*
Verkleben *n*   *s*. Verklemmen *n*
Verklemmen *n*   *des Ventilkolbens* hydraulic lock, pressure freeze, binding, gumming   *of the valve spool*

verlegen   *Oberflächenfilter* clog *v*, plug *v*
Verlegung *f*   *bei Oberflächenfiltern* clogging, plugging
Verlust *m*   loss   *wenn unbestimmt, meist* losses
Verlustbeiwert *m*   *THEOR* flow (discharge, loss) coefficient, flow resistance value
Verlustdruck *m*   loss pressure
verlustfrei   lossless
verlustlos   *s*. verlustfrei
Verlust[volumen]strom *m*   leakage flow [rate], leakage rate
Verpreßmaschine *f*   *Schlauchverbindungen* crimping device, crimper   *hose fittings*
Verpreßverbindung *f*   *für Schläuche* crimped-on (crimp, swaged-on) fitting   *for hoses*
Verriegelungsventil *n*   lock-out valve
Verriegelungszylinder *m*   locking cylinder
verrohren   pipe *v*
Verrohrung *f*   *Rohrnetz* piping, pipework
Versagen *n* **der Hydraulik**   hydraulic failure
verschieben/sich   *Kolben* stroke *v* *piston*; *Ventilkolben* shift *v*   valve spool
Verschlauchungsbedarf *m*   number of hoses needed
Verschleiß *m*   *des Öls* oil wear
Verschleißinhibitor *m*   *FLÜSS* anti-wear additive
Verschleißschutzöl *n*   anti-wear oil
verschließen   *z. B. Behälter* seal *v*   *eg tank*; *Öffnung* plug *v*   *port*
Verschlußkappe *f*   *druckdicht* pressure cap

Verschlußschraube *f*   threaded plug
verschmutzen   contaminate *v*
verschmutzt   *FLÜSS* contamination-loaded, dirty
Verschmutzung *f*   *Vorgang* contamination; *Schmutz* contaminant, contamination, dirt
   Verschmutzung *f* **durch Wasser**   *FLÜSS* water contamination
Verschmutzungsanzeige *f*   *FILTER* clogging (dirt, service) indicator, service warning device
Verschmutzungsanzeiger *m*   *FLÜSS* contamination monitor
verschmutzungsempfindlich   contaminant- (dirt-)sensitive, sensitive to contamination
Verschmutzungsempfindlichkeit *f*   contaminant (dirt) sensitivity
Verschmutzungsempfindlichkeitsklasse *f*   *eines Geräts* contaminant- (dirt-)sensitivity grade (class) *of a component*
verschmutzungsfrei   *FLÜSS* free of contamination, contaminant-free
Verschmutzungsgrad *m*   *FLÜSS* contamination (contaminant, dirt) level
Verschmutzungskennwert *m*   *FLÜSS* contamination (dirt) characteristic
Verschmutzungsklasse *f*   *FLÜSS* contamination (dirt) class
Verschmutzungskonzentration *f*   contaminant (dirt) concentration
Verschmutzungsquelle *f*   *FLÜSS* contamination source
Verschmutzungsteilchen *n*   contaminant (dirt) particle
Verschraubung *f*   *s*. Rohrverschraubung *f*; *s*. Schlauchverschraubung *f*
Verseifungszahl *f*   *FLÜSS* saponification value

versorgen   deliver *v*, supply *v*, provide *v*
Versorgung *f*   supply, admission, delivery
Versorgungsdruck *m*   supply (feed, input, inlet) pressure
Versorgungsdüse *f*   *LOGIKEL* supply nozzle (tube)
Versorgungsstrahl *m*   *LOGIKEL* supply jet
Verspannungsprüfstand *n*   closed-loop tester   *torque and power*
versperren   *z. B. eine Leitung* block *v* [off], isolate *v*   *eg a line*
verstärken   *Druck* intensify *v*, boost *v*, amplify *v*   *pressure*; *Schlauchwandung* reinforce *v*   *hose wall*
verstärkte Bördelverbindung *f*   *s*. Bördelverbindung *f* mit Klemmring
verstärkte Dichtung *f*   reinforced seal
Verstärkung *f*   *Schlauchverstärkung* hose reinforcement (support)
Verstärkungslage *f*   *LEIT* reinforcement layer
verstellbare Drossel *f*   *STROMV* variable (adjustable) restrictor (restriction)
verstellen   *PU/MOT* control *v*, stroke *v*; *steuern, betätigen STELL* operate *v*, actuate *v*, control *v*
Verstellkraft *f*   *STELL* operating (actuating, control, stroke) force
Verstellmoment *n*   *STELL* operating (actuating, control) torque
Verstellmotor *m*   variable-displacement (-volume) motor
Verstellpumpe *f*   variable-displacement (-volume, -delivery) pump
Verstellsicherung *f*   *STELL* locking mechanism

Verstellwegbegrenzer *m* einer Pumpenstelleinheit stroke limiter *of a pump control*
verstopfen, zusetzen *FILTER* clog *v*, plug *v*, *small orifices also* silt *v*; *eine Öffnung* plug *v* *a port*
Verstopfungsalarm *m* *s.* Warnfilter *m,n*
Verstopfungsanzeige *f* *FILTER* clogging (dirt, service) indicator, warning device
Verstopfungsempfindlichkeit *f* *STROMV* clogging sensitivity
Versuchsstand *m* test bench (stand)
Verteerungszahl *f* *FLÜSS* tar number
Verteilerleitung *f* header [line]
Verteilerverschraubung *f* header
verunreinigen *FLÜSS* contaminate *v*
Verweilzeit *f* *des Öls im Behälter* dwell (rest) time *of the fluid in the tank*
Verzögerungsmittel *n* *FLÜSS* inhibitor, inhibiter
Verzögerungsventil *n* *STROMV* deceleration valve
Verzweigungsverlust *m* branching loss
VGe *s.* Viskositäts-Gradexponent *m*
VI *m* *Viskositätsindex FLÜSS* viscosity index, VI
Vickers-Pumpentest *m* *FLÜSS* Vickers (vane) pump test *an anti-wear test*
Vielröhrchendüse *f* multiple-hole exhaust
vierfach-spiralarmierter Schlauch *m* 4-spiral wire wrap hose
Vierkantenfühler *m* *STELL* spool-type servovalve with four metering orifices
Vierkantenschieber *m* *WEGEV* valve spool with four metering edges

Vierkugelapparat-Test *m* *FLÜSS* four ball [wear] test
Vierlochflansch *m* four-bolt flange
Vierstellungsventil *n* four-position (4-position) valve
Vierwegeventil *n* four-way (4-way) valve
viskose Reibung *f* *THEOR* viscous (fluid) friction
viskoser Dämpfer *m* fluid absorber, viscous damper
Viskosimeter *n* visco[si]meter
Viskosimetrie *f* visco[si]metry
Viskosität *f* *FLÜSS* viscosity
Viskositäts-Druck-Verhalten *n* *FLÜSS* viscosity-pressure characteristics
viskositätserhöhendes Additiv *n* *FLÜSS* viscosity improver, thickener
Viskositätsfühler *m* viscosity detector
Viskositätsgrad *m* *SAE* viscosity (oil) grade *SAE*
Viskositäts-Gradexponent *m* *FLÜSS* viscosity grade index
Viskositätsindex *m* viscosity index, VI
Viskositätsindexverbesserer *m* viscosity index improver, VI improver
Viskositätsklasse *f* *SAE* *s.* Viskositätsgrad *m*
Viskositätskraft *f* *THEOR* viscosity force
Viskositätsmessung *f* *s.* Viskosimetrie *f*
Viskositätspol *m* viscosity pole
Viskositätspolhöhe *f* viscosity pole height
Viskositäts-Richtungskonstante *f* *Viskositäts-Temperatur-Verhalten* [viscosity] slope coefficient *viscosity-temperature characteristics*
Viskositäts-Temperatur-Blatt *n*

viscosity-temperature diagram (chart), VT diagram (chart)
Viskositäts-Temperatur-Verhalten *n* FLÜSS viscosity- temperature (VT) characteristics
VI-Verbesserer *m*   *s.* Viskositätsindexverbesserer *m*
VKA-Test *m*   *s.* Vierkugelapparat-Test *m*
Vliesfilter *m,n*   bonded fabric filter
Vollförderstellung *f*   PU/MOT full-displacement position
Vollförderung *f*   PU/MOT full displacement
vollhydraulisch   all-hydraulic, purely hydraulic
vollkommenes Fluid *n*   THEOR ideal (perfect) fluid
vollpneumatisch   all-pneumatic, purely pneumatic
vollständiges Schaltzeichen (Symbol) *n*   complete (full) symbol
Vollstromfilter *m,n*   full-flow filter
Vollstromfilterung *f*   full-flow filtration
Vollverdrängungsvolumen *n*   PU/MOT full displacement
Volumenelastizitätsmodul *m*   FLÜSS bulk modulus [of elasticity]
Volumenstrom *m*   oft nur Durchfluß flow [rate], *in a more exact context* volumetric flow rate, volume rate of flow
Volumenstrom *m* **bei Nullast** PU/MOT no-load (zero-load) flow
Volumenstrombereich *m*   range of flow rate, flow range; PU/MOT delivery (discharge, output flow) range
Volumenstromdiagramm *n* *s.* Volumenstrom-Weg-Diagramm *n*

Volumenstrommesser *m*   flowmeter
Volumenstrommessung *f*   flow [rate] measurement
Volumenstromquelle *f*   constant-flow supply (source)
Volumenstromschreiber *m*   flow recorder
Volumenstromsensor *m*   flow sensor
Volumenstromsteuerung *f* flow control circuit (system, flow control)
Volumenstromverhältnis *n*   LOGIKEL turndown ratio, TDR
Volumenstromwandler *m*   flow transducer
Volumenstrom-Weg-Diagramm *n* flow-displacement profile plot, flow [pattern] plot
Volumenstrom-Zyklusprofil *n* *s.* Volumenstrom-Weg-Diagramm *n*
volumetrische Dosierung (Zuteilung) *f*   PU/MOT volume-controlled (positive) metering
volumetrische Verluste *mpl* PU/MOT volumetric losses
volumetrischer Wirkungsgrad *m* PU/MOT volumetric efficiency
vor *einem Element*   *s.* vorgeschaltete Leitung *f*
Vorauslaß *m*   *bei entsperrbaren Rückschlagventilen* decompression *of a pilot-operated check valve*
Vorauslaßventil *n*   decompression valve
Vordrossel *f*   STROMV upstream orifice
Vorfilter *m,n*   pre-filter
Vorfilterung *f*   prefiltration
Vorfüllbehälter *m*   prefill tank
Vorfülleitung *f*   filling (loading, charging) line

vorfüllen *Pumpe* prime *v*, prefill *v*, supercharge *v*, boost *v* *pump*
Vorfüllkreislauf *m* *bei Umformmaschinen* prefill circuit *of metalforming machines*
Vorfüllpumpe *f* charge (charging, booster, supercharge, prefill) pump
Vorfüllung *f* **eines Zylinderraumes** *PU/MOT* piston priming
Vorfüllventil *n* *WEGEV* prefill valve
vorgeformte Dichtung *f* moulded (preformed) seal
vorgeformte Lippe *DICHT* moulded (preformed) lip
vorgeschaltete Leitung *f* upstream line
vorgespannte Dichtung *f* [pre]loaded (semi-automatic, installation-actuated) seal
vorgespannter Behälter *m* pressurized reservoir (tank)
vorgesteuerter Nullhubregler *m* *PU/MOT* two-stage pressure compensator
vorgesteuertes Druckbegrenzungsventil *n* pilot-operated (piloted) relief valve
vorgesteuertes Ventil *n* pilot-operated (-actuated) valve, piloted valve
Vorhub *m* *ZYL* extend (out, extension, outward) stroke
Vorkompression *f* precompression
Vorlauf *m* *ZYL* s. Vorhub *m*
Voröffnung *f* *bei entsperrbaren Rückschlagventilen* decompression *of a pilot-operated check*
Vorschubkreislauf *m* *ANWEND* feed circuit
Vorschubzylinder *m* *ANWEND* feed[ing] cylinder

Vorschweißflansch *m* *VERBIND* weld (welding neck) flange
Vorschweißverschraubung *f* butt welding fitting
Vorschweißwinkel *m* butt welding elbow
Vorspanndruck *m* *einer Dichtung* preloading (interference, fitting) pressure *of a seal*; *Fülldruck eines Speichers* precharge (charging, preload, inflation) pressure
Vorspannpumpe *f* *ANWEND* pressurizing pump
Vorspannventil *n* *DRUCKV* counterbalance (back-pressure) valve
Vorsteueranschluß *m* pilot port
Vorsteuerdruck *m* **pilot pressure**
Vorsteuerflüssigkeit *f* pilot fluid
Vorsteuerkolben *m* pilot piston (spool)
Vorsteuerkreislauf *m* pilot circuit
Vorsteuerleitung *f* pilot control (pressure) line, pilot line
vorsteuern pilot-operate *v* (-actuate) *v*, pilot *v*
Vorsteuerschieber *m* pilot piston (spool)
Vorsteuerstufe *f* pilot stage
Vorsteuerung *f* pilot actuation
Vorsteuerventil *n* pilot [valve]
Vortexdiode *f* *LOGIKEL* vortex diode
Vortexelement *n* *LOGIKEL* vortex element
Vortexverstärker *m* *LOGIKEL* vortex amplifier
Vorverdichtung *f* precompression
vorwärmen pre-heat *v*
Vorwärmer *m* heater, pre-heater
Vp-Verhalten *n* *Viskositäts-Druck-Verhalten FLÜSS* viscosity-pressure characteristics

VPI-Stoff *m über die Dampfphase wirkender Korrosionsinhibitor FLÜSS* vapour-phase inhibitor, VPI *to prevent corrosion of metals in contact with fluid vapour*
VRk *f s.* Viskositäts-Richtungskonstante *f*
VT-Blatt *n Viskositäts-Temperatur-Blatt* viscosity-temperature diagram (chart), VT diagram (chart)
VT-Verhalten *n Viskositäts-Temperatur-Verhalten FLÜSS* viscosity-temperature (VT) characteristics
VZ *f Verseifungszahl FLÜSS* saponification value

W

W-Ring *m DICHT* W-ring
wahrer Kompressionsmodul *m FLÜSS* tangent bulk modulus
Walkverdrängerpumpe *f* squeegee pump
Wanddicke *f* wall thickness
Wand[haft]effekt *m LOGIKEL* Coanda (wall-attachment) effect
Wandrauheit *f* wall roughness
Wandreibung *f* wall friction
Wandstrahl *m LOGIKEL* wall-attached jet
Wandstrahlelement *n LOGIKEL* wall-attachment (Coanda effect) element
Wärmeaustauscher *m s.* Wärmeübertrager *m*
Wärmebeständigkeit *f FLÜSS* thermal stability
Wärmebilanz *f* heat balance
Wärmeleitfähigkeit *f FLÜSS* thermal (heat) conductivity
Wärmetauscher *m s.* Wärmeübertrager *m*
Wärmeübertrager *m* heat exchanger
Warmregenerierung *f von Trockenmittel* heat regeneration *of a desiccant*
Warnfilter *m,n mit Verstopfungsalarm* self-warning filter
Wartungseinheit *f s.* Druckluft-Wartungseinheit *f*
Wasserablaß *m FILTER* water drain; *LEIT* drain (line, pipe) trap
Wasserabscheider *m AUFBER* water (moisture) separator
Wasserabscheidefilter *m,n* water-removing filter
Wasserabscheidevermögen *n FLÜSS* water demulsibility
Wasseraufbereitung *f* water preparation (conditioning, treatment)
Wasseraufnahmefähigkeit *f FLÜSS* water-carrying capacity
wasserbasische Flüssigkeit *f* aqueous (water-base) fluid
Wasserbehandlung *f s.* Wasseraufbereitung *f*
Wasserdampfemulsionszahl *f Maß der Emulgierbarkeit FLÜSS* steam-emulsion number, SEN *measure of emulsibility*
wassereingespritzter Kompressor *m* water-injected compressor
Wasserfang *m* drain (line, pipe) trap
wasserfreie Flüssigkeit *f* non-aqueous (non-water, waterfree, anhydrous) fluid
Wassergehalt *m FLÜSS* water content
wassergekühlt water-cooled
wassergekühlter Ölkühler *m* water-cooled (oil-to-water) oil cooler
Wasser-Glycol-Lösung *f* polyglycol solution, water-glycol [fluid]

wasserhaltige Flüssigkeit *f*   aqueous (water-base) fluid
Wasserhydraulik *f*   water hydraulics
Wasserhydraulikanlage *f*   water [hydraulic] system, water hydraulic
wasserhydraulische Presse *f*   *ANWEND* water hydraulic press
Wasser-in-Öl-Emulsion *f*   *FLÜSS* water-in-oil (invert) emulsion
Wasserkühlung *f*   water cooling
Wasser-Öl-Kühler *m*   water-cooled (oil-to-water) cooler
Wasserpumpe *f*   water pump
Wasserringverdichter *m*   water-ring compressor
Wasserschlag *m*   *s.* hydraulischer Stoß *m*
Wassertragvermögen *n*   *FLÜSS* water-carrying capacity
Wasserverschmutzung *f*   *s.* wäßrige Verschmutzung *f*
wäßrige Flüssigkeit *f*   aqueous (water-base) fluid
wäßrige Verschmutzung *f*   *Verschmutzung durch Wasser FLÜSS* water contamination
Wechselfilter *m,n*   spin-on (removable) filter
wechselseitig entsperrbares Rückschlagventil *n*   double (dual) check valve
Wechselstromhydraulik *f*   alternating-fluid (AF) hydraulics
Wechselstrommagnet *m*   *STELL* alternating current (A. C.) solenoid
Wechselventil *f*   *WEGEV* shuttle valve
2-Wege-Einbauventil *n*   *s.* Wegesitzventil *n*
wegegesteuerte Axialkolbenpumpe *f*   port-plate (valve-plate, flat-valve) axial piston pump
wegegesteuerte Pumpe *f*   port (valve) plate controlled pump
wegegesteuerte Radialkolbenpumpe *f*   pintle-valve (pintle- ported, valve-spindle) radial-piston pump
2-Wege-Kugelhahn *m*   *WEGEV* two-way (2-way) ball valve
3-Wege-Kugelhahn *m*   *WEGEV* three-way (3-way) ball valve
Wege-Proportionalventil *n*   *WEGEV* proportional directional control valve
Wegesitzventil *n*   cartridge (hydraulic) logic valve, logic cartridge valve, logic element,   *but s ED* cartridge valve
Wegesteuerkreislauf *m*   directional control circuit (system)
Wegesteuerung *f*   directional control
Wegeventil *n*   directional [control] valve,   *compounds s also with* Ventil *n*
***n*-Wegeventil** *n*   *n*-way valve
Wegeventil *n* **ohne selbsttätige Zentrierung**   noncentering directional valve
Wegwerffilterelement *n*   disposable (throw-away) filter element, disposable
Weichdichtung *f*   compression seal
weichschaltendes Wegeventil *n*   soft-shift directional valve
Weichsitz *m*   *VENTILE* soft seat
Weichsitzventil *n*   soft-seat (soft-seated, resilient-seal) valve
Wellendichtung *f*   [rotary] shaft seal
Wellengleichung *f*   *THEOR* wave equation
Wellenwiderstand *m*   *THEOR* wave transmission resistance
Wellmembrandruckwandler *m*   convoluted diaphragm pressure transducer
Wellrohr *n*   *LEIT* corrugated tube
Wellrohrschlauch *m*   corrugated metal hose

Werksdruckluftnetz *n*   shop (plant) air mains
Werkstattdruckluft *f*   shop (plant) air
Werkstoffverträglichkeitsprüfung *f*   FILTER material compatibility test
werkverpreßte Verbindung *f*   factory-attached [hose] fitting
Werkzeugverschraubung *f*   [compressed-] air-tool fitting
Widerstand *m*   *ohmscher Widerstand, Eigenschaft* resistance
Widerstand *m* **gegen Schaumbildung**   FLÜSS foaming resistance, resistance to foaming
Widerstandsbeiwert *m*   THEOR flow (discharge, loss) coefficient, flow resistance value
Widerstandskraft *f*   THEOR resisting force; viscous damping (drag) force
Widerstandslast *f*   resistive load
Widerstandsthermometer *n*   resistance thermometer
Widerstandsventil *n*   DRUCKV counterbalance (back-pressure) valve
Widerstandszahl *f*   *s.* Widerstandsbeiwert *m*
Wiederaufbereitung *f*   *von Öl* oil reclamation (purification)
wiederverwendbare Verbindung *f*   reusable fitting
Windkessel *m*   receiver
Winkelanschlußverschraubung *f*   elbow port fitting
Winkelbefestigung *f*   ZYL angle mount[ing]
Winkelendstück *n*   elbow end fitting (connection), elbow connector
Winkel-Schottverschraubung *f*   bulkhead elbow
Winkelverschraubung *f*   elbow [fitting], ell

Winkelverschraubung *f* **mit zweiseitigem Rohranschluß**   union elbow
45°-Winkelverschraubung *f*   45° elbow
Winteröl *n*   winter-grade oil
Wirbel *m*   vortex (*pl:* vortices *or* vortexes)
Wirbelelement *n* **mit rotierender Wirbelkammer**   LOGIKEL rate-sensitive vortex diode
Wirbelgeräusch *n*   vortex noise
Wirbelkammer *f*   LOGIKEL vortex chamber
Wirbelkammerdiode *f*   LOGIKEL vortex diode
Wirbelkammerelement *n*   LOGIKEL vortex element
Wirbelkammerverstärker *m*   LOGIKEL vortex amplifier
Wirbellärm *m*   vortex noise
Wirbelsensor *m*   LOGIKEL vortex sensor
Wirbelströmung *f*   vortex flow
wirken   *Druck, auf eine Fläche* act *v* on (upon)   *pressure, on an area*
Wirkkammer *f*   LOGIKEL interaction chamber
wirkliches Fluid *n*   THEOR real (true) fluid
Wirkraum *m*   LOGIKEL interaction chamber
wirksame Fläche   ZYL effective (net, exposed) area
Wirkstoff *m*   FLÜSS additive, agent
Wirkungsgrad *m* **bei Nennbetriebsbedingungen**   PU/MOT running efficiency
Woltmannflügel *m*   *s.* Flügelrad-Durchflußmesser *m*
Wulstnippel *m*   VERBIND beaded insert, barbed nipple
Wulstnippelverbindung *f*   barb fitting

X

X-Ring *m*   *DICHT* X-ring

Y

Y-Verschraubung *f*   Y-fitting

Z

Zähigkeit *f*   *FLÜSS* viscosity, *compounds* s *with* Viskosität *f*
Zahneingriffsleckverlust *m*   [tooth]-mesh leakage
Zähnezahl *f*   number of teeth, tooth number
Zahnkammer *f*   *in einer Zahnradpumpe* tooth space *in a gear pump*
Zahnspiel *n*   *in einer Zahnradpumpe* gear-tip clearance *in a gear pump*
Zahnkopfverlust *m*   *in einer Zahnradpumpe* peripheral leakage *in a gear pump*
Zahnradeinheit *f*   gear pump *or* motor
Zahnradmotor *m*   gear motor
Zahnradpumpe *f*   gear pump
  Zahnradpumpe *f* **mit axialem Spielausgleich (mit druckabhängigem Axialspalt)**   gear pump with pressurized side-plates (with pressure-dependent axial clearance, with pressure loading)
  Zahnradpumpe *f* **mit festem Seitenspiel**   fixed axial clearance gear pump
Zahnringpumpe *f*   *trochoidenverzahnt* progressing-tooth gear pump
zahnstangenbetätigtes Ventil *n*   rack-and-pinion operated (actuated) valve
Zahnstangenschwenkmotor *m*   rack-and-pinion (piston rack) rotary actuator
Zapfenbefestigung *f*   *ZYL* trunnion (stud) mount
Zapfstelle *f*   tap, tapping (take-off) point
Zehnkantenschieber *m*   *WEGEV* valve spool with ten metering edges
Zeichenschablone *f*   *für Schaltzeichen* circuit tool
zeitgesteuerter Ablaß *m*   *FILTER* timer-actuated (interval) drain
Zeitventil *n*   *STROMV* time-delay (timing) valve
Zelle *f*   *einer Flügelzellpumpe* chamber *of a vane pump*
Zellenverdichter *m*   rotary vane (sliding-vane, vane) compressor
Zellulosefilter *m,n*   cellulose filter
zentrale Druckluftversorgung *f*   central compressed-air source
Zentralhydraulik *f*   *ANWEND* central hydraulic system
Zentrierfeder *f*   *STELL* centering spring
Zentriernut *f*   *WEGEV* balancing (centering) groove
Zentrifugalabscheider *m*   centrifugal dryer
Zentrifugaldämpfer *m*   involute damper
Zentrifugalfilter *m,n*   centrifugal filter
Zentrifugalpumpe *f*   centrifugal (rotodynamic) pump
Zentrifugaltrockner *m*   centrifugal dryer
zentrisch angeordneter Kolben *m*   centered spool

Zerfallstemperatur *f* *FLÜSS* decomposition temperature
ziehende Last *f* overrunning load
Zinkdialkyldithiophosphat *n* *FLÜSS* zinc dialkyldithiophosphate, ZDDP
Zinkdithiophosphat *n* *FLÜSS* zinc dithiophosphate, ZDP
zirkulieren *Flüssigkeit* circulate *v* *fluid*
Zubehör *n* accessories *pl*
Zufluß *m* input, inlet, intake, inflow, *compounds s with* Zulauf *m*
Zufluß *m*: **im Zufluß steuern** *STROMV* meter-in *v*
Zuförderdruck *m* input (inlet, supply, feed) pressure
Zuförderleitung *f* filling (loading, charging) line
Zuförderpumpe *f* charge (charging, booster, supercharge, prefill) pump
Zufuhr *f* supply, admission, delivery
zuführen *Druck* pressurize *v*, apply *v* pressure to *a component*, pressure-load *v*, charge *v* with pressure, expose *v* to a pressure
Zuführung *f* input, inlet, intake, inflow, *compounds s with* Zulauf *m*
Zugankerzylinder *m* tie-rod cylinder
Zugmagnet *m* *STELL* pull-type (tractive) solenoid (magnet)
Zugseilzylinder *m* cable cylinder
Zughub *m* *ZYL* pull stroke
Zugstangenbefestigung *f* *ZYL* tie-rod (extended tie-rods) mount[ing]
Zugzylinder *m* pull-action cylinder
Zulauf *m* input, inlet, intake, inflow
Zulauf *m*: **im Zulauf steuern** *STROMV* meter-in *v*
Zulaufanschluß *m* input (inlet, intake, supply) port, *also, at a pump* suction port, *also, at a motor* pressure port
Zulaufdrosselung *f* *STROMV* meter-in flow control, metering-in
Zulaufdruck *m* input (inlet) pressure, *also, at a pump* suction pressure, *also, at a motor or valve actuator* supply (feed) pressure
zulaufgesteuert metered-in, *also as a verb:* meter-in *v*
Zulaufkammer *f* input (inlet, intake) chamber, *also, in a pump* suction chamber, *also, in a motor* pressure chamber
Zulaufkanal *m* input (inlet, intake) channel (duct), *also, in a pump* suction channel, *also, in a motor* pressure channel
Zulaufleitung *f* input (inlet, intake) line, *also, to a pump* suction line, *also, to a motor* supply (feed, pressure) line
Zulaufplatte *f* *einer Ventilbatterie* inlet (intake) section *of ganged valves*
Zulaufseite *f* input (inlet, intake) side, *also, of a pump* suction side, *also, of a motor* pressure side
Zulaufsteuerung *f* *STROMV* meter-in flow control, metering-in
Zulaufsteuerventil *n* *STROMV* meter-in valve
Zulauf[volumen]strom *m* input (inlet, intake) flow [rate], *also, into a pump* suction flow
Zuleitung *f* *s.* Zulaufleitung *f*
Zuluft *f* input (intake) air
Zuluftdrosselung *f* *s.* Zuluftsteuerung *f*
Zuluftsteuerung *f* intake air-metering, metering-in

zumessen *im Zulauf steuern* meter-in *v*
Zumeßkreislauf *m* STROMV meter-in circuit
Zumessung *f* *s.* Zulaufsteuerung *f*
Zumeßventil *n* *s.* Zulaufsteuerventil *n*
zurückfließen return *vi*
  zurückfließen lassen return *vt*
zurückleiten *einen Strom* return *vt*
zurückströmen return *vi*
  zurückströmen lassen return *vt*
zurückverstellen *Verstellpumpe* destroke *vi and vt*, *variable pump*
zurückziehen[/sich] ZYL retract *vi and vt*, withdraw *vi and vt*
Zusammenbrech-/Berstfestigkeitsprüfung *f* FILTER collapse-burst test
Zusammenbrechdruck *m* FILTER collapse pressure
zusammendrückbar FLÜSS compressible
Zusammendrückbarkeit *f* FLÜSS compressibility
zusammengebrochenes Filterelement *n* collapsed filter element
Zusatz *m* FLÜSS additive, agent
Zusatzbetätigung *f* STELL auxiliary (override) control
Zusatzkreislauffilterung *f* off-line filtration
Zuschaltventil *n* DRUCKV sequence valve
Zusetzempfindlichkeit *f* STROMV clogging sensitivity
zusetzen[/sich] *enge Drosselquerschnitte* silt *vi and vt*, clog *vi and vt*, plug *vi and vt* *small orifice*
Zusetzen *n* FILTER clogging, plugging

Zusteuerventil *n* *s.* Zulaufsteuerventil *n*
Zustrom *m* *s.* Zulaufstrom *m*
Zutritt *m* *s.* Zulauf *m*
zuwachsen *s.* zusetzen/sich
Zwanglauffilter *m,n* full-flow filter
Zwanglauffilterung *f* full-flow filtration
zweibasische Flüssigkeit *f* double-base fluid
zweidimensionales Kopieren *n* ANWEND two-dimensional copying
Zweidruckkreislauf *m* KREISL high-low (hi-lo, dual pressure) circuit
zweidüsiges Prallplattenventil *n* STELL double-nozzle (two-jet, double) flapper valve
Zweiebenen-Rohrgelenk *n* double-plane swivel joint
Zweiflügelschwenkmotor *m* double-vane rotary actuator
zweiflutige Schraubenpumpe *f* double-suction (double-inlet) screw pump
Zweigkreislauf *m* branch circuit
Zweig[volumen]strom *m* branch flow [rate]
Zweikantenfühler *m* STELL spool-type servovalve with two metering orifices
Zweikantenschieber *m* WEGEV valve spool with two metering edges
Zweikreispumpe *f* tandem (dual, double, split-flow) pump
Zweikugelventil *n* SPERRV two ball valve
Zweilippenabstreifer *m* double-lip wiper
Zweilippenring *m* U-seal (-cup), double-lip seal

Zweipunkt-Stelleinheit *f* *PU/MOT* two-position control[ler] (stroker)
Zweipunktverstellung *f* *PU/MOT* two-position control
Zweispindelschraubenpumpe *f* two-screw (double-screw) pump
Zweistellungsventil *n* *WEGEV* two-position (2-position) valve
Zweistoffdichtung *f* composite (combined, dual-material) seal
Zweistoff-Flüssigkeit *f* double-base fluid
Zweistrangabsperrkupplung *f* *VERBIND* two-way (double shut-off, double-valved) coupling
Zweistrangabsperrung *f* *VERBIND* two-way (double) shut-off
Zweistrangsitzkupplung *f* *VERBIND* double poppet coupling
Zweistrompumpe *f* tandem (dual, double, split-flow) pump
Zweistufenfilter *m,n* two-stage filter
Zweistufenpumpe *f* two-stage pump
zweistufiger Nullhubregler *m* *PU/MOT* two-stage pressure compensator
zweistufiges Servoventil *n* two-stage servovalve
zweistufiges Ventil *n* two-stage valve
Zweiwegeabsperrkupplung *f* s. Zweistrangabsperrkupplung *f*
Zweiwegeabsperrung *f* s. Zweistrangabsperrung *f*
Zweiwege-Einbauventil *n* cartridge (hydraulic) logic valve, logic element (cartridge valve)
Zweiwegekugelhahn *m* *WEGEV* two-way (2-way) ball valve
Zweiwege-Strombegrenzungsventil *n* s. Zweiwege- Stromregelventil *n*

Zweiwege-Stromregelventil *n* series flow-control valve (flow regulator)
Zweiwegeventil *n* *WEGEV* two-way (2-way) valve
Zwillingsfilter *m,n* duplex (twin) filter
Zwillingspumpe *f* tandem (dual, double, split-flow) pump
Zwillingsrückschlagventil *n* *wechselseitig entsperrbar* double (dual) check valve
Zwischenheizer *m* *für Druckluft* reheater *for compressed air*
Zwischenkühler *m* interstage cooler, intercooler
Zwischenkühlung *f* interstage cooling, intercooling
Zwischenplatte *f* *VENTILE* interconnecting plate
Zwischenstellung *f* *WEGEV* intermediate position; *des Kolbens im Zylinder s ED* midstroke stopping
Zyklusdiagramm *n* cycle plot (diagram)
Zylinder *m* cylinder
Zylinder *m* **in Sonderausführung** specialty (special, custom) cylinder
Zylinder *m* **mit beid[er]seitiger (doppelseitiger) Kolbenstange** double-rod [end] cylinder
Zylinder *m* **mit einseitiger Kolbenstange** single-rod [end] cylinder
Zylinder *m* **mit Federrückzug** spring-return cylinder
Zylinder *m* **mit feststehendem Zylinderkörper** stationary-body cylinder
Zylinder *m* **mit feststehender Kolbenstange** stationary-rod cylinder
Zylinder *m* **mit geführtem Kabel** guided cable cylinder
Zylinder *m* **mit großem Innendurchmesser** large-bore cylinder

Zylinder *m* **mit kleinem Innendurchmesser**   small-bore cylinder
Zylinder *m* **mit nicht lösbar verbundenen Deckeln**   one-piece (solid end) cylinder
Zylinder *m* **mit Quadratkörper**   square-bodied cylinder
Zylinder *m* **mit Scheibenkolben**   piston-type cylinder
Zylinder *m* **mit Schwerkraftrückzug**   gravity-return (weight-returned) cylinder
Zylinder *m* **mit seitlicher Befestigung**   side mount[ed] cylinder
Zylinder *m* **mit Tauchkolben**   ram (plunger) cylinder
Zylinder *m* **mit Teleskopkolben**   telescopic (telescoping) cylinder
Zylinder *m* **mit überlangem Hub**   long-stroke cylinder
Zylinder *m* **mit verstellbarem Anschlag**   adjustable-stroke cylinder
Zylinder *m* **ohne Endlagenbremsung**   non-cushioned cylinder
Zylinder *m* **ohne Kolbenstange**   rodless cylinder
Zylinderanschluß *m*   *WEGEV* cylinder port (connection)
Zylinderantrieb *m*   linear actuator
Zylinderausrichtung *f*   cylinder alignment
Zylinderbauform *f*   cylinder design (type)
Zylinderbefestigung *f*   cylinder mount[ing]
Zylinderblock *m*   *PU/MOT* cylinder block
Zylinderboden *m*   cylinder [end] cover (cap, closure);   *Zylinderfuß* cylinder bottom
Zylinderbohrung *f*   *Innendurchmesser* cylinder bore [size], cylinder inner (inside) diameter, cylinder ID;   *Innenwand* cylinder bore (inner wall)
Zylinderdämpfung *f*   s. *Endlagenbremsung f*
Zylinderdeckel *m*   cylinder [end] cover (cap, closure)
Zylinderdruckanschluß *m*   *WEGEV* cylinder pressure port
Zylinderenddichtung *f*   cylinder end (head) seal
Zylinderende *n*   cylinder end (head)
Zylinderendkappe *f*   s. *Zylinderdeckel m*
Zylinderfläche *f*   cylinder [bore] area
Zylinderflansch *m*   cylinder flange
Zylinderfuß *m*   cylinder bottom
Zylinderinnendurchmesser *m*   cylinder bore [size], cylinder inner (inside) diameter, cylinder ID
Zylinderinnenwand *f*   cylinder bore (inner wall)
Zylinderkopf *m*   cylinder head
Zylinderkopf[deckel] *m*   cylinder head (front) cap (cover)
Zylinderkopfdichtung *f*   cylinder end (head) seal
Zylinderkörper *m*   *PU/MOT* cylinder block;   *ZYL* cylinder body
Zylindermantel *m*   cylinder barrel
Zylindermantelschlitz *m*   barrel slot
Zylinderquerschnittsfläche *f*   cylinder [bore] area
Zylinderraum *m*   cylinder chamber
Zylinderrohr *n*   s. *Zylindermantel m*
Zylinderrücklaufanschluß *m*   *WEGEV* cylinder return port
Zylindersensor *m*   position (displacement, piston, cylinder) sensor

Zylinderstern *m* *Radialkolbenmaschinen* cylinder block
Zylindertrommel *f* *Axialkolbenmaschinen* cylinder block
Zylindertyp *m* cylinder design (type)
Zylinderwand *f* cylinder wall, barrel wall
Zylinderzulaufanschluß *m* . *WEGEV* cylinder pressure port
zylindrischer Bremsansatz *m* *ZYL* straight spear